Developments and Applications of Advanced Engineering Ceramics and Composites

Developments and Applications of Advanced Engineering Ceramics and Composites

Ceramic Transactions Series, Volume 192

Proceedings of the 6th Pacific Rim Conference on Ceramic and Glass Technology (PacRim6); September 11–16, 2005; Maui, Hawaii

Edited by
Mrityunjay Singh
and
Hua-Tay Lin

The American Ceramic Society

WILEY-INTERSCIENCE
A JOHN WILEY & SONS, INC., PUBLICATION

Published by John Wiley & Sons, Inc., Hoboken, New Jersey
Published simultaneously in Canada.

For general information on our other products and services please contact our Customer Care Department within the U.S. at 877-762-2974, outside the U.S. at 317-572-3993 or fax 317-572-4002.

Wiley also publishes its books in a variety of electronic formats. Some content that appears in print, however, may not be available in electronic format.

Library of Congress Cataloging-in-Publication Data is available.

ISBN-13 978-0-470-08290-4
ISBN-10 0-470-08290-9

Printed in the United States of America.

10 9 8 7 6 5 4 3 2 1

Contents

Processing and Fabrication—Non-Oxide Systems

Microstructural and Thermal Characterization

Thermomechanical Characterization

Preface

Advanced ceramics and composites materials have seen tremendous growth worldwide over the last few decades. These engineered ceramics and composites are enabling materials for a number of demanding applications in aerospace, power generation, ground transportation, nuclear and chemical industries. A number of these applications require high performance and long life at elevated temperature in aggressive environments. Substantial progress has been made globally in the technology development, manufacturing, commercialization, and applications of these materials over the last two decades. However, significant challenges still remain for their wide scale commercial applications worldwide.

This volume contains 28 selected papers from two international symposia held in the area of advanced ceramics and composites during the sixth international conference of Pacific Rim Ceramic Societies (PACRM-6), Maui, Hawaii, September 11-16, 2005. These symposia covered all areas related to science, technology, and applications of advanced ceramics and ceramic matrix composite materials. Special emphasis was placed to invite eminent ceramic scientists and engineers from various backgrounds and disciplines (academia, government laboratory, research institutes, and industries) from all over the world to discuss the advances in engineering ceramics and composites worldwide.

We would like to thank The American Ceramic Society staff for their help during the organization of the symposium and publication of this volume. We would also like to thank all the conference sponsors. The excellent quality conference would not have been possible without the dedication and diligence of all the invited, contributed, and poster presenters from all over the world.

We hope that the papers in this volume will provide an open forum for the vigorous discussion of issues and opportunities in the area of composite materials. It is the earnest hope of the editors that this volume will contribute to the growth and popularization of advanced engineering ceramics and composite technology all over the world. These enabling technologies will contribute to better future through more efficient, economical, and environmentally benign systems for the benefit of mankind.

MRITYUNJAY SINGH
Ohio Aerospace Institute
NASA Glenn Research Center

HUA-TAY LIN
Metals and Ceramics Division
Oak Ridge National Laboratory

Processing and Fabrication—Oxide Systems

PROCESSING AND PROPERTIES OF Ti-BASE INTERMETALLIC PARTICLES DISPERSED TETRAGONAL ZIRCONIA

Norihito Tanaka, Tohru Sekino, Takafumi Kusunose
The Institute of Scientific and Industrial Research, Osaka University,
Mihogaoka 8-1, Ibaraki, Osaka, 567-0047, Japan

Tadachika Nakayama, Koichi Niihara
Extreme Energy-Density Research Institute, Nagaoka University of Technology,
Kamitomioka 1603-1, Nagaoka, Niigata, 940-2188, Japan

ABSTRACT

Continuous three step heating process consisted of decomposition, hydrogen reduction and hot-pressing of 3Y-TZP/NiO(or CoO)/TiH_2 powder mixtures has been attempted to obtain 3Y-TZP/NiTi and 3Y-TZP/CoTi composites. Intermetallic compounds Ti-Ni or Ti-Co binary systems were homogeneously dispersed in Y-TZP matrix. Fracture strength of 3Y-TZP/30vol%NiTi composite sintered at 1200°C has reached 680MPa that is much higher than the monolithic 3Y-TZP sintered for the same condition. The fracture strength, however, decreased with the increase in sintering temperature above 1300°C. For 5vol% CoTi reinforced 3Y-TZP composite, the fracture strength has increased up to 800MPa. However, when carbon nanotubes (CNT) mixed with this composite system the fracture strength did not improve, because of incomplete reaction and the residual amounts of Co and Ti particles.

INTRODUCTION

It is widely known that zirconia (ZrO_2) is stabilized to tetragonal phase by the solid solution with the addition of compounds such as yttria or ceria. Especially Y_2O_3 doped tetragonal zironia polycrystal (Y-TZP) has attracted special attention and been expected as promising structural material because of their excellent mechanical properties owning to stress induced martensitic phase transformation from tetragonal to monoclinic (t-m) phase. Y-TZP monolith is thus widely used in structural parts, connection part of optical fibers, and/or smart media. Therefore, a great deal of research has been conducted to improve the mechanical properties in order to spread the industrial use of this material.

Compositing of tetragonal zirconia with a second phase is one way to achieve the enhancement of mechanical properties. Y-TZP/Ni nanocomposites reported by Kondo *et al.*[1,2] exhibited excellent fracture strength of over 1 GPa. Recent investigation in ceramic matrix composites revealed that the dispersion of intermetallic particulates into ceramic matrix lead interpenetrating network microstructure.[3] For example, Y-TZP/Ni_3Al composites have been reported to possess excellent mechanical properties.[4] Thus, a new ceramic/metal composite system with superior properties can be developed by appropriate combination of ceramic and intermetallic materials.

On the other hand, NiTi intermetallic compound is well known to have excellent functions such as pseudoelasticity, shape memory effect, good corrosion resistance and so on due to its martensitic transformation characteristics,[5] and these good mechanical properties have made NiTi attractive for many industrial applications. Thus, NiTi-based intermetallics are considered as attractive second phase dispersion into ceramic materials. However there are few reports on the fabrication of NiTi intermetallic particles dispersed Y-TZP composites and their mechanical properties.

Therefore, we have focused on the tetragonal-zirconia (Y-TZP)/NiTi intermetallic composite system. Two kinds of Ti-based intermetallics NiTi and CoTi have been chosen as a second phase dispersion, because TiNi and CoTi (both B2-type titanides) generally exhibits an attractive combination of physical and mechanical properties. The authors have previously reported that 3Y-TZP/NiTi could be successfully prepared by the reduction and hot-pressing process at relatively low sintering temperature (1200°C) below the melting point of NiTi (1240°C).[6] However, the effect of higher sintering temperature above liquidus of NiTi has not been investigated yet. Therefore, the effect of higher sintering temperature on the microstructure and mechanical properties has been investigated for the 3Y-TZP/NiTi composites in this study. In addition, CoTi dispersed 3Y-TZP composites have also been fabricated with and without the addition of carbon nanotubes (CNT). Carbon nanotube is considered to be suitable strengthening dispersion, was recently reported for the alumina/CNT system,[7] because of its small volume and high specific strength as well as good elasticity. Incorporation of this kind of low-dimensional electrical conductive materials is expected to realize electrically functionalized composite owing to its unique 3-dimensional conductive pathway in bulk materials, which might be, for instance, used as probe to detect in-situ change of electrical properties of dispersed intermetallic particulates.

EXPERIMENTAL PROCEDURE

Zirconia powder containing 3 mol% of Y_2O_3 (Sumitomo Osaka Cement Co., Ltd., Osaka, Japan) was used as a ceramic matrix. NiO (Nakalai Tesque Inc., Kyoto, Japan), CoO (Wako Pure Chemical Industries, Ltd., Osaka, Japan) and TiH_2 (Kojundo Chemical Lab Co., Ltd, Saitama, Japan) powders were selected as source materials for dispersed intermetallic compounds. Y-TZP, NiO (or CoO) and TiH_2 powders were mixed by conventional wet ball-milling for 24 h in ethanol using zirconia balls. The mixed powders were dried and then dry ball-milled for 24 h to crush the hard agglomerate of the powders. For the 3Y-TZP/NiTi system, NiTi content in the final sintered composite was adjusted to be 30vol%. In the case of 3Y-TZP/CoTi/CNT composite, multi-walled carbon nanotube (MWCNT) having diameter of around 30 nm were dispersed in ethanol with a surfactant. Then 3Y-TZP, CoO, and TiH_2 powders were mixed with MWCNT/ethanol slurry, and wet and dry ball-milled for 48h and 24h, respectively. The content of CoTi was varied from 0, 5, 15, 30 and 40 vol% while the MWCNT was fixed at 0.75%.

The powder mixtures were packed into a 44 mm diameter graphite die, placed into a multipurpose high-temperature furnace (Fujidenpa Kogyo Co., Ltd., Osaka, Japan), and subjected to the sintering process. The sintering process basically consisted of dehydrogenation of TiH_2 to metallic Ti, reduction of NiO to Ni or CoO to Co and hot-press sintering process at 1200 to1600°C under an applied pressure of 30 MPa for 1h in an Ar atmosphere. To fabricate 3Y-TZP/CoTi and 3Y-TZP/CoTi/CNT composites, however, dehydrogenation process at 500°C was skipped because it was found that the heating at 700°C in H_2 atmosphere could provide metallic-Ti. The sintered samples were then cooled in the furnace, and the applied pressure was released at 500 °C.

The sintered samples were cut with a diamond saw to obtain rectangular bars of 3x4x37mm for the mechanical properties investigation. They were grounded with a diamond whetstone, and polished with 0.5 μm diamond slurry to get a fine surface. Fracture strength was measured by the 3-point bending method with a span of 30mm at a crosshead speed of 0.5mm/min using a universal testing machine (Autograph, Shimazu Co. Ltd., Japan) for all samples. At least five specimens were tested to determine the bending strength.

X-ray diffractometer (XRD, RAD-C system, Rigaku Co., Tokyo, Japan) with CuKα radiation was used to identify crystalline phases of sintered specimens. In order to investigate the microstructure such as grain size and/or distribution of second phase dispersoid, fracture surfaces were observed by using scanning electron microscopy (SEM, model S-5000, Hitachi Co., Ltd., Hitachi, Japan).

RESULTS AND DISCUSSION

a) 30vol%NiTi / 3Y-TZP composites:

TEM observation and EDX analysis of the prepared powder revealed that ZrO_2 particles (approximately 50 to 100 nm) dispersed around large TiH_2 particles (0.5 to 1 μm) and NiO particles (100 to 200 nm).[6] We have successfully prepared this NiTi/3Y-TZP composite by a mixed route synthesis. We have observed in the XRD profile that TiH_2 completely transformed to Ti by dehydrogenation at 500°C, and NiO completely transformed to Ni by H_2 reduction at 700°C. After sintering above 1200°C, only the constituent intermetallic compounds of Ni_3Ti, Ni_2Ti, Ni_4Ti_3 and NiTi among Ni-Ti binary system as well as tetragonal zirconia were observed, as shown in Fig.1. Equiatomic NiTi phase was, however, minor phase even though the nominal composition of prepared powder was adjusted to form Ni:Ti atomic ratio as 50:50. It was reported that the titanium tend to solid solute into ZrO_2 phase when ZrO_2/Ti mixture was sintered.[8] In this research, clear evidence showing the solid solution of Ti into 3Y-TZP was not observed, however, it was suspected that such a reaction might caused the actual reduction of Ti fraction, which would result in the formation of various Ni-rich intermetallic compounds. Hereafter, the composite is abbreviated as 3Y-TZP/Ni-Ti due to the multi-phase dispersoid among the binary system. During this process, neither reaction phases such as TiO_x nor reduction of Y-TZP was observed.

By comparing the fracture strength of 3Y-TZP/30vol%Ni-Ti composites hot-pressed at different sintering temperatures, the strength of specimen sintered at 1200°C was much higher (around 680 MPa) than that sintered at 1300 to 1600°C (around 200 to 350 MPa). According to the binary phase diagram of Ni-Ti system, liquid phase is expected to form above 1300°C for the composition of Ni:Ti=50:50.[9] In usual, a liquidus sintering is one of the most useful method to fabricate dense sintered bodies which indicates enhanced mechanical properties. In the present case, however, liquidus sintering (i.e., higher temperature sintering above 1300°C) in 3Y-TZP/Ni-Ti system do not have good influence on their mechanical properties.

Figure 2 shows typical SEM morphologies of fractured surface of the 3Y-TZP/30vol%Ni-Ti composites hot-pressed at various temperatures. Ni-Ti intermetallic particles were observed and they were dispersed with spherical shape among 3Y-TZP matrix. It should be noted that the 3Y-TZP/30vol%Ni-Ti composite sintered at 1200°C exhibited fairly high fracture strength of around 680MPa than that of 3Y-TZP monolith (300MPa) hot-pressed at the same temperature.[6] SEM investigation (Fig.2(a)) showed that the microstructure of this sample was fine and dense. These facts indicate that the dispersed Ni-Ti compounds seem to accelerate sintering of zirconia, and resulted in the enhancement of the fracture strength. In the case of the composites sintered at higher temperature (1300 ~ 1600°C), grain sizes of both t-ZrO_2 matrix and dispersed Ti-Ni particles were much larger than those sintered at 1200°C. This grain growth of both t-ZrO_2 and dispersed Ni-Ti inatermetallics are supposed to be the reason behind the decrease in the fracture strength down to 200~350 MPa. Hence, it is concluded that hot-pressing at 1200°C was the appropriate sintering temperature to fabricate 3Y-TZP/Ni-Ti composite system.

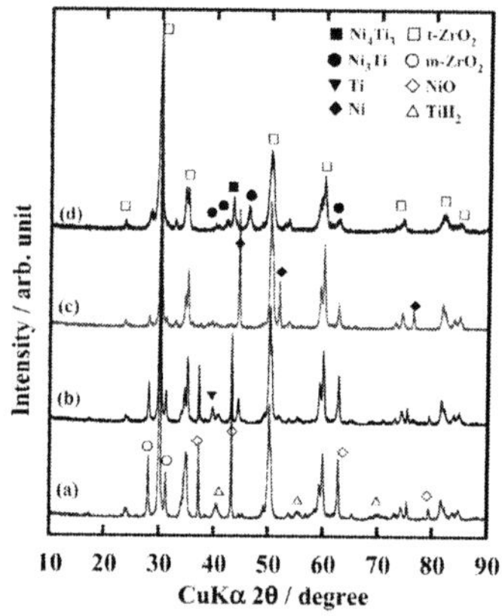

Fig.1 XRD profiles of (a)3Y-TZP/TiH_2/NiO mixture,(b) the powder after heating at 500°C, for 1h in vacuum, (c) after hydrogen reduction at 700°C for 1h, and (d) after sintering at 1200°C for Y-TZP/30vol%Ni-Ti composite.

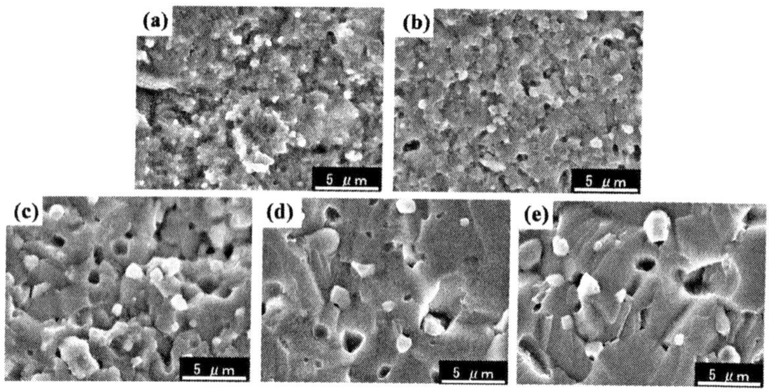

Fig.2 Typical SEM morphologies of fractured surfaces for the 3Y-TZP/ 30vol%TiNi composites hot-pressed at (a)1200, (b)1300, (c)1400, (d)1500, and (e)1600°C.

b) 3Y-TZP/CoTi and 3Y-TZP/CoTi/CNT composites:

In the case of 3Y-TZP/Co-Ti systems, heating process was slightly changed from that of the 3Y-TZP/Ti-Ni system as is mentioned in the experimental section. As a result shown in Fig.3, complete dehydrogenation of TiH_2 as well as reduction of CoO could be accomplished by the heating at 700°C in H_2 atmosphere. Sintered body was found to contain Co_3Ti and CoTi phases that indicated the formed intermetallic phase was lesser than the case for 3Y-TZP/Ni-Ti system. It is probably due to the stability of CoTi compound than that of NiTi phase.

On the other hand, as shown in Fig.4, XRD analysis of the 3Y-TZP/0~40vol%CoTi/0.75vol%CNT composites revealed that the constituent intermetallic phase was Co_2Ti, and un-reacted Co and Ti were identified. It implied that CNT had an influence on the formation behavior of CoTi intermetallics during sintering.

In the case of CoTi/CNT composite without 3Y-TZP was prepared to know the phase development behavior, where a small amount of Ti_2O_3 and CoO was found as seen in Fig.4(f). It was considered that Ti_2O_3 formed by oxidation of Ti metal with CoO after decomposition of TiH_2 at around 500°C. In addition, Titanium has high affinity with oxygen and hence its oxide is thermodynamically stable than the CoO, which might result in remaining of Ti_2O_3 after hydrogen reduction. On the contrary, CoO was thought to be the residual oxide that could not be completely reduced probably due to the existence of oxide agglomerates and resultant inhibition of reduction by the rate controlling of gas diffusion within the oxide agglomerates. This kind of phenomenon was pointed out for the reduction of agglomerates of nanocrystalline NiO.[10]

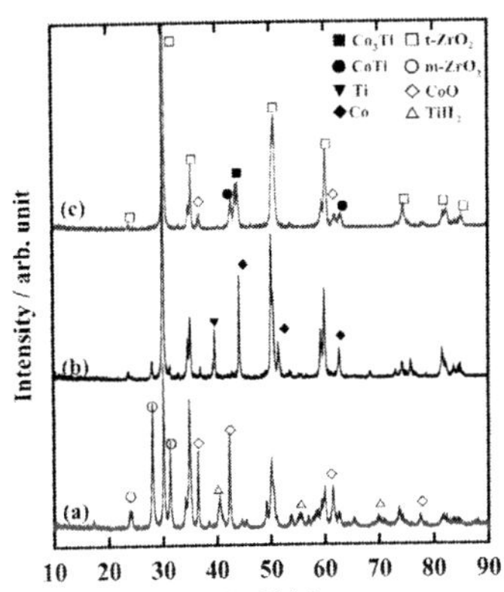

Fig.3 XRD profiles of (a) 3Y-TZP/CoO/TiH_2 mixture, (b) reduced t-ZrO_2/Co/Ti composite powder at 700°C, and (c) hot-pressed 3Y-TZP/ 40vol%Co-Ti composite.

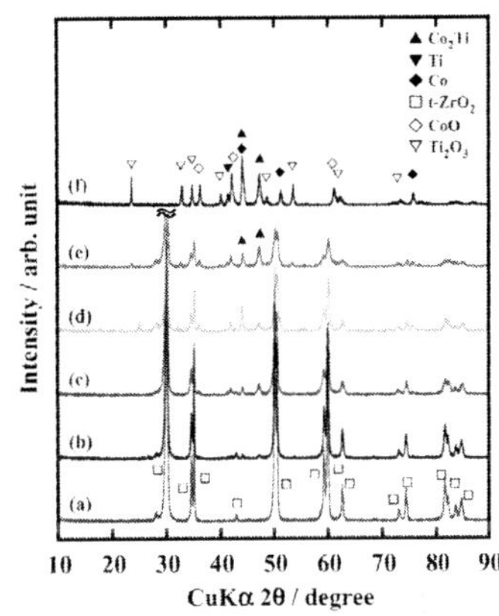

Fig.4 XRD profiles of Y-TZP/Co-Ti/CNT composites sintered at 1200°C for 1h. (a) 3Y-TZP/CNT, (b) 5vol%CoTi, (c) 15vol%CoTi, (d)30vol%CoTi, and (e) 40vol%CoTi containing 3Y-TZP/CNT, and (f) CoTi/CNT. All the composites contain 0.75vol% of CNT.

c) Mechanical properties of 3Y-TZP/Co-Ti and 3Y-TZP/Co-Ti/CNT composites:

Figure 5 shows variation of fracture strength of 3Y-TZP/Co-Ti composites and 3Y-TZP/Co-Ti/0.75vol%CNT composites as a function of CoTi content. In spite of relatively low sintering temperature of 1200°C, 3Y-TZP/Co-Ti revealed fairly high strength of over 500MPa, which value is higher than the 3Y-TZP monolith (~300MPa) sintered at the same temperature (1200°C). Maximum strength of around 800MPa was achieved for the 3Y-TZP/5vol%CoTi sample. It indicates that the 3Y-TZP/Co-Ti intermetallic composite system as well as present fabrication method is advantageous to realize high strength composite materials which can be prepared at lower temperature.

Fracture strength of the composites containing CNT was not improved much as seen in Fig.5 except the composite of 3Y-TZP/15vol%Co-Ti/0.75vol%CNT, which was approximately 200MPa higher than the 3Y-TZP/15vol%CoTi (without CNT). However large degradation of strength for the sample containing 40vol%CoTi compared with that of composite without CNT was confirmed. The CNT content for all the composites is 0.75vol% and this seems to be too small amount to effect on the degradation of fracture strength. As is shown in Fig.4, the composites exhibit a residual Ti and Co. These residual elements may be one of the reasons of degradation of fracture strength. The other reason may be the microstructure of this composite (Fig.6). As is seen from this micrograph, some large shapeless matter was often observed at the fractured surface together with added CNTs (not shown here). Detailed formation mechanism of this obstacle have not been clarified yet, however, it is thought to be formed due to the addition of CNT or converted from organic surfactant which has been used to disperse the CNT. Detailed analysis is under investigation, and will be reported near future.

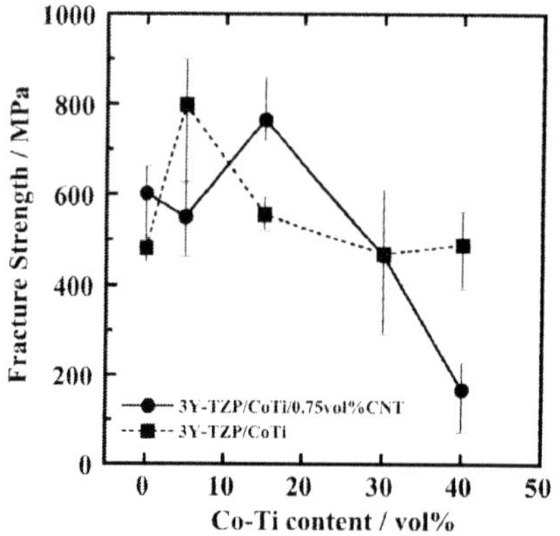

Fig.5 Fracture strength of 3Y-TZP/Co-Ti/0.75 vol%CNT composites.

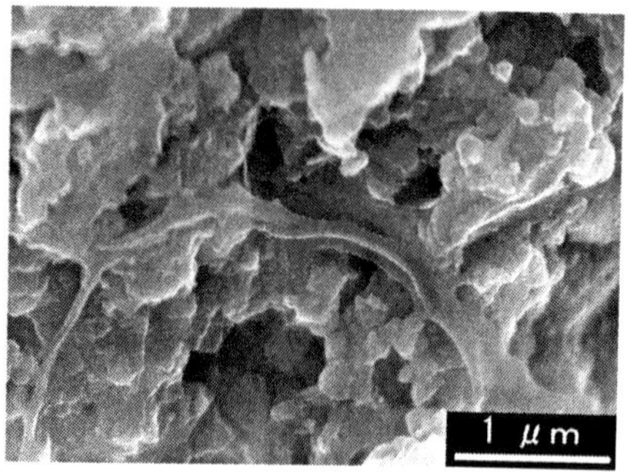

Fig.6 Fracture surfaces of hot-pressed 3Y-TZP/40 vol%Co-Ti/0.75vol%CNT composites.

CONCLUSIONS

Ti-based intermetallic compounds dispersed 3Y-TZP composites have been successfully fabricated by reductive sintering process including dehydrogenation, reduction, reaction and hot-pressing sequences. 3Y-TZP/NiTi composites consisted of various Ni-rich binary compounds among Ti-Ni system. The fracture strength of 3Y-TZP/30vol%Ni-Ti composite sintered at 1200°C was higher than the monolithic 3Y-TZP sintered at the same condition, while it decreased with increasing sintering temperature.

On the other hand, composite of 3Y-TZP and Co-Ti system contained only 2 phases of intermetallic compounds, and the maximum strength of 800MPa was successfully achieved for the 3Y-TZP/5vol%CoTi composite. From this research, it is considered that multi-step heating that allowed in-situ reaction and sintering has some advantage on the fabrication of Ti-based intermetallic compound dispersed ceramic composites.

Unlike expectation, the addition of carbon nanotubes could not enhance mechanical properties of the 3Y-TZP/Co-Ti composite except some case because of the inhibition of formation of intermetallic phase and resultant residuals of fairly large monolithic metal particles such as Co and Ti, and also because of formation of large shapeless matter which might correspond to the fracture origin.

REFERENCES

[1]H. Kondo, T. Sekino, N. Tanaka, T. Nakayama, T. Kusunose, and K. Niihara, "Mechanical and Magnetic Properties of Novel Yttria-Stabilized Tetragonal Zirconia/Ni Nanocomposite Prepared by the Modified Internal Reduction Method", *J. Am. Ceram. Soc.*, **88**[6], 1468-1473 (2005).

[2]H. Kondo, T. Sekino, Y.-H. Choa, T. Kusunose, T. Nakayama, M. Wada, T. Adachi, and K. Niihara, "Mechanical and Magnetic Properties of nickel Dispersed Tetragonal Zirconia Nanocomposites, " *J. Nanosci. Nanotechnol.*, **2**, 485-90 (2002).

[3]S. Skirl and R. Krause, " Processing and Mechanical Properties of Al_2O_3/Ni_3Al Composites with Interpenetrating Network Microstructure, "*J. Am. Ceram. Soc.*, **84**[9], 2034-40 (2001).

[4]Jia Li, Yan-Sheng Yin, Xun-Yan Tan, and Jin-De Zhang, "A ZrO_2(3Y) Matrix Composite Toughened with Fe_3Al Intermetallic, "*J. Am. Ceram. Soc.*, **88**[1], 235-238 (2005).

[5]G.F. Bastin and D. Rieck, "Diffusion in the titanium-nickel system: 1. Occurrence and growth of the various intermetallic compounds," *Metall. Trans.*, **5**, 181-1827(1974).

[6]N. Tanaka, T. Sekino, T. Nakayama, Y. Hayashi, T. Kusunose and K. Niihara, "Mechanical and Electrical Properties of Y-TZP Containing TiNi-base Intermetallic Compounds, " *Materials Science Forum*, **486-487**, 366-369 (2005).

[7]R.W. Siegel, S. K. Chang, B. J. Ash, J. Stone, P. M. Ajayan, R. W. Doremus, "Mechanical behavior of polymer and ceramic matrix nanocomposites," *Scripta Mater*., **44**, 2061-2064 (2001).

[8]L. D. Teng, F. M. Wang, W. C. Li, "Thermodynamics and microstructure of Ti-ZrO2 metal-ceramic functionally graded materials," *Mater. Sci. & Eng.*, **A293**, 130-136(2000).

[9]J. L. Murray, in *Phase Diagrams of Binary Nickel Alloys*, (Ed. by P. Nash, ASM, Materials Park, Ohio, USA), 1991, p.342.

[10]B. S. Kim, J. S. Lee, T. Sekino, Y.-H. Choa , K. Niihara, "Hydrogen reduction behavior of NiO dispersoid during processing of Al_2O_3/Ni nanocomposites," *Scripta Mater.*, **44**, 2121-2125(2001).

IN-SITU SYNTHESIS OF OXIDE/OXIDE COMPOSITES

Zhong-Chun Chen, Toshihiko Okazawa, and Keisuke Ikeda
Department of Materials Science and Engineering, School of Engineering, Tohoku University
Aoba-yama 6-6-02, Aoba-ku
Sendai 980-8579, Japan

ABSTRACT

Alumina-based composites reinforced with barium-β-alumina and zirconia have been synthesized via reactive sintering of alumina and barium zirconate binary powder mixtures. The synthesis and densification were performed by conventional pressureless sintering, hot pressing, and spark plasma sintering (SPS) techniques. The in-situ synthesized composites exhibited a microstructural feature of the elongated barium-β-alumina platelets and equiaxed zirconia particles, which were distributed in alumina matrix homogeneously. Pressureless sintering of alumina and barium zirconate powder mixtures resulted in formation of barium-β-alumina platelets with larger aspect ratios, compared to hot pressing and SPS techniques. SPS is an effective means for synthesis and simultaneous densification of the alumina-based in-situ composites under a relatively low temperature and short holding time. However, the orientational growth of barium-β-alumina phase is not well developed due to a lower temperature and shorter time used in the SPS process. The heat treatment of densified composites at a higher temperature can promote the preferential diffusion and growth, thus improving the aspect ratios of barium-β-alumina platelets.

INTRODUCTION

All-oxide ceramic matrix composites (CMCs) consisting of oxide matrix and oxide reinforcements are attracting much attention for high temperature structural applications (such as gas turbine and aerospace engines, heat exchangers, and hot gas filters), because of their high strength and inherent stability in oxidizing atmospheres at high temperatures.[1] In recent years, considerable efforts have been made to highlight oxide fiber-reinforced oxide matrix composites.[2-8] In these oxide/oxide composites, a weak fiber/matrix interfacial layer, i.e., an interphase between fibers and surrounding matrix is required to allow fiber/matrix interfacial debonding and subsequent fiber pullout during fracture.[9] The incorporation of reinforcements with large aspect ratios (such as fibers, whiskers, and platelets) may cause some difficulties in achieving full density, due to the constraint of network formed by the reinforcements. Accordingly, pressure-assisted sintering, typically, hot-pressing technique is often used to obtain fully dense composites. Moreover, applications of these reinforcements lead to high cost and difficulties in handling and in distributing them into a matrix homogeneously.

The above-mentioned problems may be solved by in-situ synthesis, which is an attractive and cost-effective processing means to obtain ceramic matrix composites with high performances. In the last decade, in-situ toughening approach has been used to enhance the mechanical properties of ceramic materials. Second phases or grains with elongated morphology (platelet, rodlike or needlelike) are formed in situ during materials processing, thus resulting in an improvement in fracture toughness. For example, aluminate compounds, called as hexaluminates (e.g., $LaAl_{11}O_{18}$, $SrAl_{12}O_{19}$, and $CaAl_{12}O_{19}$), were incorporated into an alumina matrix to form alumina-based oxide/oxide composites.[10-14] These hexaluminates have an elongated morphology with β-alumina, magnetoplumbite (MP) or related layered structure,[15] hence leading to microstructural toughening.

Recently, the authors have examined the interfacial reactions between alumina fibers and barium zirconate ($BaZrO_3$) coatings during heat treatment or hot pressing.[16,17] Besides, an attempt has been made to synthesize an alumina matrix composite by reactive sintering of Al_2O_3 and $BaZrO_3$ powder mixtures.[18] It has been demonstrated that the presence of an elongated Ba-β-Al_2O_3 phase, formed through in-situ reactions in Al_2O_3 matrix, caused crack deflection and crack bridging and thus improvement in mechanical properties. The Al_2O_3-based composite containing 30% Ba-β-Al_2O_3 and ZrO_2 showed an optimum combination of mechanical properties: high flexural strength, large work of fracture, and improved fracture toughness.[18]

In the current work, all-oxide composites consisting of Ba-β-Al_2O_3 and ZrO_2 (reinforcing phases), and Al_2O_3 (matrix) have been synthesized by conventional pressureless sintering, hot pressing, and spark plasma sintering (SPS) techniques. SPS is an innovative technique and has been successfully used to rapidly sinter or synthesize a large number of materials including metals, polymers, ceramics, and various composites. The objective of the present investigation was to examine the effects of powder particle size and reactive-sintering conditions on densification behavior, formation of reinforcements, and microstructural development during consolidation.

EXPERIMENTAL PROCEDURE

Two types of commercially available α-Al_2O_3 powders with average particle sizes of 0.16μm (A-1) and 8.8μm (A-2), and $BaZrO_3$ powder with an average particle size of 1.5μm were used as the starting materials. The Al_2O_3 and $BaZrO_3$ powders were ball milled with high-purity alumina balls in ethanol for 24 h and then dried. In this work, Al_2O_3 matrix composites were designed to possess nominal compositions with volume fractions (denoted as V_r) of reinforcements (Ba-β-Al_2O_3 and ZrO_2) of 0-0.4 after complete reactive sintering.

Three different densification methods: (a) pressureless sintering (PLS), (b) uniaxial hot pressing (HP), and (c) spark plasma sintering (SPS), were used in this experiments. For pressureless sintering, the powder mixtures were uniaxially pressed at 100MPa, followed by cold-isostatic pressing (CIP) at 300 MPa. The green compacts were sintered in air at 1500-1600°C for 2 h. The hot pressing was carried out under vacuum at 1400°C for 1 h at a pressure of 35 MPa. With regard to spark plasma sintering, the powders were consolidated in a graphite die by passing a large pulsed current through powders, die, and punches under applied pressure to achieve rapid sintering. The densification was performed in vacuum in a temperature range of 1200-1400°C for 10 min at a pressure of 50 MPa. The heating rates for PLS, HP, and SPS were 10°C/min, 15°C/min, and 100°C/min, respectively. Moreover, some sintered specimens by SPS technique were heat-treated in air at 1500°C for 5 h.

The bulk density of reactive-sintered specimens was determined by the Archimedes' method. The crystalline phases present in the mixed powder and sintered samples were identified by X-ray diffraction (XRD) with CuK_α radiation. Microstructural characterization was performed by scanning electron microscopy (SEM), and energy dispersive X-ray spectroscopy (EDS) was used for compositional analysis.

RESULTS AND DISCUSSION

Densification behavior

Fig.1 shows the relative density plotted as a function of volume fraction of the reinforcements (V_r) for composites fabricated by pressureless sintering and hot pressing. The hot-pressed composites exhibited a high density level close to the theoretical densities (TD), regardless of composition of the composites. In the case of pressureless sintering, however, the relative density decreased as V_r increased. It was noted that the density reduction with V_r was much smaller when finer Al_2O_3 powder (A-1) was used, in comparison with coarser Al_2O_3 powder (A-2). Furthermore, the composites prepared from A-1 powder (0.16μm) showed higher densities (more than 97%TD) at ≥1500°C, whereas those from A-2 powder (8.8μm) had very lower density values, showing that coarse A-2 powder has poor densification behavior. This reveals that the particle size has a significant effect on densification behavior of pressureless sintered composites. When an Al_2O_3 powder with a smaller average particle size is used as one of the starting powders, on the one hand, the densification of Al_2O_3 matrix easily occurs due to a large specific surface area (hence large driving force for sintering). On the other hand, a smaller particle size gives rise to a decrease in diffusion distance, accelerating the chemical reactions between Al_2O_3 and $BaZrO_3$ powders. Moreover, the composites, fabricated from A-1 powder at 1500°C and 1600°C, showed almost the same density values at a given composition. On the contrary, in the case of A-2 powder, the composites sintered at 1500°C had ~10% lower density values than those sintered at 1600°C.

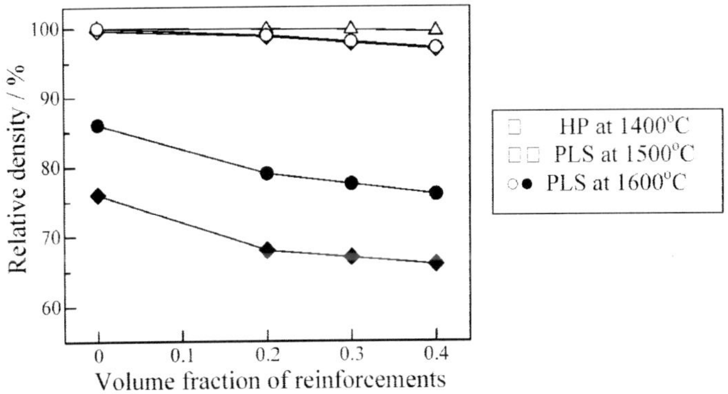

Fig.1 Relative density of the composites consolidated by pressureless sintering (PLS) and hot pressing (HP) as a function of volume fraction of reinforcements. Open and filled symbols stand for A-1/$BaZrO_3$ and A-2/$BaZrO_3$ powder systems, respectively.

The dependence of relative density of the composites (V_r = 0.3) consolidated via SPS technique on sintering temperature is shown in Fig.2. The density increased with increasing sintering temperature, and finer A-1 powder exhibited a higher density at a given sintering temperature than A-2. Such densification behavior is similar to that in conventional sintering process. However, the densification temperature was shifted to a lower one under SPS conditions, compared to pressureless sintering. For example, nearly fully dense composite could be obtained at 1250°C for A-1 powder, and 99.4%TD was attained at 1350°C even for coarse A-2 powder. Besides, the densification could be completed in minutes. These unique features result from applications of a mechanical pressure, a rapid heating rate, and pulsed direct current.[19] The above results indicate that SPS technique is also effective for synthesis and simultaneous densification of composites including in-situ chemical reactions between powder components.

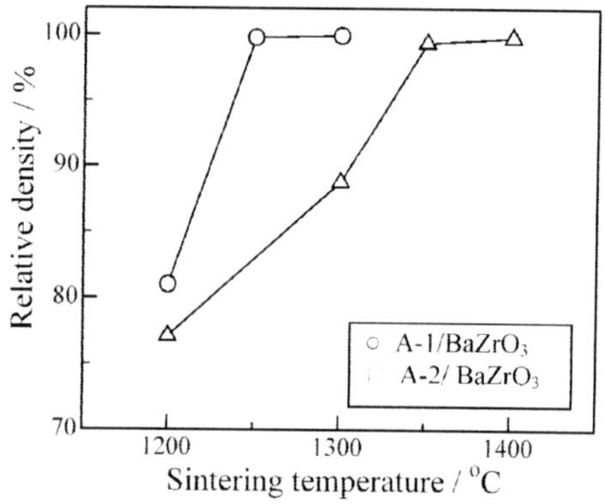

Fig.2 Dependence of relative density of the composites (V_r=0.3) reactive-sintered via SPS technique on sintering temperature.

X-ray diffraction analysis

The XRD patterns for mixed powder and sintered specimens (V_r=0.3) by SPS at different temperatures are shown in Fig.3. For A-1/$BaZrO_3$ powder mixture, the pattern showed well-defined peaks of α-Al_2O_3 and $BaZrO_3$, as expected (Fig.3(a)). After sintering at 1200°C, the intensity of $BaZrO_3$ peaks was greatly reduced, while some small peaks of ZrO_2 (monoclinic) and barium monoaluminate $BaAl_2O_4$ ($BaO{\cdot}Al_2O_3$) phases appeared. At 1250°C, almost no trace of $BaO{\cdot}Al_2O_3$ could be found from the XRD pattern, with the dominant phases being a barium hexaluminate $BaAl_{14.66}O_{23}$ (i.e., $BaO{\cdot}7.3Al_2O_3$) and ZrO_2. Moreover, some small peaks corresponding to $BaZrO_3$ were still detected, indicating the presence of remnant $BaZrO_3$. When sintering temperature was raised to 1300°C, no significant changes were observed from the pattern (Fig.3(d)) except for relative intensity of peaks, and the XRD spectrum presented a feature similar to that at 1250°C.

The results shown in Fig.3 reveal that the chemical reactions between $BaZrO_3$ and Al_2O_3 powders do occur during sintering process, concurrent with occurrence of sintering shrinkage. The predominant reaction equation can be written in the following general form

$$BaZrO_3 + x\,Al_2O_3 \rightarrow BaO \cdot x\,Al_2O_3 + ZrO_2 \qquad (1)$$

In this work, since Al_2O_3 is designed as a matrix of the composites, there exists excessive Al_2O_3 in Al_2O_3-$BaZrO_3$ system with respect to stoichiometric compositions (Eq.(1)). The x value in Eq.(1) depends mainly on sintering temperature. When reactive sintering was conducted at 1200°C, x=1, that is, $BaO \cdot Al_2O_3$ was formed. At higher temperatures more than 1250°C, however, the barium hexaluminate was present in the form of $BaO \cdot 7.3Al_2O_3$ (i.e., x=7.3 in Eq.(1) or $BaAl_{14.66}O_{23}$ or $Ba_{0.75}Al_{11}O_{17.25}$), one of the nonstoichiometric hexaluminates or Ba-β-Al_2O_3. These results, obtained in the composites densified by SPS, are in good agreement with our previous investigations on interfacial reactions between $BaZrO_3$ coatings and Al_2O_3 fibers.[16]

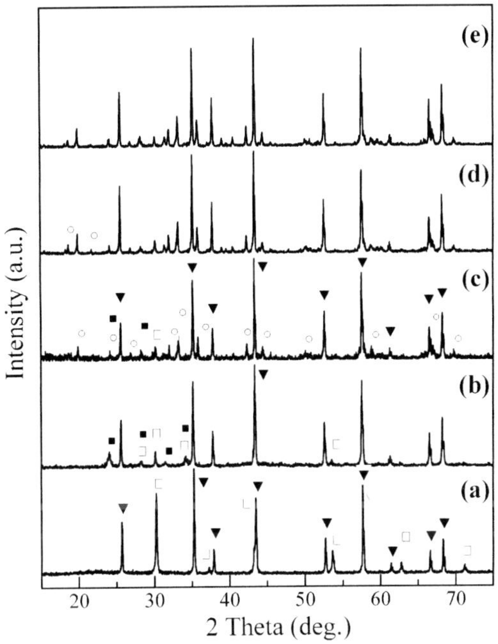

Fig.3 XRD patterns of (a) A-1/$BaZrO_3$ powder mixture, (b-d) sintered, and (e) sintered (1200°C) and then heat-treated (1500°C) specimens (V_r=0.3). Reactive sintering was conducted via SPS technique at (b) 1200°C, (c) 1250°C, and (d) 1300°C, respectively. (□) $BaZrO_3$, (▼) α-Al_2O_3, (□) $BaO \cdot Al_2O_3$, (■) monoclinic ZrO_2, and (○) $BaO \cdot 7.3Al_2O_3$.

Note that the Ba-β-Al_2O_3 phase has a composition close to $BaAl_{14.66}O_{23}$ from the XRD analysis in this work. It should be pointed out that Ba-β-Al_2O_3 includes a number of nonstoichiometric compounds with a wide Al/Ba ratio in the range of 9.1-14.6.[20] It has been well established that stoichiometric barium hexaluminate ($BaAl_{12}O_{19}$ or $BaO \cdot 6Al_2O_3$) is actually constituted by two nonstoichiometric phases, referred to as phase I and phase II, with compositions $BaAl_{14.66}O_{23}$ and $BaAl_{9.15}O_{14.73}$, respectively.[21,22] On the other hand, according to the phase diagram of Al_2O_3-$BaAl_2O_4$ system,[21] only phase I exists together with Al_2O_3, provided the molar fraction of BaO is lower, for instance, <10%. In the current work, a small amount of $BaZrO_3$ powder reacts with excessive Al_2O_3 (Al-rich) to form monophasic $BaO \cdot 7.3Al_2O_3$ in the Al_2O_3 matrix. This is consistent with the results reported by Kimura et al.[21]

To examine the effect of heat treatment on in-situ reactions and phase evolution, some specimens, consolidated by SPS, were subjected to further heat treatment at a higher temperature in air. As an example, Fig.3(e) shows the XRD pattern of a sintered specimen at 1200°C by SPS, followed by heat treatment at 1500°C. The pattern showed a feature extremely similar to that sintered at 1300°C (Fig.3(d)). This suggests that the chemical reactions between $BaZrO_3$ and Al_2O_3 as well as between $BaAl_2O_4$ and Al_2O_3 occur during the heat treatment at a higher temperature, even though the reaction is not complete in a sintered specimen due to a lower sintering temperature. Furthermore, this result implies that barium monoaluminate $BaAl_2O_4$ is a transient phase, which can continue to react with Al_2O_3 at higher temperatures to form more stable Al-rich $BaAl_{14.66}O_{23}$ phase.

Fig.4 illustrates the XRD patterns of the composites prepared by pressureless sintering in air and hot pressing in vacuum, respectively. All the patterns were similar to each other. It should be noted that the relative intensity of $BaZrO_3$ seemed to be higher for hot-pressed composite (Fig.4(c)) than those for pressureless-sintered ones (Figs.4(a) and (b)). The reason for this is not very clear, further investigations are necessary. In addition, the results shown in Figs.3 and 4 were based on A-1/$BaZrO_3$ mixed powder. When A-2 powder was used, the resulting composites have been confirmed to have similar phase evolution characteristics.

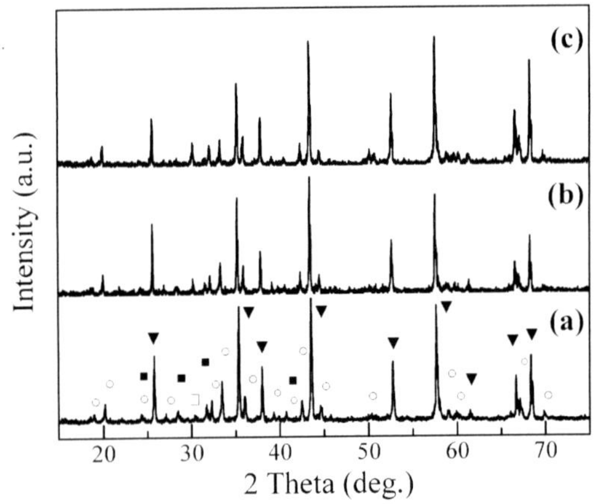

Fig.4 XRD patterns of sintered composites by pressureless sintering at (a) 1500°C and (b) 1600°C, and (c) by hot pressing at 1400°C. (□) $BaZrO_3$, (▼) α-Al_2O_3, (■) monoclinic ZrO_2, and (○) $BaO \cdot 7.3Al_2O_3$.

Microstructures

The microstructures of the composites consolidated from an A-1/$BaZrO_3$ powder mixture (V_r = 0.3) by different sintering techniques are shown in Fig.5. In these micrographs, there exist

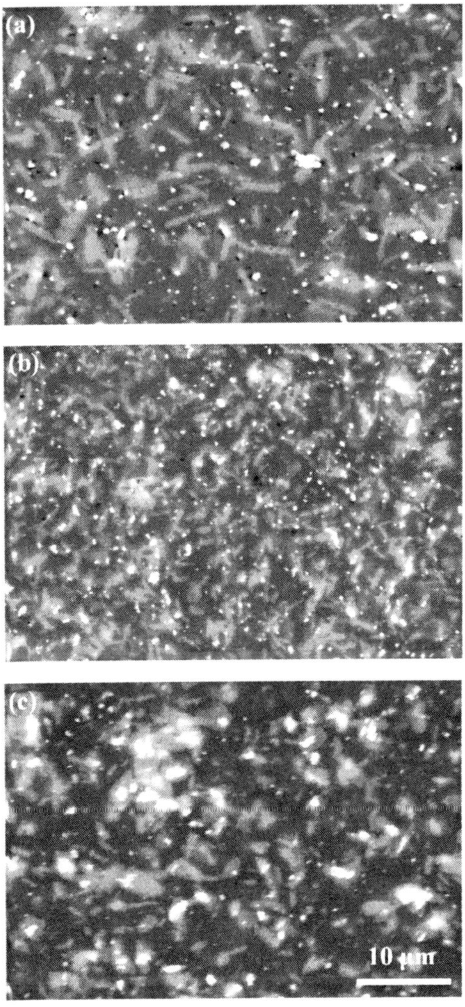

Fig.5 SEM micrographs (backscattered-electron images) showing the microstructures of composites (V_r=0.3) consolidated by (a) pressureless sintering at 1500°C, (b) hot pressing at 1400°C, and (c) SPS at 1300°C.

white, light gray, and dark gray regions. EDS analyses have demonstrated that the light gray regions contained Al, Ba, and O, revealing to be Ba-β-Al_2O_3, in conjunction with the XRD results. On the other hand, the EDS spectra for the white regions (particles) consisted of O and intense Zr peaks. In addition to O and Zr, our previous work[18] has confirmed the presence of Ba in some white particles, in particular, in the central regions of those particles with large particle sizes. This indicates that the white regions in Fig.5 primarily correspond to ZrO_2 and, possibly, with a trace amount of $BaZrO_3$, which remains in the specimen after sintering. This has also been verified by XRD analysis, which clearly showed the existence of $BaZrO_3$ phase in the sintered composites (Figs.3 and 4). Besides, the dark gray regions in the microstructures shown in Fig.5 were associated with Al_2O_3, and the dark points corresponded to pores. As shown in Fig.5, the composites prepared by hot pressing and SPS methods were very dense, whereas some small pores were involved in pressureless-sintered composite. These pores are considered to result from (i) normal sintering pores and (ii) pores due to solid-state diffusion and reactions.

For pressureless-sintered composite (Fig.5(a)), as a result of reactions between $BaZrO_3$ and Al_2O_3 powders during sintering, the microstructure had a feature of rodlike Ba-β-Al_2O_3 (light gray phase) and quasi-spherical or equiaxed ZrO_2 (perhaps containing $BaZrO_3$) particles, which were distributed in Al_2O_3 matrix homogeneously. The Ba-β-Al_2O_3 platelets were measured to have an average aspect ratio of ~3.5. The microstructural feature of elongated morphology of Ba-β-Al_2O_3 phase has been found in other alkaline-earth and rare-earth hexaluminates.[4,23]

The mechanism for formation of Ba-β-Al_2O_3 and ZrO_2 phases during reactive sintering is attributed to diffusion of Ba cations. Due to the diffusion of Ba cations from $BaZrO_3$ to surrounding Al_2O_3, ZrO_2 is formed at those positions where the $BaZrO_3$ particles are originally present. In addition, evaporation of BaO during sintering may lead to the formation of ZrO_2 phase because of a high BaO partial pressure at an elevated temperature.[24] Concurrent with the formation of ZrO_2, elongated Ba-β-Al_2O_3 particles are generated in the Al_2O_3 matrix, which is likely to be due to the oriented growth or an anisotropy in their growth kinetics. Moreover, coalescence of neighboring Ba-β-Al_2O_3 can further promote the development of elongated Ba-β-Al_2O_3 phase.[18]

With regard to the composite fabricated by hot pressing at 1400°C, the microstructure (Fig.5(b)) was much finer, compared to that shown in Fig.5(a). Although Ba-β-Al_2O_3 particles displayed a platelet shape, their length and width became smaller. When Al_2O_3-$BaZrO_3$ binary powder mixture was consolidated by SPS technique (Fig.5(c)), a small amount of Ba-β-Al_2O_3 platelets were observed, and most of Ba-β-Al_2O_3 exhibited an approximately equiaxed shape. In other words, the aspect ratios of Ba-β-Al_2O_3 particles were very small. This might be due to a lower sintering temperature and shorter holding time used in the SPS process. Consequently, the preferred orientational growth of Ba-β-Al_2O_3 phase is not well developed. As shown in Fig.5(c), ZrO_2 and/or unreacted $BaZrO_3$ (white particles) had larger sizes than those shown in Figs.5(a) and (b). A lot of these particles were surrounded by Ba-β-Al_2O_3. This suggests that Ba cations in $BaZrO_3$ are diffused to surrounding Al_2O_3 at nearly the same speed in different directions to form Ba-β-Al_2O_3 without a preferred orientation.

When a densified specimen by SPS is heat-treated at a higher temperature, the microstructure, especially the morphology of Ba-β-Al_2O_3 phase may be changed. As an example, Fig.6 shows the microstructure of the composite consolidated from A-2/$BaZrO_3$ mixture by SPS at 1350°C and subsequently heat-treated in air at 1500°C. A number of Ba-β-Al_2O_3 platelets with an elongated morphology were distributed in the Al_2O_3 matrix, together with ZrO_2 particles. The formation of Ba-β-Al_2O_3 platelets might be ascribed to the following two reasons. (i) Some Ba-β-Al_2O_3 particles with nearly equiaxed shape in the sintered specimen are transformed to platelets with large aspect ratios, due to diffusion of cations or atoms along preferred directions during the heat treatment. (ii) Some unreacted $BaZrO_3$ reacts with Al_2O_3 to form Ba-β-Al_2O_3 platelets. As described previously, due to a lower sintering temperature and shorter holding time, there exists unreacted $BaZrO_3$ in the densified specimen by SPS. These $BaZrO_3$ can continue to react with Al_2O_3 to form Ba-β-Al_2O_3. It can be found from Fig.6 that some pores were generated as a result of solid-state diffusion and reactions.

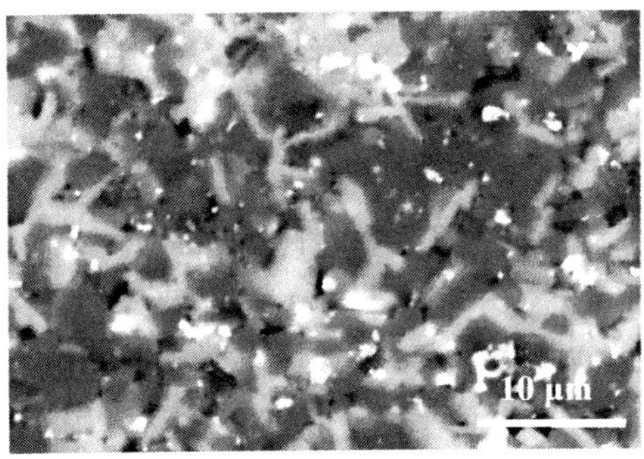

Fig.6 Microstructure of the composite (V_r=0.3) consolidated by SPS at 1350°C, followed by heat treatment at 1500°C.

CONCLUSIONS

All-oxide composites consisting of Al_2O_3 (matrix), Ba-β-Al_2O_3 and ZrO_2 (reinforcing phases) have been synthesized via reactive sintering of Al_2O_3 and $BaZrO_3$ powder mixtures. The effects of powder particle size and reactive-sintering conditions on densification behavior, formation of reinforcements, and microstructural development during consolidation have been investigated. The in-situ synthesized composites showed a microstructural feature of elongated Ba-β-Al_2O_3 platelets and equiaxed ZrO_2 particles, which were distributed in Al_2O_3 matrix homogeneously. Pressureless sintering of Al_2O_3 and $BaZrO_3$ powder mixtures resulted in formation of Ba-β-Al_2O_3 platelets with larger aspect ratios, compared to hot pressing and SPS techniques. It has been demonstrated that powder particle size has a significant effect on densification behavior of composites during reactive sintering. To obtain denser composites, it is important to use fine powders under the conditions of pressureless sintering. SPS technique is an effective means for synthesis and simultaneous densification of the Al_2O_3-based in-situ composites under a relatively low temperature and short holding time. However, the orientational growth of Ba-β-Al_2O_3 phase is not well developed due to a lower temperature and shorter time used in the SPS process. The heat treatment of densified composites at a higher temperature can promote the preferential diffusion and growth, thus improving the aspect ratios of Ba-β-Al_2O_3 platelets.

REFERENCES

[1]K. K. Chawla, C. Coffin, and Z. R. Xu, "Interface Engineering in Oxide Fibre/Oxide Matrix Composites," *Inter. Mater. Rev.*, **45**, 165-89 (2000).

[2]P. E. D. Morgan and D. B. Marshall, "Functional Interfaces for Oxide/Oxide Composites," *Mater. Sci. Eng. A*, **162**, 15-25 (1993).

[3]M. K. Cinibulk, "Magnetoplumbite Composites as a Fiber Coating in Oxide/Oxide Composites," *Ceram. Eng. Sci. Proc.*, **15**, 721-28 (1994).

[4]M. K. Cinibulk and R. S. Hay, "Textured Magnetoplumbite Fiber-Matrix Interphase Derived from Sol-Gel Fiber Coatings," *J. Am. Ceram. Soc.*, **79**, 1233-46 (1996).

[5]D.-H. Kuo, W. M. Kriven, and T. J. Mackin, "Control of Interfacial Properties through Fiber Coatings: Monazite Coatings in Oxide-Oxide Composites," *J. Am. Ceram. Soc.*, **80**, 2987-96 (1997).

[6]T. A. Parthasarathy, E. Boakye, M. K. Cinibulk, and M. D. Petry, "Fabrication and Testing of Oxide/Oxide Microcomposites with Monazite and Hibonite as Interlayers," *J. Am. Ceram. Soc.*, **82**, 3575-83 (1999).

[7]K. K. Chawla, H. Liu, J. Janczak-Rausch, and S. Sambasivan, "Microstructure and Properties of Monazite ($LaPO_4$) Coated Saphikon Fiber/Alumina Matrix Composites," *J. Euro. Ceram. Soc.*, **20**, 551-59 (2000).

[8]M. H. Lewis, A. Tye, E. G. Butler, and P. A. Doleman, "Oxide CMCs: Interphase Synthesis and Novel Fibre Development," *J. Euro. Ceram. Soc.*, **20**, 639-44 (2000).

[9]A. G. Evans and F. W. Zok, "The Physics and Mechanics of Fiber-Reinforced Brittle-Matrix Composites," *J. Mater. Sci.*, **29**, 3857-96 (1994).

[10]K. Tsukuma and T. Takahata, "Mechnical Property and Microstructure of TZP and TZP/Al_2O_3 Composites," *Mater. Res. Soc. Symp. Proc.*, **78**, 123-35 (1987).

[11]M. Yasuoka, K. Hirao, M. E. Brito, and S. Kanzaki, "High-Strength and High-Fracture-Toughness Ceramics in the Al_2O_3/$LaAl_{11}O_{18}$," *J. Am. Ceram. Soc.*, **78**, 1853-56 (1995).

[12]R. A. Cutler, R. J. Mayhew, K. M. Prettyman, and A. V. Virkar, "High-Toughness Ce-TZP/Al_2O_3 Ceramics with Improved Hardness and Strength," *J. Am. Ceram. Soc.*, **74**, 179-86 (1991).

[13]P.-L. Chen and I-W. Chen, "In-Situ Alumina/Aluminate Platelet Composites," *J. Am. Ceram. Soc.*, **75**, 2610-12 (1992).

[14]L. An and H.M. Chan, "R-Curve Behavior of In-Situ-Toughened Al_2O_3:$CaAl_{12}O_{19}$ Ceramic Composites," *J. Am. Ceram. Soc.*, **79**, 3142-48 (1996).

[15]N. Iyi, S. Takekawa, and S. Kimura, "Crystal Chemistry of Hexaaluminates: β-Alumina and Magnetoplumbite Structures," *J. Solid-State Chem.*, **83**, 8-19 (1989).

[16]Z. C. Chen, S. Duncan, K. K. Chawla, M. Koopman, and G. M. Janowski, "Characterization of Interfacial Reaction Products in Alumina Fiber/Barium Zirconate Coating/Alumina Matrix Composite," *Mater.* Character., **48**, 305-14 (2002).

[17]Z. C. Chen, R. Kulkarni, K. K. Chawla, M. Koopman, and K. Ikeda, "Processing and Microstructure of an All-Oxide Ceramic Composite," *Mater. Sci. Forum*, **475-479**, 1301-304 (2005).

[18]Z. C. Chen, K. K. Chawla, and M. Koopman, "Microstructure and Mechanical Properties of In Situ Synthesized Alumina/Ba-β-Alumina/Zirconia Composites," *Mater. Sci. Eng. A*, **367**, 24-32 (2004).

[19]Z. Shen, M. Johnsson, Z. Zhao, and M. Nygren, "Spark Plasma Sintering of Alumina," *J. Am. Ceram. Soc.*, **85**, 1921-27 (2002).

[20]G. Groppi, C. Cristiani, and P. Forzatti, "Phase Composition and Mechanism of Formation of Ba-β-Alumina-Type Systems of Catalytic Combustion Prepared by Precipitation," *J. Mater. Sci.*, **29**, 3441-50 (1994).

[21]S. Kimura, E. Bannai, and I. Shindo, "Phase Relations Relevant to Hexagonal Barium Aluminates," *Mater. Res. Bull.*, **17**, 209-15 (1982).

[22]N. Iyi, S. Takekawa, Y. Bando, and S Kimura, "Electron-Microscopic Study of Barium Hexaluminates," *J. Solid State Chem.*, **47**, 34-40 (1983).

[23]M. G. Cain, R. L.Cain, M.H. Lewis, and J. Gent, "In Situ Reacted Rare-Earth Hexaaluminate Interphases," *J. Am. Ceram. Soc.*, **80**, 1873-76 (1997).

[24]R. Odoj and K. Hilpert, "Evaporation and Standard Enthalpy of Formation of $BaZrO_3$(s)," *Z. Phys. Chem.-Frank.*, **102**, 191-201 (1976).

FABRICATION AND CHARACTERIZATION OF ZIRCONIA-BASED COMPOSITES CONTAINING DISPERSED MONAZITE PARTICLES

A. Hirvonen,[1,*] R. Ramaseshan,[1] T. Kusunose,[1] T. Sekino[1] and K. Niihara[1,2]

[1]The Institute of Scientific and Industrial Research, Osaka University, 8-1, Osaka 567-0047, Japan

[2]National University Corporation, Nagaoka University of Technology Director Extreme Energy Density Research Institute, 1603-1 Kamitomioka, Nagaoka, Niigata 940-2188, Japan

ABSTRACT

The performance and characteristics of ceramic components are considerably influenced by the characteristics of precursor powder used to prepare the materials. The outstanding properties possessed by advanced ceramics are achieved through exceptional composition and microstructure that require careful control throughout the successive stages of the applied processing. The present research aimed originally to obtain and describe the new composite systems in yttrium stabilized tetragonal-zirconia (Y-TZP) combined with monazite (LaPO|*bsub*|4|*esub*|) that possesses very low coefficient of thermal conductivity with maintaining its good mechanical properties. This class of composite might be used for novel high-temperature structural application such as thermal barrier coating and/or mechanical parts. Monazite has been synthesized using wet chemical processing and heat-treatment method. Mixed raw powders have been consolidated by pulse electric current sintering (PECS, also called as SPS) method, which could provide dense materials within a shot sintering time. By using this novel method, reaction between zirconia and monazite was successfully inhibited. Fabrication and sintering characteristics as well as microstructure will be presented in relation to their mechanical and thermal properties.

INTRODUCTION

In recent times, a new family of high-temperature ceramic composites was identified to be $LaPO_4$ composites with Al_2O_3 or ZrO_2. Refractory nature of $LaPO_4$ and compatibility with the above compounds are characteristically pointed towards to structural ceramics. Other benefit such as low fracture energy and stability in oxidizing environments including good machinability because of their weak interfaces makes this kind of combination of ceramics suitable for high-temperature oxide/oxide composites[1-3].

PECS (Pulse Electric Current Sintering) is one of the newest developed sintering processes, which can sinter ceramic powders quickly to its full density at a relatively lower temperature compared with other conventional sintering methods. It enables a powder compact to be sintered within minute's trough rapid heating[4-9].

• Corresponding author.
E-mail address: ari15@sanken.osaka-u.ac.jp
On leave from Materials Science & Engineering, Helsinki University of Technology, Helsinki, Finland

Monazite is a promising candidate for high temperature structural ceramics because of its excellent creep resistance, low thermal expansion and good thermo chemical stability[10-11]. This class of materials is recently projected for a range of high temperature advanced applications and it is often referred to as monazite ceramics. Before Morgan et al. reported that the bulk of monazite/alumina interface was stable in air for 24 h at 1600 °C, which makes the structure very stabile[12]. Lanthanum phosphate functions have a weak interphase in ceramic matrix composites [13-14] and its room temperature mechanical properties are weak compared for example with zirconia or alumina, but a weak interface can change the fracture mode from brittle to non-brittle. Combination which consists of $LaPO_4$-ZrO_2 composite was reported to be thermally stable and machinable[15-17]. Lanthanum phosphate ($LaPO_4$) has been found to be a suitable and effective deboding material for high-temperature oxide/oxide composites[3, 18-19].

Cubic zirconia stabilized with 8mol% yttria (8YSZ) has the advantages like it is stable in oxidizing, reducing environment and it does not degrade readily from chemical reactions at the electrode materials[19-20]. However, the mechanical strength of 8YSZ is poor, compared to tetragonal zirconia stabilized with 3 mol% Y_2O_3 (Y-TZP), which has a higher bending strength at room temperature[21]. However 8YSZ is easier to handle and it has better ionic conductivity compared to 3Y-TZP zirconia[21].

In the present study, monazite type of $LaPO_4$ was synthesized via the wet precipitation and mechanochemical reaction method and used to reinforce the 8YSZ. The guideline for designing this kind of composite was that, in a fully dense, fine-grained, two-phase mixture, easy material removal should occur by formation and linking of cracks at the weak interfaces between two phases like shown earlier by Davis et al.[22]. By using PECS sintering method it is possible to tailor the fine microstructure and thermal conductivity beyond the ranges attainable with other machinable ceramics, properties which are essential and important for the advanced ceramics.

EXPERIMENTAL PROCEDURE

Characterization of $LaPO_4$ powders

In the present work as a first stage was used Lanthanum oxide (La_2O_3) from Ishizu Seiyaku Ltd. (Osaka, Japan), and HCl from Wako Pure Chemical Ind. Inc. (Japan). La_2O_3-powder and hydrogen acid were synthesized by mixing in a glass beaker. Latter acid was added until La_2O_3 was dissolved. Obtained $LaCl_3$ solution was mixed with phosphoric acid at the La to P of 1:1, while stirring. Ammonia water was added to the precipitate to adjust the pH and set for 1 day. After that the precipitate solution was washed several times with de-ionized water and then with acetone, and finally dried at room temperature for 1 day. In order to achieve $LaPO_4$ as a the final product, the powder were dry ball-milled at room temperature for 1 day. Following that, the powders were calcined at 600 °C for 2 h to convert the rhabdophane to monazite.

Fabrication of 8YSZ-$LaPO_4$ composites

Commercially available 8YSZ powder (TZ-8YZ, Tosoh Corp.) with average grain size of ca. 0.1 μm was used as the starting material. The investigated contents of $LaPO_4$ within the initial powder mixture varied from 5 to 20 volume percent (vol. %), while the used powder mixture contained always 8 mol % of the yttria stabilizer. The powders were grounded after mixing together by means of wet-ball-milling in ethanol for 24 h using zirconia balls diameter of 5 mm and then the slurry was dried by using rotary vaporizer. For preparing the samples for the mechanical strength measurements, ca. 35 g of 8YSZ-$LaPO_4$ powder mixture were placed in a Ø

35 mm graphite die and then an electric current of 3000-5000 A was applied. The sintering temperature rate was 100 °C/min, from room temperature until the sintering value was 1300°C. The sintering time was 5 minutes at reached temperature in Argon atmosphere, while the pressure was 30 MPa. After sintering the current was switched off, pressure was released, and the sample was cooled to the room temperature. During the sintering process, the temperature was measured by means of an optical pyrometer, what was focused on the die hole surface, which was centred on the sintering sample.

Characteristics of sintered 8YSZ-$LaPO_4$

The PECS-sintered compacts were analyzed by X-ray diffraction technique (XRD, $CuK\alpha$ V=-50kV-150mA, RU-200B, Rigaku Co. Ltd., Tokyo, Japan) to identify the phase structure caused by the applied processing. The relative density of the bulk composites was determined from the ratio of bulk and true densities. The true density was measured by using the Archimedes' method with distilled water at room temperature.

The microstructure of the obtained materials was observed by means of scanning electron microscopy (SEM). Sintered specimens were polished with diamond pastes (grade of 1 μm) and finally the procedure applied exclusively to the samples with a perfect, mirror surface. For thermal etching procedure the specimens were additionally annealed for 20 min at the temperatures that were 1250 °C, while the applied heating and cooling rates were equalled 10°C/min.

Evaluation of the coefficient of thermal conductivity was measured at room temperature by the laser flash thermal analyzer (ULVAC-RICO, TC 7000) of the sintered materials. Ground and polished samples dimensions were pellets 12 and 1 mm in diameter and thickness, respectively.

It is worth to emphasize that the rectangular specimens (3 x 4 x 40 mm) were grounded and polished prior to mechanical testing, with fine diamond paste (grade of 2 and 0.5 μm), to avoid the surface effect on the strength measurement, and through this to obtain reliable results. Strength of the newly fabricated composites was measured by using the three-point bending method with a span of 30 mm and a crosshead speed of 0.5 mm/min, while the nominal applied load was 20N.

Thermogravimetry was used to monitor any weight changes on reoxidation of the pellets prepared by the PECS method. The thermal analysis was carried out in air atmosphere up to 1400°C with the applied heating rate equalled 10 deg/min, while annealing was for 3 hours. In particular, the oxidation of composite containing 10 and 20 vol. % of $LaPO_4$ was analysed in detail using DTA and thermogravimetry (TG) method (equipment: Bruker AXS, TG-DTA 2000SA).

RESULTS AND DISCUSSION

XRD-patterns of PECS-sintered pellets are shown in figure 1. In case of 100% 8YSZ-pellet sample is indexed as a cubic unit cell structure. Similar results are reported previously by Takeuchi et al. 19). In case of pure $LaPO_4$ sample the data is showing clear structure of monazite. Moreover, after adding monazite to zirconia matrix the cubic structure is changing to tetragonal face following the monazite content. Clear microstructural distinction between zirconia matrix and monazite was not obvious.

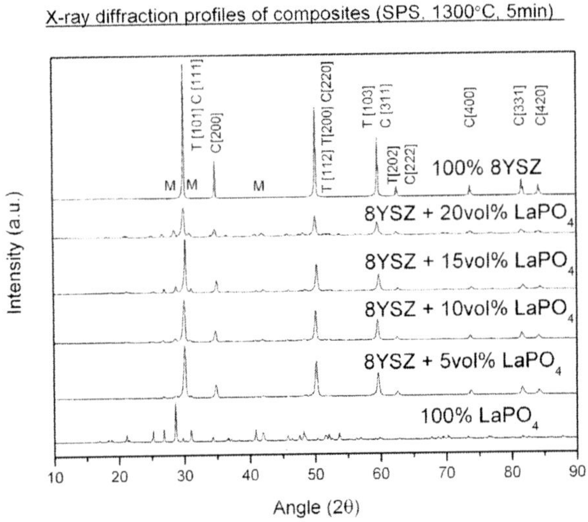

Fig. 1. X-ray diffraction patterns of 8YSZ-based composites.

Relative densities are shown in figure 2, what indicates that composite density systematically decreases with the content of monazite in the sintered body. In case of simple 8YSZ pellet, density value is 99% of the theoretical density value.

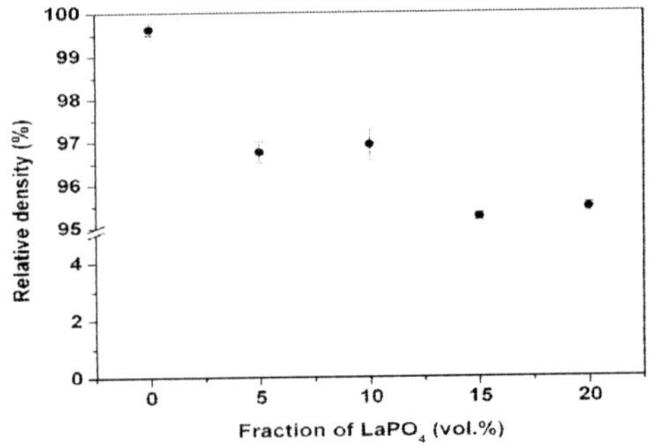

Fig. 2. The relative density of the 8YSZ-based composite with the different level of monazite.

In figure 3 shows the polished cross-section of PECS sintered 8YSZ/monazite composites after thermal etching. The SEM micrographs pellet of the structure of the obtained nanocomposites observed using by high-resolution scanning electron microscopy of the fractured surfaces. In contrast to the sintered pellets. the average grain sizes are in samples 300 nm, as reported previously for dense 8YSZ ceramics. For the composite which contains 5 vol.% of monazite is showing more irregular uniform structure compared with higher monazite sample. In case of 15 vol.% of monazite amount the granular grains with homogenous distribution can be observed all over of the matrix[23]. In figure 3 B shows the monazite pinning effect around zirconia grains and that phenomenon at high temperatures should play a significant role against grain growth of zirconia.

Fig. 3. SEM-micrographs of the polished and thermally etched surfaces of nanocomposites obtained by sintering at 1300°C the nanopowder mixture containing 5vol.% (a) and 15 vol.% (b) of monazite.

To our satisfaction, the coefficient of thermal conductivity (CTC) measured for the obtained 8YSZ-based composites with monazite and fabricated from the powder including also high content of monazite[20] vol.%, is lower than the one evaluated for pure zirconia (figure 4). This kind of phenomena is contrary compared with results made by Min et al.[1] One of the reasons for this kind of trend is higher yttria content, which makes structure for pure cubic face as shown is picture 1. The rule of mixture for YSZ and $LaPO_4$, which can be used in TBC systems, are supporting this by lower thermal conductivity value of monazite[23,24]. The 15 vol.% monazite reinforced composite has slightly higher k value than the pure 8YSZ, which is due to system error ±10%. This kind of results registered effect clearly points towards higher contents of monazite powders as the recommended initial materials for fabrication of thermal barrier coatings. Moreover, the output of the research agrees with the recent findings by Khor *et al.* who lately elaborated on the influence of the thermal conductivity of plasma sprayed 7wt.% Y_2O_3 stabilized ZrO_2 for solid oxide fuel cell (SOFC) applications[25].

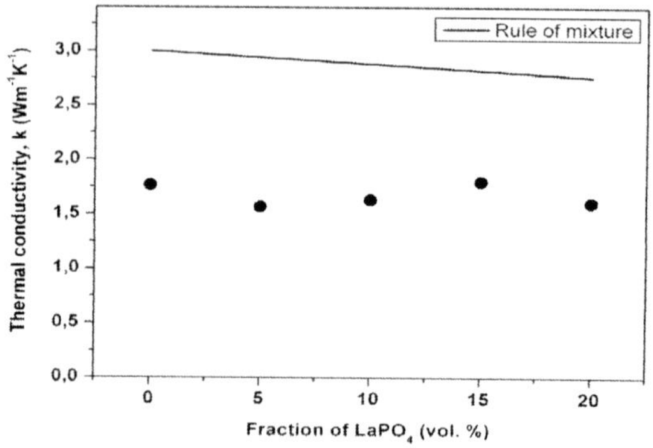

Fig. 4. The coefficient of thermal conductivity measured for the 8YSZ-based nanocomposites sintered from the powders with different initial content of the monazite.

Figure 5 shows the three-point bending strength for the sintered ceramics. The 8YSZ composite is comparable to those reported previously[23]. The toughness strength of the obtained zirconia-monazite composites depend on the weight percent of monazite in the initial powder mixture. The real benefit of monazite content in composite is expected through the monazite pinning effect (figure 3b) at high temperatures applications.

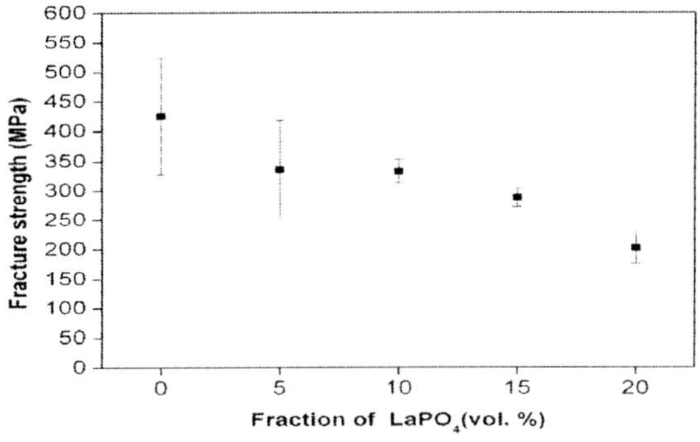

Fig. 5. Fracture strength as a function of monazite content measured by three-point bending technique for the 8YSZ composites.

The oxidation of composite containing 10 and 20 vol.% of $LaPO_4$ was analyzed in detail using thermogravimetry (TG) method at 1400 °C for 3 hours in air atmosphere. From data which can be seen in figure 6, cannot be seen any significant weight loss. Thus this kind of material has a high potential for high temperature applications.

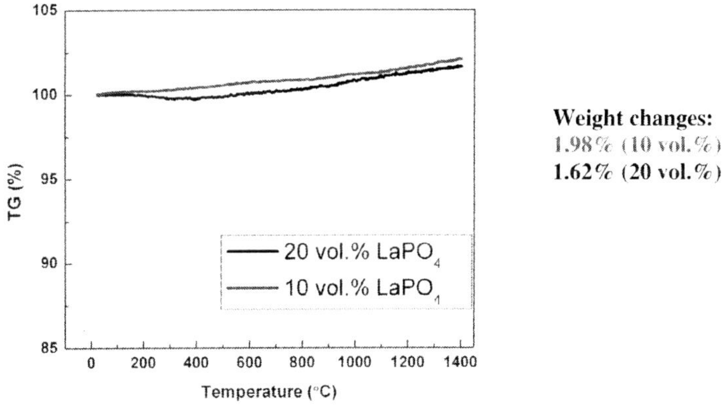

Fig. 6. TG and data registered during three hours annealing (1400°C) of the 8YSZ based composite that contained 10 vol. % of monazite.

CONCLUSIONS

The present paper elaborates on fabrication and characterization of 8YSZ-Monazite composites prepared by Pulse Electric Current Sintering (PECS) of powders of zirconia and monazite, while the initial content of the latter component varied from 5 to 20 vol.%. PECS process permits lower sintering time and temperature compared with conventional sintering method and it seems to be promising fabrication method to produce dense 8YSZ-$LaPO_4$ ceramic nano-composite with improved mechanical properties. The observed microstructure reached 99% of its theoretical density and has fine grained nano structure.

PECS is found to be an effective method for obtain dense zirconia based composite as well they provide evidence that the density of final composite increases with the content of zirconia in the sintered body. In addition of monazite content is successfully manufactured by colloidal method and added to zirconia matrix. These results strongly suggest that this kind of composite is suitable candidate for higher temperature applications.

REFERENCES

1. W. Min, K. Daimon, T. Matsubara and Y. Hikichi, "Thermal and mechanical properties of sintered machinable $LaPO_4$–ZrO_2 composites," *Mat. Res. Bull.*, **37**, 1107-1115 (2002).
2. D.B. Marshall, P.E.D Morgan and R.M. Housley, "Microstructure/Properties Relations of Advanced Materials Debonding in Multilayered Composites of Zirconia and $LaPO_4$," *J. Am. Ceram. Soc.*, **80** (7), 1677-1683 (1997).
3. Parthasaranthy, T.A., Boakye, E., Cinibulk, M.K. and Perry, M.K., "Fabrication and Testing of Oxide/Oxide Microcomposites with Monazite and Hibonite as Interlayers," *J. Am. Ceram. Soc.*, **82**, 3575-3658 (1999).
4. D. L. Jonson, "Microwave and plasma sintering of ceramics," *Ceram. Int.*, **17,** 295-300 (1991).
5. S. H. Risbud and C-H. Shan, "Fast consolidation of ceramic powders," *Mater. Sci. Eng.*, **A 204,** 146-151 (1995).
6. Su, H., Johnson, D. L., *J. Am. Ceram. Soc.*, Vol.**79,** pp.3199-3210 (1996).
7. J. Hong, L. Gao, S. D. D. L. Torre, H. Miyamoto and K. Miyamoto, "Spark plasma sintering and mechanical properties of $ZrO_2(Y_2O_3)$–Al_2O_3 composites," *Mater. Lett.*, **43,** 27-31 (2000).
8. C-H. Shan, , S.H. Risbud and K.S. Yamazaki, "Rapid consolidation of Bi-Pb-Sr-Ca-Cu-O powders by a plasma activated sintering process," *Mater. Sci. Eng.*, **B26,** 55-60 (1994).
9. L. Gao, H.Z. Wang and J.S. Hong, "SiC-ZrO_2(3Y)-Al_2O_3 nanocomposites superfast densified by spark plasma sintering, " *Nanostruct. Mater.*, **11,** 43-49 (1999).
10. Hikichi, Y., Nomura, T., Tanimura, Y., Suzuki, S. and Miyamoto, M., *J. Am. Ceram. Soc.*, Vol.**73**, pp.3594 (1990).
11. Hikichi, Y. and Nomura, T., *J. Am. Ceram. Soc.*, Vol.**70**, pp.c-252-c253 (1987).
12. P.E.D Morgan, D.B.Marshall, and R.M. Housley, "High-temperature stability of monazite-alumina composites" *Mater. Sci. Eng.*, **195,** 215-20 (1995).
13. Morgan, P.E.D and Marshall, D.B., *J. Am. Ceram. Soc.*, Vol.**78**, pp.1553 (1995).
14. J.B. Davis, D.B. Marshall, and P.E.D. Morgan, "Monazite-containing oxide/oxide composites, " *J. Eur. Ceram. Soc.*, **20,** 583-87 (2000).

15 Marshall, D.B., Morgan, P.E.D. and Housley R.M., *J. Am. Ceram. Soc.*, Vol.**78** pp. 677 (1992).

16 W. Min, and K. Daimon, T. Matsubara and Y. Hikichi, "Thermal and mechanical properties of sintered machinable $LaPO_4$–ZrO_2 composites," *Mater. Res. Bull.*, **37**, 1107-1115 (2002).

17 Hikichi, Y., Ota, T. and Hattori, T., *Mineralogical Journal*, Vol.**19**, pp.123-130 (1997).

18 B.F. Sorensen and R. Talreja, "Toughness of damage tolerant continuous fibre reinforced ceramic matrix composites, " *J. Eur. Ceram. Soc.*, **15**, 1047-1059 (1995).

19 T. Takeuchi, I. Kondoh, , N. Tamari, N. Balakrishnan, K. Nomura, H. Kageyama, and Y. Takeda, "Improvement of Mechanical Strength of 8 mol % Yttria-Stabilized Zirconia Ceramics by Spark-Plasma Sintering, " *J. Electrochemical Soc.*, **149**, A455-A461 (2002).

20 K. A. Khor, L-G Yu, S.H. Chan and X.J. Chen, "Densification of plasma sprayed YSZ electrolytes by spark plasma sintering (SPS), " *J. Eur. Ceram. Soc.*,**23,** 1855-1863 (2003).

21 T.K. Gupta, F.F. Lange and J.H. Bechtold, "Microstructurally developed toughening mechanisms in ceramics: Transformation toughening in the Al2O3/ZrO2 composite system, " *J. Mater. Sci.*, **13**, 1464-1470 (1978).

22 J.B. Davis, D.B. Marshall, R.M. Housley and P.E.D. Morgan, "Machinable Ceramics Containing Rare-Earth Phosphates, " *J. Am. Ceram. Soc.*, **81,** 2169-75 (1998).

23 Sundre, O., Cheung, J., Marshall, D., Morgan, P. and Levi, C. G., The 25th Annual International Conference on composites, Advanced Ceramics, Materials, and Structures: B (Cocoa Beach of Florida, Jan. 2001), *J. Am. Ceram. Soc.*, pp. 367-374 (2001).

24 X.Q. Cao, R. Vassen and D. Stoever, "Ceramic materials for thermal barrier coatings, " *J. Eur. Ceram. Soc.*,**24**, 1-10 (2004).

25 K. A. Khor, L-G. Yu, S.H. Chan and X.J. Chen, "Densification of plasma sprayed YSZ electrolytes by spark plasma sintering (SPS)," *J. Euro. Ceram. Soc.*, **23**, 1855-1860 (2003).

MONAZITE FIBER COATINGS FOR OXIDE COMPOSITES

Emmanuel E. Boakye*, Pavel Mogilevsky*
Randall S. Hay, and Michael K. Cinibulk
Materials and Manufacturing Directorate, Air Force Research Laboratory,
Wright-Patterson Air Force Base, OH 45433-7817

ABSTRACT

Recent work on oxide fiber coatings for ceramic matrix composites is reviewed. Attempts to make hermetic coatings and avoid stress corrosion of the fiber are emphasized. The effect of heat-treatment temperature and time on coating density is discussed.

1. INTRODUCTION

Monazite fiber coatings have proven to promote graceful failure in ceramic matrix composites (CMCs).[1-3] To maintain high strength and toughness during service, CMCs require the maximum possible fiber strength, so fiber strength degradation during coating or composite processing should be avoided. Previous work on sol- and solution-derived monazite fiber coatings found that fiber strength degradation correlated with weight loss at high temperature from coating decomposition products. Stress corrosion was suggested to be the degradation mechanism.[4-11] Corrosive species can also be expected during CMC matrix processing or during CMC use, so a hermetic fiber coating that seals the fiber surface from these species is desirable.

Current monazite coatings made from rod-shaped precursors are porous unless heat-treated at temperatures at which the currently available oxide fibers exhibit a loss of strength.[6,9,12,13] The porosity of the coatings is partially due to poor packing of the rod-shaped particles.[12] Spherical particles pack to higher green densities and will require lower temperatures to densify.[14,15] In this paper, work on fiber coatings from five different monazite precursors is reviewed. Strength degradation mechanisms are discussed. Preliminary results for densification of fiber coatings made from spherical monazite particles are presented.

2. EXPERIMENTS

2.1 Fiber Coatings and Characterizations

All fiber coatings were applied using a continuous vertical fiber coater.[4,16,17] Nextel 720, 610, Tyranno-ZX, Tyranno-SA, Sylramic, Hi-Nicalon and Hi-Nicalon S fibers were desized prior to coating by passing them through a furnace held at 1100°C/3 cms^{-1}. Typically, fiber tows were coated with sols or solutions with a monazite yield of 20 g/L. The coatings were cured by passing them through a furnace at 900° to 1300°C with a speed of 3cms^{-1}. The furnace hot zone was ~8 cm in length so the fibers were held at maximum temperature for 3s. The coating process was repeated up to 7 times to increase the coating thickness.

The fiber coatings were examined by scanning electron microscopy SEM (Model 360 FE, Ernst Leitz, Western, Germany) equipped with energy dispersive spectroscopy (EDS) operating at 10-15 kV. Transmission electron microscopy TEM (Model CM 200, Philips, Almelo, the Netherlands) thin foils were made from coated fiber cross-sections, as described elsewhere.[18,19] Single filament tensile strengths and Weibull modulii were measured using a 2.54 cm gauge

---

* UES, Inc., Dayton, OH. Work performed under USAF Contract #: F33615-01-C-5214.

length. A minimum of 50 tests were done for each data point.[20] The effect of coating temperature and number of coatings on fiber strength was evaluated.

3. RESULTS AND DISCUSSION

3.1 Prior Coatings

Precursor #1 consisted of a solution of lanthanum nitrate (LAN) and trimethyl phosphate (TMP).[4,10] The La:P molar ratio was 1:1 and 1:1.18. After heat treatment at 800°C[4], X-ray diffraction (XRD) showed a pure monazite phase for the 1:1 molar ratio and mixtures of monazite and LaP_3O_9 for the 1:1.8 molar ratios.[4] Electron diffraction studies of coated fibers showed that monazite formed at >900°C.[21] The coatings were dense equiaxed grains, and bubbles were formed on fiber surfaces (Fig. 1). Bubble formation was attributed to out-gassing of decomposition products from the precursor. $AlPO_4$ formed at the fiber coating interface.

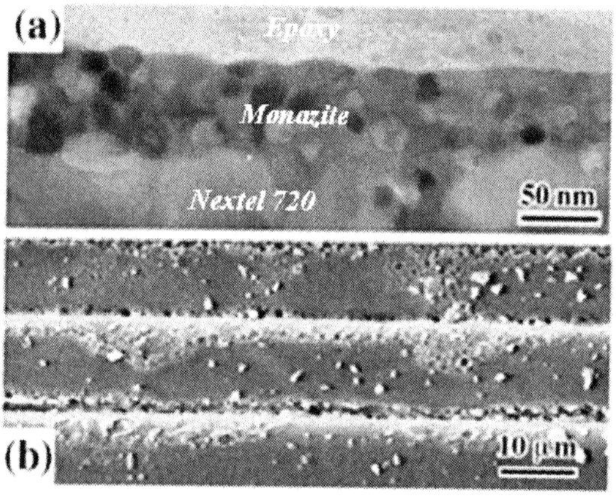

Fig. 1 $La(NO_3)_3/(OCH_3)_3P{=}O$ derived coating showed the highest strength loss. Bubbles indicate out-gassing.

Precursors # 2-4 formed rod-shaped La- and Ce- rhabdophane.[6,13] They were made from $La(NO_3)/H_3PO_4$, $La(NO_3)/(NH_4)_2HPO_4$ and $Ce(NO_3)_3/H_3PO_4$ (Eqs. 1-3). Precipitates from reactions 1-3 corresponded to the rhabdophane crystalline structure with a water of hydration of ~0.5.[6,9,13] XRD pattern of powders heat treated at 200° - 600°C was consistent to rhabdophane and were the same for Ce- and La- derived precursors.[6,13] Monazite formed at 750°C.[6,13]

Precursor #2 : $La(NO_3)_3 + (NH4)_2HPO_4 \rightarrow LaPO_4{\cdot}xH_2O + 2NH_4NO_3 + HNO_3$ [1]

Precursor #3 : $La(NO_3)_3 + H_3PO_4 \rightarrow LaPO_4{\cdot}xH_2O + 3HNO_3$ [2]

Precursor #4 : $Ce(NO_3)_3 + H_3PO_4 \rightarrow CePO_4.xH_2O + 3HNO_3$ [3]

The monazite morphology and crystalline structure were the same for La- and Ce-PO_4 derived monazite coatings (Fig. 2). Both formed porous coatings of rod-shaped (~15 x 150 nm) monazite particles. Also, both La- and Ce-PO_4 derived monazite had a crystallographic texture with preferred double orientation.[12] Some grains were aligned along the Nextel™ 720 fiber axis, while others were parallel to the circumference of the fiber axis.

3.2 Current Coating (Precursor # 5)

To aid in the densification of monazite at low temperatures, a nanometer sized spherical monazite precursor was made.[22] Rhabdophane was made by controlled precipitation from lanthanum-citrate chelate (La-Cit) and phosphoric acid (PA). La-Cit chelate was made by dissolving lanthanum nitrate and citric acid in deionized water and stirring for ~1h (Eq. 4).

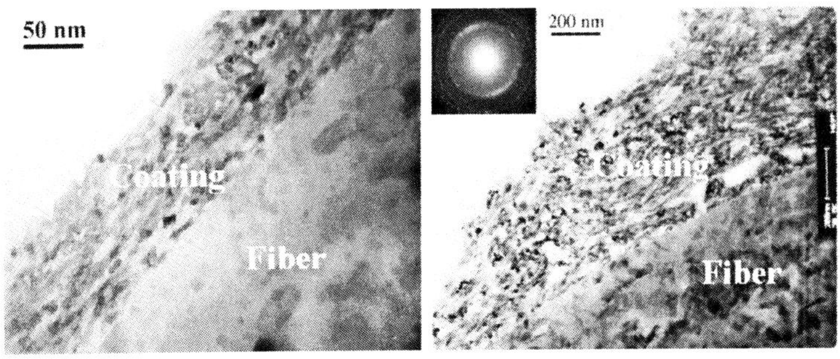

Fig. 2 TEM micrographs of Nextel 720 fiber coated with (a) $LaPO_4$ and (b) $CePO_4$.

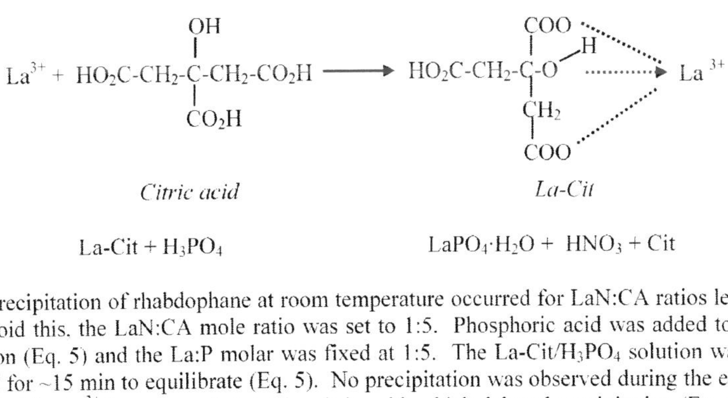

[4]

Citric acid *La-Cit*

$La\text{-}Cit + H_3PO_4 \quad LaPO_4{\cdot}H_2O + HNO_3 + Cit$ [5]

Fast precipitation of rhabdophane at room temperature occurred for LaN:CA ratios less than 1:5. To avoid this, the LaN:CA mole ratio was set to 1:5. Phosphoric acid was added to the La-Cit solution (Eq. 5) and the La:P molar was fixed at 1:5. The La-Cit/H_3PO_4 solution was stirred at ~22°C for ~15 min to equilibrate (Eq. 5). No precipitation was observed during the equilibration period. The La^{3+} ions were chelated by citric acid, which delayed precipitation (Eq. 4). Heating the solution to >30°C destabilized the La-Cit and caused precipitation of rhabdophane particles (Eq. 5).

The particles were washed to remove residual nitrates and citric acid prior to fiber coating. The washing was repeated until the sol acidity decreased from a pH of <1 to a pH of ~3.

Similar to the results of the rod-shaped particles the XRD patterns of the spherical powders heat treated at 200- 600°C were consistent with rhabdophane, and monazite formed at 750°C.[22]

Rhabdophane particles precipitated with the La-Cit/H_3PO_4 precursor were spherical with a primary particle size of 10 nm. The monazite grains in as-deposited coatings were ~15 nm in size and electron diffraction patterns showed no evidence of texture (Fig. 3). SEM visual inspection showed higher packing density for the coatings derived from spherical particles so they were used for the densification studies (Fig. 4).

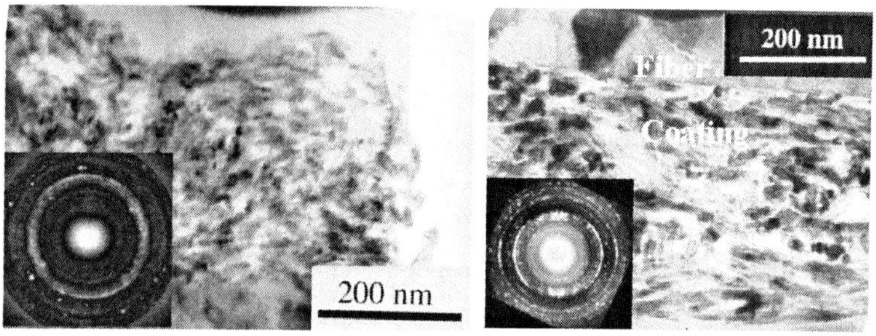

Fig. 3 TEM micrographs of coating from spherical and rod-shaped monazite; (a) spherical and (b) rod-shaped.

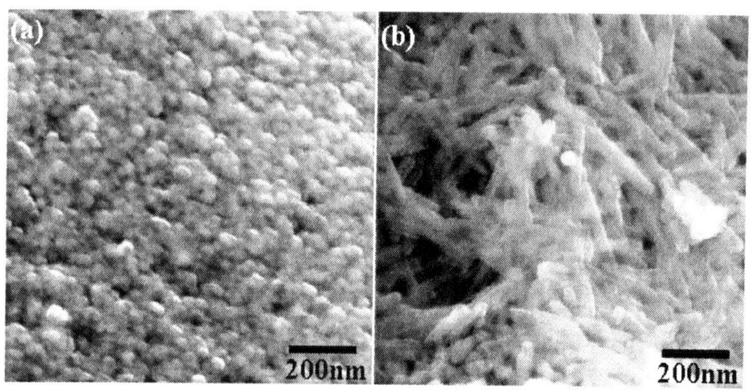

Fig. 4 SEM micrographs of monazite coating; (a) spherical particles and (b) rod-shaped particles.

3.3 Densification of Monazite Coatings

The fiber coatings were densified by heat-treatment at 1000°, 1100°, 1200° and 1300°C for 1, 10 and 100 h. Coatings thicker than ~500 nm were cracked and layered (Fig. 5). The outer portion of the coating was dense, while the inner portion was porous. Cracking of coatings may be due to weight loss from the La-Cit^{+}/H_3PO_4 precursor at high heating rates and the residual stress from the thermal expansion mismatch between the coating and the fiber. The average thermal expansion coefficient of monazite (9.6×10^{-6} °C^{-1})[23] is significantly higher than that of Nextel^{TM}720 fiber (6.0×10^{6}).[24]

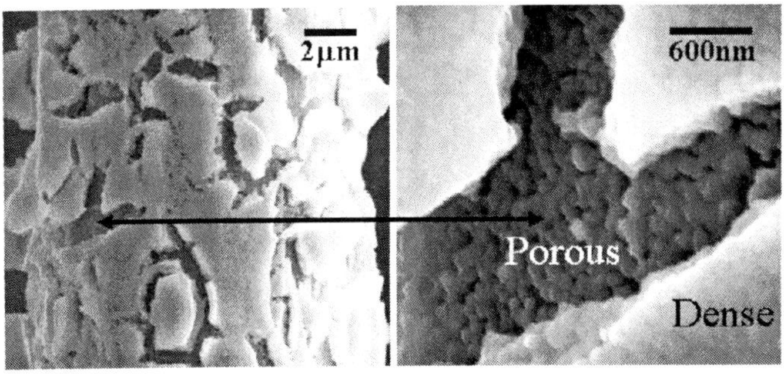

Fig. 5 SEM micrographs of monazite coating heat treated at 1100°C/100h.

Inference from grain growth studies[25] and TEM (Fig. 6) showed that coatings heat-treated at >1100°C were at least locally hermetic if they were uncracked. Heat treatments of coatings thinner than ~100 nm at >1200°C, formed monazite spheroids and left voids between the spheres of monazite particles (Fig. 7). Therefore, crack formation and spheroidization limited the thickness range to 100 – 500 nm.

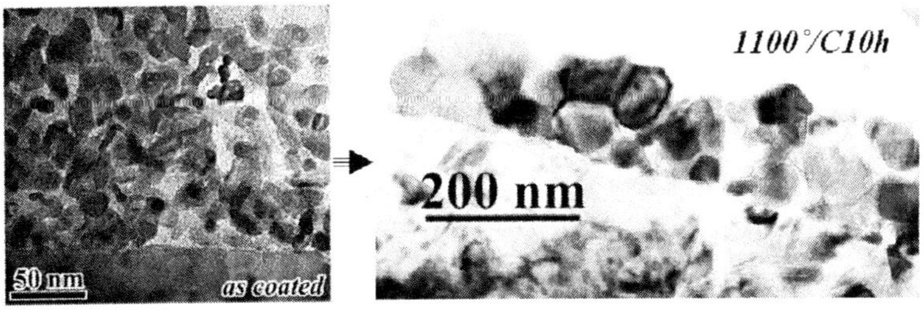

Fig. 6 TEM micrographs of monazite fiber coating derived from spherical particles.

3.4 Effect of Coating Temperature on Fiber Strength

Previous work showed that large variations in strength of monazite coated fibers were precursor dependent.[4] Evidence of stress corrosion was inferred from a comparison of TGA weight loss for the different monazite precursors (Fig. 8) and the strength of fibers coated with these precursors (Fig. 9).[5,6] Fibers coated with unwashed precursors with high weight loss above the coating temperature had low strength (Precursors #1 and #2) and fibers coated with the washed precursors #2 and #3 with almost no weight loss above the coating temperature had the highest filament strength (Fig. 9). Figure 10 is a plot of precursor weight loss versus filament strength for fibers coated at 900°, 1100° and 1300°C. A strong correlation between fiber strength and precursor weight loss above the coating temperature is observed, implying a dependence on the amount of residual precursor decomposition left in the coating.

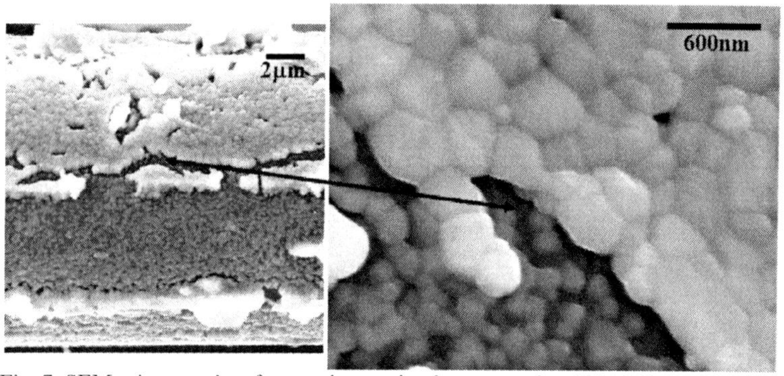

Fig. 7 SEM micrographs of monazite coating heat treated at 1300/10h showing hermetic and spheroidized coating.

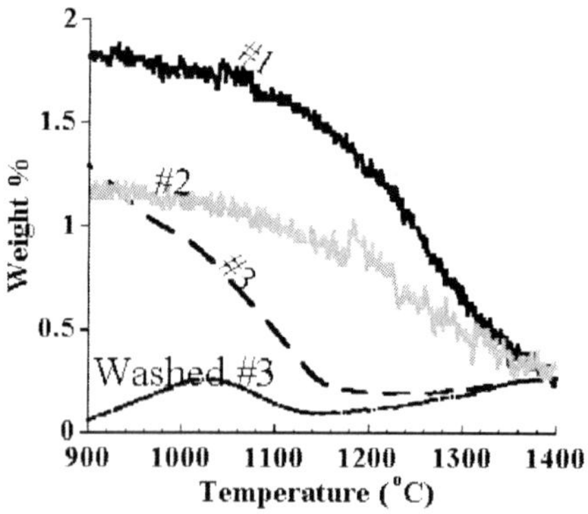

Fig. 8 TGA weight loss from different precursors.

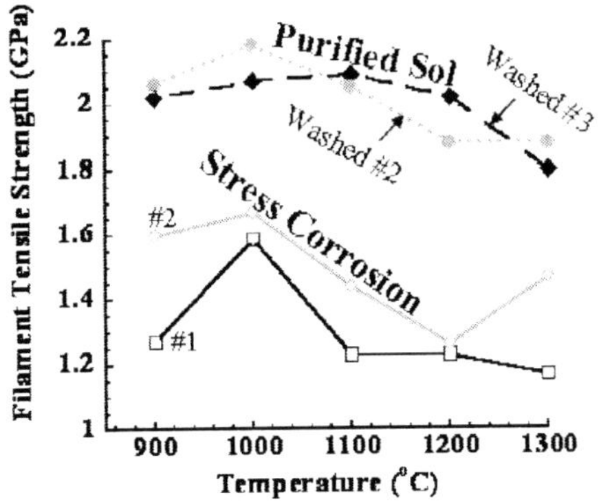

Fig. 9 Effect of monazite precursor on fiber strength. Purified rhabdophane precursors showed the highest strength retention.

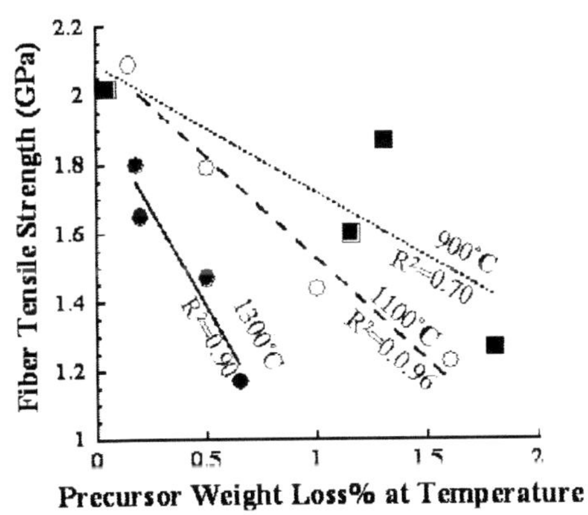

Fig. 10 TGA weight loss versus fiber strength for different monazite precursors.

3.5 Effects of Number of Coats

Oxide and SiC fibers coated multiple times with the washed monazite precursor retained their strengths (Fig. 11 and Table 1). The filament strength of monazite coated Nextel™ 720 and 610 fibers were ~2 GPa and 3 GPa respectively, for two or more coats and changed little with the number of coats. The filament strengths of desized Nextel™ 720 and 610 were 2.0 GPa and 3.0 GPa, respectively. The tensile filament strengths of Tyranno-SA, Tyranno-ZX, Sylramic, Hi-Nicalon and H-Nicalon S fibers were all slightly higher than their desized strengths after 5 coats with the washed monazite precursor (Table 1). However, preliminary results of strength retention following extended heat treatments of coated SiC fibers depends strongly on the environment. Strength degradation may occur at very low oxygen partial pressures, where monazite is reduced and reacts with the SiC fiber.[26]

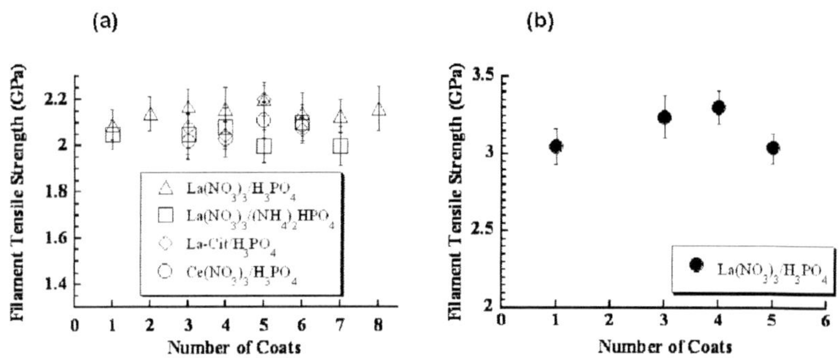

Fig. 11 Effect of number of coats on filament strength of monazite coated fiber; (a) Nextel 720 and (b) Nextel 610.

Table1. Effect of monazite coating on SiC fiber strength.

Fiber	Desized Filament Strength (GPa)	Monazite Coated Filament Strength (GPa)
Tyranno-SA	2.3	2.6
Tyranno-ZX	2.4	2.7
Sylramic	3.1	3.3
Hi-Nicalon	2.9	3.1
Hi-Nicalon S	2.8	2.9

4. DISCUSSION AND CONCLUSIONS

Fibers coated with solutions and sols containing nitric acid and ammonium nitrate have relatively low strengths, but fibers coated with washed sols that eliminate nearly all the nitric acid or ammonium nitrate retain their original strength. The observed fiber strength retention was independent of 1) the type of rare earth metal; lanthanum and cerium derived monazite coated fibers retained strength (Fig. 11), 2) phosphorus source; H_3PO_4 and $(NH_4)_2HPO_4$ derived monazite precursors retained strength 3) monazite morphology; fibers with coatings derived from precursors with either spherical or rod-shaped monazite both retained strength 4) fiber; Nextel™ 610, 720, (Fig. 11) Tyranno-ZX, Tyranno-SA, Hi-Nicalon, Hi-Nicalon-S and Sylramic fibers retained strength (Table 1).

Prior and current work shows lack of evidence of strength degradation by solid state reaction between fiber and impure monazite.[5,6,9,11] The only solid state product observed was $AlPO_4$ at the fiber/coating interface. However, monazite coated fibers, where the precursor was slightly phosphate rich, were stronger than the corresponding fiber with stoichiometric monazite coating.[4] Surface flaw healing was proposed as the strength enhancement mechanism.[4,10] Also, no correlation between fiber strength and small amounts of grain growth at the fiber surface was observed.[9]

Degradation from stress corrosion at a particular temperature was found to correlate with the amount of gas-phase precursor decomposition products that remain in the coating at temperatures > 900°C (Figure 10).[9] All the precursors used $La(NO_3)_3$ as the lanthanum source. Above the coating temperature (900°C), mass spectroscopy study showed N_2, H_2O and N_2O as the only detectable decomposition products.[8] The identity of the corrosive species is unknown but it is suspected to be a surface active metastable species, which is difficult to detect.[11] Partial pressures of precursor decomposition products in coating pores may be significantly higher than atmospheric pressure, and therefore enhance corrosive species activity if the coating is hermetic. Strength degradation of Nextel™720 from zirconia-carbon correlated with the use of the nitrogen containing precursor; zirconium oxynitrate.[27]

REFERENCES

1. Keller, K. A. et al. Effectiveness of Monazite Coatings in Oxide/Oxide Composites after Long-Term Exposure at High-Temperature. *J. Am. Ceram. Soc.* **86**, 325-332 (2003).
2. Kaya, C., Butler, E. G., Selcuk, A., Boccaccini, A. R. & Lewis, M. H. Mullite (NextelTM 720) Fibre-Reinforced Mullite Matrix Composites Exhibiting Favourable Thermomechanical Properties. *J. Eur. Ceram. Soc.* **22**, 2333-2342 (2002).
3. Lee, P.-Y., Imai, M. & Yano, T. Effects of Sintering Condition on the Mechanical Properties of Monazite-Coated Alumina Fiber/Alumina-Matrix Composites Fabricated by Hot-Pressing. *J. Ceram. Soc. Japan* **112**, 29-34 (2004).
4. Boakye, E., Hay, R. S. & Petry, M. D. Continuous Coating of Oxide Fiber Tows Using Liquid Precursors: Monazite Coatings on Nextel 720™. *J. Am. Cer. Soc.* **82**, 2321-2331 (1999).
5. Boakye, E. E., Petry, M. D., Hay, R. S. & Douglas, L. M. Monazite Coatings on Nextel 720TM, 610TM, and Tyranno-SA Fiber Tows: Effects of Precursors on Fiber Strength. *Ceram. Eng. Sci. Proc.* **21**, 229-236 (2000).
6. Boakye, E. E., Hay, R. S., Mogilevsky, P. & Douglas, L. M. Monazite Coatings on Fibers: II, Coating Without Strength Degradation. *J. Am. Cer. Soc.* **84**, 2793-2801 (2001).
7. Hay, R. S., Petry, M. D. & Boakye, E. in *20th Annual Conference on Composites and Advanced Ceramics* (Am. Cer. Soc., Cocoa Beach, 1996).
8. Hay, R. S., Boakye, E. & Petry, M. D. Effect of Coating Deposition Temperature on Monazite Coated Fiber. *J. Eur. Ceram. Soc.* **20**, 589-597 (2000).
9. Hay, R. S. & Boakye, E. E. Monazite Coatings on Fibers: I, Effect of Temperature and Alumina Doping on Coated-Fiber Tensile Strength. *J. Am. Ceram. Soc.* **84**, 2783-2792 (2001).
10. Boakye, E., Petry, M. D. & Hay, R. S. Porous Aluminum Oxide and Lanthanum Phosphate Fiber Coatings. *Cer. Eng. Sci. Proc.* **17**, 53-60 (1996).

11. Hay, R. S., Boakye, E. E. & Mogilevsky, P. in *Corrosion/2002, Denver Colorado* 1-6 (NACE International, Denver Colorado, 2002).
12. Mogilevsky, P., Hay, R. S., Boakye, E. E. & Keller, K. A. Texture of Rhabdophane Derived Monazite Coatings. *J. Am. Ceram. Soc.* . **86**, 1767-1772 (2003).
13. Boakye, E. E. & Mgilevsky, P. Fiber Strength Retention of La and Ce Coated Nextel 720. *J Am. Ceram. Soc.* **87**, 314-316 (2004).
14. Cocks, A. C. F. The structure of Constitutive Laws for the Sintering of Fine Grained Materials. *Acta Metall. Mater.* **42**, 2191-2210 (1994).
15. Yeh, T.-S. & Sacks, M. D. Effect of Particle Size Distribution on the Sintering of Alumina. *J. Am . Ceram. Soc.* **71**, C484-C487 (1988).
16. Hay, R. S. & Hermes, E. E. Sol-Gel Coatings on Continuous Ceramic Fibers. *Cer. Eng. Sci. Proc.* **11**, 1526-1532 (1990).
17. Hay, R. S. Sol-Gel Coating of Fiber Tows. *Cer. Eng. Sci. Proc.* **12**, 1064 (1991).
18. Hay, R. S., Welch, J. R. & Cinibulk, M. K. TEM Specimen Preparation and Characterization of Ceramic Coatings on Fiber Tows. *Thin Solid Films* **308-309**, 389-392 (1997).
19. Cinibulk, M. K., Welch, J. R. & Hay, R. S. Method for Preparation of TEM Specimens of Coated Fibers. *J. Am. Ceram. Soc.* **79**, 2481-2484 (1996).
20. Petry, M. D., Mah, T. & Kerans, R. J. Validity of UsingAverage Diameter for Determination of Tensile Strength and Weibull Modulus of Ceramic filaments. *J. Am . Ceram. Soc.* **80**, 2741-2744 (1997).
21. Boakye, E., Hay, R. S. & Petry, M. D. (eds. Voigt, J. A., Wood, T. E., Bunker, B. C., Casey, W. H. & Crossey, L. J.) 363-368 (Materials Research Society, San Francisco, California, U.S.A., 1996).
22. Boakye, E. E., Mogilevsky, P. & Hay, R. S. Synthesis of Nanosized Spherical Rhabdophane Particles. *In Press J Am. Ceram. Soc.* (2005).
23. Morgan, P. E. D. & Marshall, D. B. Ceramic Composites of Monazite and Alumina. *J. Am. Cer. Soc.* **78**, 1553-63 (1995).
24. Wilson, D. M. & Visser, L. R. High Performance Oxide Fibers for Metal and Ceramic Matrix Composites. *Composites* **A32**, 1143-1153 (2001).
25. Boakye, E. E., Mogilevsky, P., Hay, R. S. & Keller, K. in *Presented at 27th Annual International Conference on Advanced Ceramics and Composites,* (Cocoa Beach, FL, 2003).
26. Boakye, E. E., Mogilevsky, P., Welter, J., Hay, R. S. & Kerans, R. J. Monazite Coatings on SiC Fiber Tows. *In preparation* (2005).
27. Boakye, E. E., Hay, R. S., Petry, M. D. & Parthasarathy, T. A. Zirconia-Silica-Carbon Coatings on Ceramic Fibers. *J. Am. Ceram. Soc.* **87**, 1967-1976 (2004).

FABRICATION AND MECHANICAL PROPERTIES OF LEUCITE-CONTAINING ALUMINA CERAMICS

Fumiharu Sato, Shinobu Hashimoto, Sawao Honda and Hideo Awaji

Materials Science and Engineering, Nagoya Institute of Technology, Gokiso-cho, Showa-ku, Nagoya-shi, 466-8555, Japan

ABSTRACT

Powder compacts consisting of 10-40 mass% of leucite, which was synthesized from natural potash feldspar, and alumina were sintered using a PECS (pulse electric current sintering) technique. Under the condition of heating at 1400 °C and 30 MPa for 5 min, the relative density of the whole samples was above 98 %. Then, leucite and alumina in the sample did not react with each other. Subsequently, the mechanical properties of the dense sintered body were examined. In the case of 40 mass% of leucite-containing alumina sample after heating at 1400 °C and 30 MPa for 5 min, the bending strength, Vickers hardness and elastic modulus of the sample were 360 MPa, 9 GPa and 152 GPa, respectively. The values of Vickers hardness and elastic modulus brought close to those of the natural bone, and the strength retained the value over the natural bone.

INTRODUCTION

Leucite has an aesthetic property and a superior biocompatibility, so that it is expected to be used as a dental porcelain. Further, it has a high melting point (1693 °C)[1] and a high linear thermal expansion coefficient of $1\text{-}3\times10^{-5}$ / K (till 600 °C),[2] therefore the use of leucite as a reinforcement or coatings for their metal materials is also expected. Previously, authors reported the mechanical properties of monolithic dense leucite body.[3] However, the report on mechanical properties of leucite composites has been limited, excepting glass material with leucite crystal that precipitated by heat treatment.[4] Then the improved mechanical properties of the glass material is presumed to be due to prestress caused by the difference in thermal expansion between leucite and glass matrix.[2]

Alumina takes an important role as a biomaterial due to having a high biocompatibility and an excellent wear resistance, so that it is used as an artificial joint or an artificial root. However, it is clear that alumina has a hard property, so that it is difficult to make the complicated shape compared to metal material or damage the natural bone due to the stress concentration.

In this study, novel leucite-containing alumina composite materials were fabricated in order to have close hardness and elastic modulus to the values of natural bone, retaining a high bending strength using a PECS technique. Mechanical properties of these composite materials were examined.

EXPERIMENTAL

Aluminum sulfate hydrate ($Al_2(SO_4)_3\cdot14\text{-}18H_2O$), K_2SO_4 and natural potash feldspar (Sekikawa, Niigata, Japan) were used as raw materials. The chemical composition of the potash feldspar is shown as Table I. The aluminum sulfate hydrate was calcined at 300 °C for over 12 h to form $Al_2(SO_4)_3$. Starting powder mixture consisting of $Al_2(SO_4)_3$: potash feldspar : K_2SO_4 =

3 : 3 : 15 (mass%) was placed into an alumina crucible, the crucible was covered with an alumina lid and the crucible was heated in an electric furnace at 1100 °C for 3 h. After heating, the reactant mass that remaining at the bottom of the crucible was soaked in 3 % HCl at 70-80 °C for 0.5-1 h. Further, remaining reactant was washed with pure water and dried. A white powder was obtained as a final product. The crystalline phase of the final product was analyzed by X-ray diffraction (XRD; XD-D1, Shimadzu Co., Japan). The obtained leucite powder was ground by ball milling for 24 h using alumina balls of 2 mm diameter to get finer leucite particles with an average size of 0.2 μm diameter. Subsequently, the fine leucite powder was added to fine alumina powder (Taimei Co. Japan, TM-D; 0.1 μm) in the range of 10, 20, 30 and 40 mass%. The mixed powder was also mixed well by ball milling with 2 mm alumina balls and ethanol as a media for 24 h. After milling, the dried mixed powder was run through a sieve of 100 μm to form starting sample powder. The obtained starting powder was heated using the PECS technique at a temperature range 1300-1500° C under 30 MPa for 5 min. The density of the heated samples was measured by Archimedes method. The crystalline phase of the heated samples was confirmed by XRD analysis. The mechanical properties of the dense sample, such as three-point bending strength (DSS-500, Shimadzu, Japan), elastic modulus (LUM-77KS, Tungaloy Co., Japan), Vickers hardness (AKV-A, Akashi Seisakusyo, Japan), Wear rate (FPD-05VS, Tokyo Sikenki Seisakusyo, Japan) and linear thermal expansion (TMA-50, Shimadzu, Japan) were determined. In order to clarify the sintering mechanism, the microstructure of the dense sample was observed by scanning electron microscopy (SEM, JMS5200, JEOL, Japan).

Table I. Chemical composition of potash feldspar used (mass%)

SiO_2	68.0
Al_2O_3	18.0
Fe_2O_3	0.11
TiO_2	0.02
CaO	0.12
MgO	0.31
K_2O	13.0
Na_2O	1.20
Rb_2O	0.11
Ir_2O_3	0.02
PbO	0.01
SO_3	0.01
Cl	0.02
Total	100.93

RESULTS AND DISCUSSION

Figure 1 shows the change of the relative density of the leucite-containing alumina with the amount of leucite at the temperature range of 1300-1500 °C. At 1400 °C and 1500 °C, the relative density of the samples was about 98 % and the difference between the temperature conditions was little. However, the relative density of the both conditions slightly decreased with increasing the amount of leucite. On the other hand, at 1300 °C, the relative density of the samples with 10-30 mass% of leucite was about 96 %, that is lower than those of the samples fabricated at 1400 °C and 1500 °C. Furthermore, in the case of the 40 mass% leucite-containing sample, the relative density decreased to 94 %. When only alumina was heated at 1250 °C, the relative density reached 98 %. Therefore, it is thought that addition of leucite to alumina leads to decrease the relative density.

Figure 2 shows SEM photographs of the 10 and 40 mass% leucite-containing alumina obtained at 1400 °C. According to this figure, both samples were composed of fine grains under below 1μm. However, each grain was not able to distinguish alumina or leucite. Particularly, the trace of the existence of the liquid phase during heating was observed in the sample with 40 mass% of leucite.

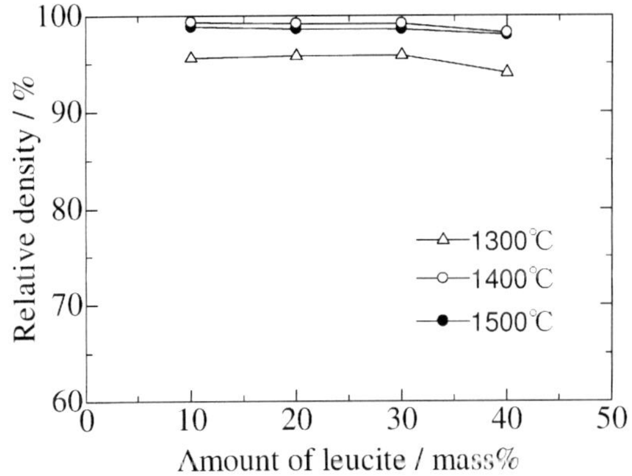

Fig. 1 The change of the relative density of the leucite-containing alumina with the amount of leucite at the temperature range of 1300-1500 ℃

Fig.2 SEM image of fracture surfaces of leucite-containing alumina after heating at 1400 °C. (a); 10 mass% leucite - containing, (b); 40 mass% leucite - containing.

Figure 3 shows XRD patterns of the leucite-containing alumina with the amount of leucite obtained at 1400 °C. From this result, monolithic alumina and leucite, those were the same with the starting sample, were detected in the samples, so that no reaction seemed to occur between alumina and leucite during heating. It is clear that no reaction occur below 1588±5 °C as shown in Figure 4 that is a phase diagram between leucite and alumina. However, in Fig. 2, the trace of the existence of the liquid phase during heating was observed in the sample with 40 mass% of leucite. In this experiment, leucite was made from natural potash feldspar, as shown in Table I. Therefore it is considered that some impurities such as Na_2O or MgO react with leucite-containing alumina to form some liquid phase during heating. This liquid phase might accelerate the sintering of the leucite-containing alumina.

Figure 5 shows Vickers hardness of the leucite-containing alumina with the amount of leucite obtained at 1400 °C. In the case of 40 mass% leucite-containing alumina, the hardness reached 9 GPa that is a half value of monolithic alumina: 18 GPa. Previously, authors have reported that the Vickers hardness of monolithic leucite that fabricated from potash feldspar has 5.3 GPa.

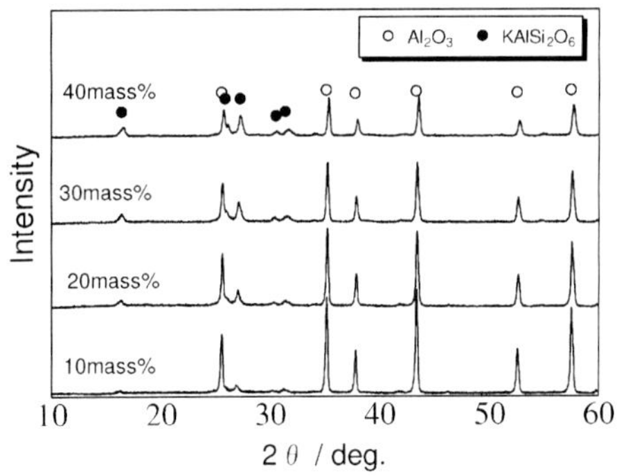

Fig.3 X-ray patterns in leucite-containing alumina with the amount of leucite after heating at 1400 °C.

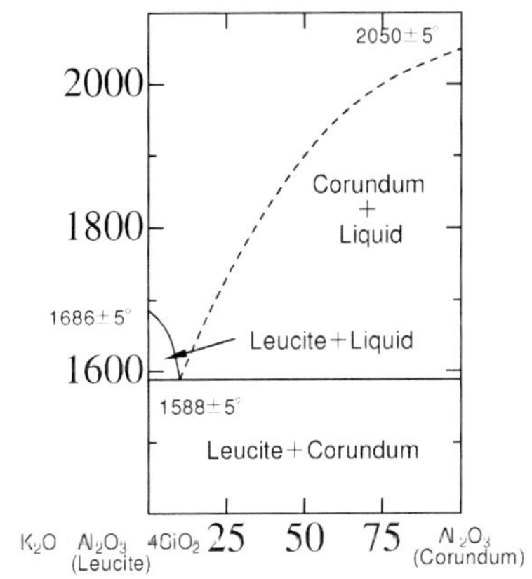

Fig.4 System $K_2O \cdot Al_2O_3 \cdot 4SiO_2$-$Al_2O_3$.

J. F. Schairer and N. L. Bowen, *Am.J. Sci*, 253 [12] 719 (1955).

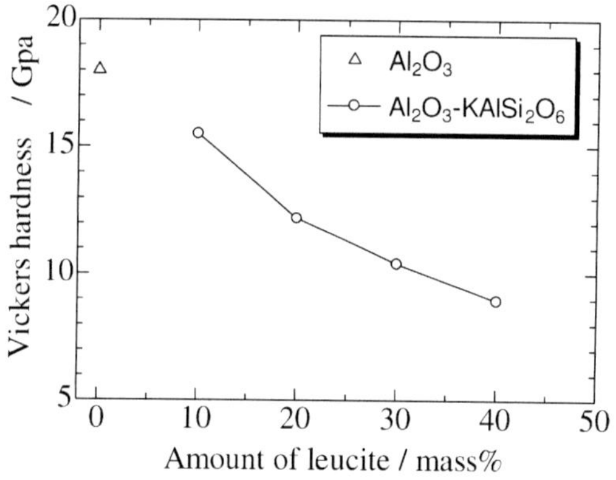

Fig.5 Vickers hardness of leucite-contain alumina with the amount of leucite after heating at 1400 ℃ and monolithic alumina.

Therefore, addition of 40 mass% of leucite to alumina have a grate influence of decreasing the hardness of alumina. When the amount of leucite added to alumina increase over 40 mass%, the Vickers hardness is thought to decrease slightly till 5.3 GPa.

Figure 6 shows elastic modulus of the leucite-containing alumina with the amount of leucite obtained at 1400 °C. The elastic modulus of the 40 mass% leucite-containing alumina was 152 GPa that was 60 % off compared to monolithic alumina, because that value of monolithic alumina is 380 GPa. Further, in the case of the 30 mass% leucite-containing alumina, the elastic modulus was 186 GPa that was a half value of monolithic alumina. Thus 40 mass% leucite-containing alumina has a good effect on decreasing of elastic modulus in the same manner as Vickers hardness. Further when the amount of leucite added to alumina increase over 40 mass%, the elastic modulus is thought to slightly decrease till 63 GPa that is the value of monolithic leucite. If the difference of elastic modulus between artificial material and natural bone (about 12 - 20 GPa of elastic modulus)[5] is high, there is a possibility that serious bone damage will occur in the natural bone due to stress concentration. Therefore as an ideal condition, the elastic modulus of the artificial material is close to that of natural bone.[6] As a result, 40 mass% leucite-containing alumina having 152 GPa of elastic modulus is expected to be used as an artificial bone material.

Figure 7 shows bending strength of the leucite-containing alumina with the amount of leucite obtained at 1400 °C. Required strength of an artificial bone is the same with or higher strength compared to that of natural bone. As shown in Fig. 7, the bending strength of the 40 mass% leucite-containing alumina was 360 MPa. This value is higher than general value of natural bone (30-200 MPa),[7] so that leucite-containing alumina is expected to be used as an artificial bone. Further, according to Fig.7, since the bending strength of 20 mass%

leucite-containing alumina was 475 MPa, this composite material is also thought to be able to use as an artificial tooth root whose requiring strength is 500 MPa.

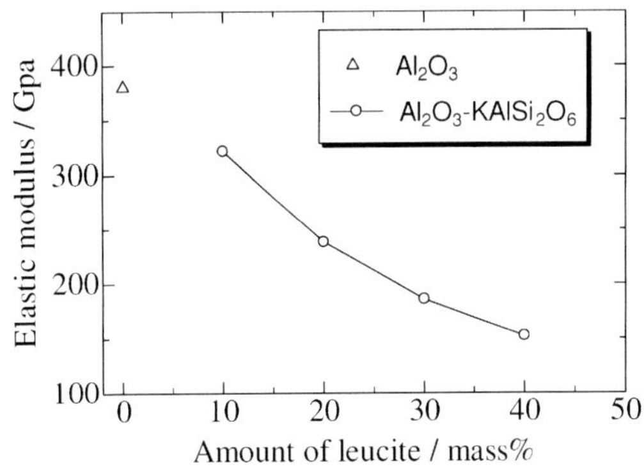

Fig.6 Elastic modulus of leucite-containin alumina with the amount of leucite after heating at 1400 ℃ and monolithic alumina.

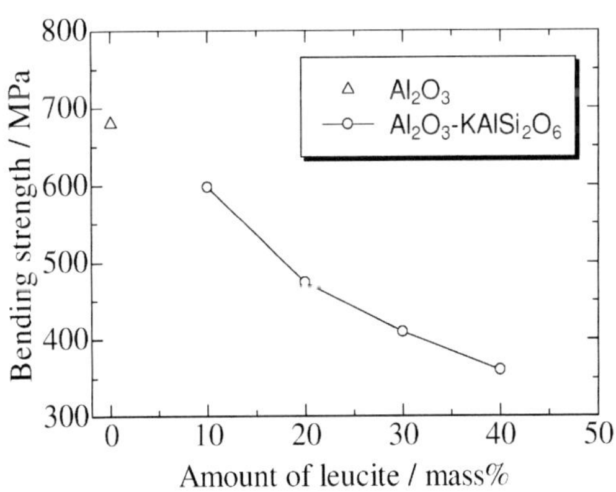

Fig.7 Bending strength of leucite-containing alumina with the amount of leucite after heating at 1400 ℃ and monolithic alumina.

Figure 8 shows wear rate of the leucite-containing alumina with the amount of leucite at 1400 °C. As a result in Fig. 8, the wear rate of 20 mass% leucite-containing alumina reached 1.5 times, and 40 mass% leucite-containing alumina reached 2 times compared to that value of monolithic alumina. Therefore mechanical workability (machinability) is thought to be improved when leucite is added to alumina. Because the wear rate of monolithic leucite (8.2×10^{-14} / $m^3\cdot m^{-1}$) is about 7 times compared to that of only alumina. When the amount of leucite in the composite increases more than 40 mass%, wear rate will be thought to drastically increase.

Finally, Figure 9 and Table II show thermal expansion behavior and thermal expansion coefficient of the leucite-containing alumina with the amount of leucite obtained at 1400 °C. From both the results, it is clear that the thermal expansion coefficient of the leucite-contaning alumina increased with increasing the amount of leucite. Particularly, in the sample with 30 mass% and 40 mass% of leucite, the similar thermal expansion behavior of monolithic leucite was observed. After 660 °C, the expansion curve of the samples had a small slope due to the transformation of leucite from the tetragonal phase to the cubic phase. The thermal expansion coefficient of the 40 mass% leucite- containing alumina was 1.3×10^{-5} / K (till 600 °C).

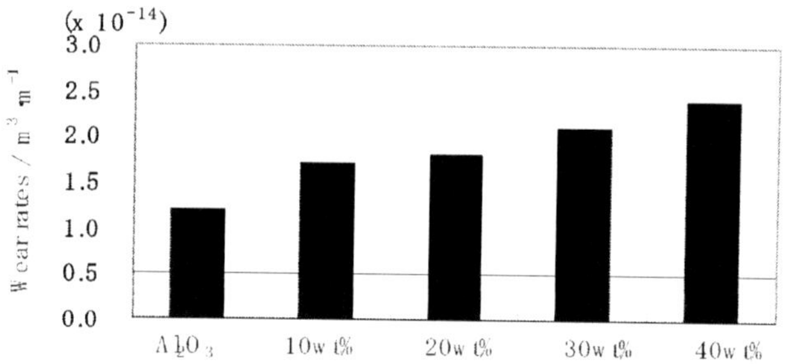

Fig.8 Wear rate of leucite-containing alumina with the amount of leucite heated at 1400 °C and monolithic alumina. Disk is SiC, Load is 4.9 N, Revolution speed is 1m / sec.

Table II. Thermal expansion coefficient of leucite-containing alumina with amount of leucite after heating 1400 °C

	10 mass%	20 mass%	30 mass%	40 mass%
RT-600°C	8.4	9.5	11.1	12.7

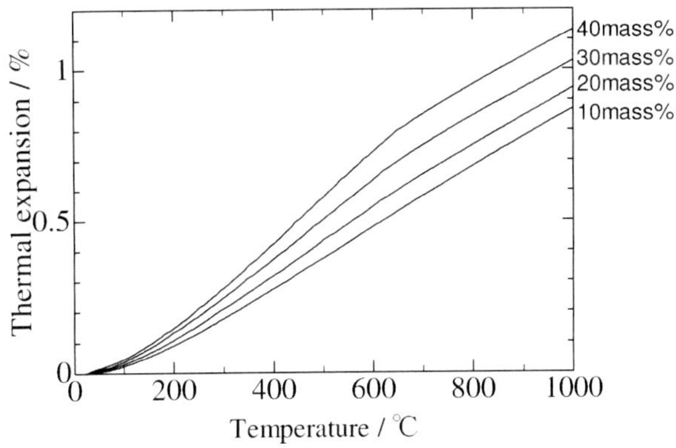

Fig.9 Linear thermal expansion curves of leucite-containing alumina with the amount of leucite after heating at 1400 ℃

CONCLUSION

Leucite-containing alumina composites were fabricated. The Vickers hardness and elastic modulus of the composites compared to monolithic alumina decreased and were close to the values of natural bone. On the other hand, bending strength of the composite keeps the required value as an artificial bone material. The bending strength, Vickers hardness and elastic modulus of the 40 mass% leucite-containing alumina were 360 MPa, 9 GPa and 152 GPa, respectively. Further, then the wear rate of the composite material was 2 times compared to that of monolithic alumina. Finally, the thermal expansion coefficient of the composites was 1.3×10^{-5} / K (till 600 °C).

REFERENCES

[1]Levin, E. M., Robbins, C. R. and McMurdie, H. F., "Phase Diagram for Ceramists," Ed. by Reser, M. K., Am. Ceram. Soc., Columbus, Ohio, USA (1967) Fig. 407.

[2]Ban, S., Matsuo, K., Mizutani, N., Kaikawa, K. and Hasegawa, J., *J. Jpan Soc. Dent. Mater. Dev.*, Vol. 19, pp. 318-325 (2000) [in Japanese].

[3]Hashimoto, S., et al., *J. Ceram. Soc. Japan*, Vol 113, pp. 488-490 (2005).

[4]M. J. Cattell. et al., Dent. Mater, Vol 17, pp. 21-33 (2001).

[5]Fujikawa, A., *J. Ceram. Soc. Japan*, Vol 38, pp. 40-43 (2003) [in Japanese].

[6]Nozue, A., *J. Ceram. Soc. Japam*, Vol 38, pp. 21-25 (2003) [in Japanese].

[7]J. D. Currey, Chinical Orthop. Related Res. 73, 210 (1970).

Mo-SiO_2 FUNCTIONALLY GRADED MATERIALS FOR HIGH INTENSITY DISCHARGE LAMP

Ayumu UMEMOTO and Koichi HAYASHI
TOTO Ltd., Advanced Ceramic Division
2-1-1 Nakashima, Kokurakita-ku, Kitakyushu 802-8601, Japan

Kunihiko NAKASHIMA, Noritaka SAITO, Kenji KANEKO and Keisaku OGI
Department of Materials Process and Engineering, Graduate School of Engineering, Kyushu University
6-10-1 Hakozaki, Higashi-ku, Fukuoka 812-8581, Japan

ABSTRACT

The conventional high intensity discharge lamps are made of Mo foil-sealing method, in which the Mo-metal of 30 μm thick is sandwiched by the fused silica envelope. The conventional high intensity discharge lamps therefore requires cooling system to reduce the thermal stress between Mo and fused silica during the switching operation and a long sealing zone is usually required to complete hermeticity, especially in high power devices, so that the production of compact high intensity discharge lamps has been a Herculean task.

Here, we show the invention of the new type of hermetically-sealed high intensity discharge lamps, made of Mo-SiO_2 functionally graded material (FGM) as an electrode and a sealing component. In the case of high intensity discharge lamps with Mo-SiO_2 FGM (FGM-Lamp), the thermal expansion coefficient between Mo and SiO_2 is functionally graded so that it tolerates a large number of heating cycles, with no cooling system are required. Furthermore lamps survive without breakage of lamp takes place. The W electrode is totally separated from the lamp envelope by the FGM, so that no leakage of the luminous elements or gases takes place, when a large gas pressure exists inside a lamp.

INTRODUCTION

The concept of the functionally graded material (FGM) has been conceived as a material in which the chemical composition and/or microstructural parameters gradually changes over the definable geometric distances, resulting in a corresponding variation in the properties of the material. Potential applications of FGMs are; high efficiency engine components, ceramic turbine components, direct metal tools for industrial use, biomaterials used in artificial human implants, drug delivery devices with controlled release rates, armor and armament components for defense and others. For these applications, material composition control is necessary for improving a variety of properties such as thermo-mechanical performance, toughness and strength, and for reducing interfacial stresses between dissimilar materials [1-3].

In general, two powders with different compositions are layered to prepare bulk FGMs [4-6], which results in macroscopically graded compositions. However, they are not perfectly graded microscopically, since there are always interfaces present between each layer. In this study, Mo-SiO_2 FGM for high intensity quartz envelopes discharge lamps was successfully achieved using a combined sedimentation and pressurized slip casting method [7-9] with uniform grading from the conductive part (Mo-rich phase) to the sealing part (fused silica).

In recent years, high intensity discharge lamps (HiDLs) have been widely used in the field of lighting vision, instead of the tungsten filament halogen lamps. Hg and metal halides are used as luminous elements. They provide a combination of a good color and a high luminous efficiency, which can be controlled by the lamp dose, the tube design and the power arc operation.

In the case of conventional HiDLs (c-HiDLs), the Mo-foil directly contacts the fused silica envelope as shown in Figure 1-b, to maintain the hermetic seal and electrical conduction at less than 10^5 Pa inside the lamp. This method (Mo-sealing method) is used to manufacture c-HiDLs throughout the world where a large thermal stress is always present between Mo and SiO_2 (Fig.1-a) [4,5], potentially resulting the thermal stresses resulting in delamination and fracture at the factory (Fig.1-c). In practice, the temperature of the sealing component drops from more than 1000 °C to approximately 300 °C, or vice versa, and the difference in the thermal expansion between Mo and SiO_2 becomes apparent (Fig.1-a) during the switching operation. Leakage of the luminous elements or gases and the breakage of the lamp (Fig.1-d) often occurs, when especially lamps are being run with short arc gaps and high pressure. Therefore, some c-HiDLs require cooling system to reduce the thermal stress. An option to improve the luminous efficacy, is to increase the pressure within the lamp part to more than 10^5 Pa, which increases the difficulty of hermetical sealing at current HiDLs. Furthermore, c-HiDLs are expected to become smaller and shorter, but there is a practical limit due to the Mo foil-sealing method. The sealing component is lengthened to maintain the same temperature suppressing the effect of thermal stress to achieve high input power. Thus the fabrication of compact conventional HiDLs has been a Herculean task.

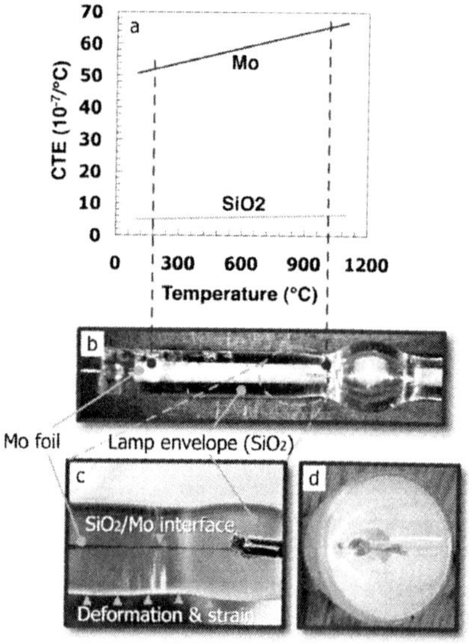

Figure 1. a) Thermal expansion coefficient for SiO_2 and Mo. b) Conventional high intensity discharge lamp. c) Sealing fraction of the lamp. d) Breakage of the lamp.

EXPERIMENTAL

Powders of Mo with 99.9% purity and SiO_2 with 99.999% purity were used in the fabrication of Mo-SiO_2 FGMs. The FGMs were fabricated using a Mo-SiO_2 mixed water suspension with 50 wt% content of solids, prepared by an ultrasonic stirring. The Mo-SiO_2 mixed slurry was placed in a vacuum desiccator to exclude air bubbles from the slurry. The FGM was prepared using a combined sedimentation and pressurized slip casting method [7-9]. The Mo-SiO_2 mixed slurry was cast in a plastic mold on a porous ceramic block, under a controlled pressure. Cylindrical green compacts were cast in this manner. After drying the samples at room temperature in vacuum for 24 hours. the samples were sintered in vacuum at the temperature just below the melting temperature of SiO_2 for fixed time.

The microstructure of the $Mo-SiO_2$ FGMs was observed using reflected light optical microscopy and by scanning electron microscopy (SEM). To analyze the element of the FGM along the gradient, thin slices (about 1mm thick) were cut perpendicular to it. The average Mo content of each slice was measured using calibrated EPMA of both faces.

The $Mo-SiO_2$ FGMs was joined with an electrode, and then the fused SiO_2 lump bulb. A relative spectral power distribution of the FGM-lamp was measured using spectroradiometer.

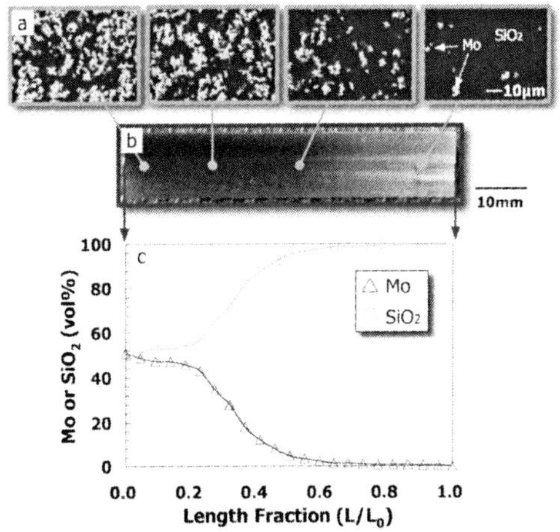

Figure 2. Characterization of $Mo-SiO_2$ functionally graded materials. a) SEM photographs of different regions of $Mo-SiO_2$ FGM. b) Optical photograph of $Mo-SiO_2$ FGM c) Compositional gradients of $Mo-SiO_2$ FGM.

RESULTS AND DISCUSSION

Figure 2 shows the microscopic characterization of $Mo-SiO_2$ FGM feedthrough for the lamp. All SEM micrographs represent the typical images of different sections taken from the FGM sample (Fig.2-a). The appearance of the FGM sample taken with a optical microscope (Fig.2-b) and the compositional gradients of the FGM (Fig.2-c) are shown. As can be seen from Fig.2-a, the Mo particles are homogeneously distributed and their distributions and sizes are gradually varied, which resulted in a microscopic compositional gradient as shown in Fig.2-c. The maximum Mo content is almost 48vol% at the front (L/L_0 nearly equal to 0) of the total FGM length, and one-fourth of the sample is practically pure silica. In addition, the composition gradient of $Mo-SiO_2$ FGM can easily be controlled by using this combined sedimentation and pressurized slip casting method.

Here we show the invention of the new type of hermetically-sealed HiDLs, using $Mo-SiO_2$ FGM ($Mo-SiO_2$

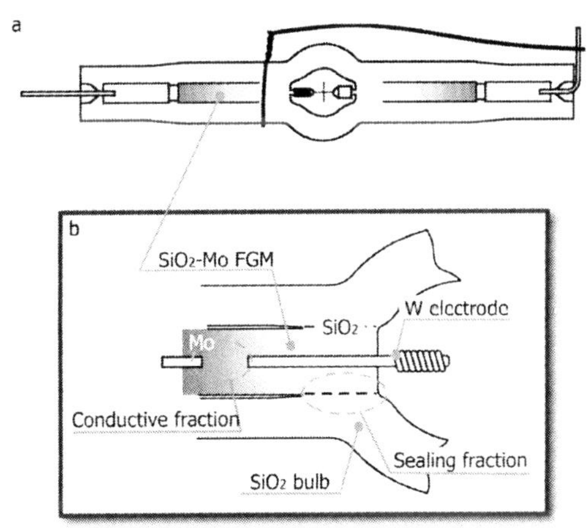

Figure 3. Schematic illustrations of a proposed FGM sealed high intensity discharge lamp. a) Structure of FGM high intensity discharge lamp. b) Sealing and the conductive fractions of lamp.

FGM-HiDL), as an electrode and a sealing component. The schematic diagrams of a proposed Mo-SiO$_2$ FGM-HiDL, and that of the sealing part and the conductive part are shown Figure 3-a, b. In the case of Mo-SiO$_2$ FGM-HiDL, the thermal expansion coefficient between Mo and SiO$_2$ is functionally graded so that it tolerates a large number of heating cycles. Namely, a part of FGM should have low resistivity in order to ensure a good electrical conductivity with a tungsten electrode, whereas the other part should be insulating and composed of pure SiO$_2$, so that this part can directly be contacted with the lamp envelope with a hermetic sealing (Fig.3-b). There is no difference in the thermal expansion coefficient between the SiO$_2$ part of Mo-SiO$_2$ FGM-HiDL and the lamp envelope, so that hermetic sealing can be achieved. Furthermore, the W electrode is totally separated from the lamp envelope by Mo-SiO$_2$ FGM, so that no cooling system is required. Lamps constructed in the manner show no leakage of the luminous elements or gases and no breakage of lamp take place, even when a high gas pressure is present inside the lamp. This enables us to fabricate compact HiDLs without the development of catastrophic thermal stresses.

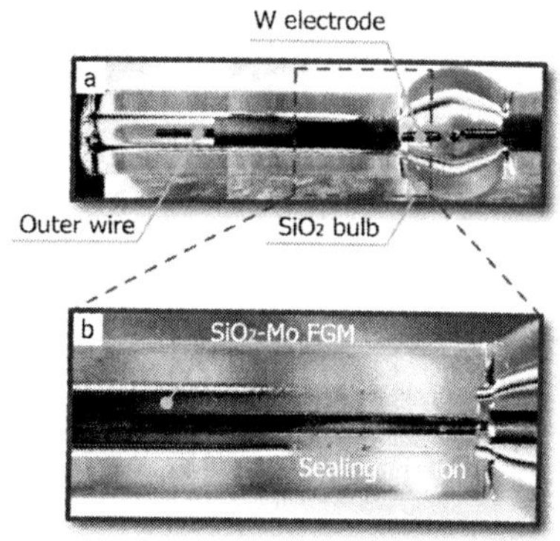

Figure 4. Structure of FGM sealed high intensity discharge lamp. a) FGM lamp structure. b) Sealing fraction of FGM lamp.

The structure and the sealing component of Mo-SiO$_2$ FGM-HiDL are as shown in Figure 4-a, b. The sealing of Mo-SiO$_2$ FGM-HiDL (Fig.4-b) is much easier than that of c-HiDLs, because the insulating part of Mo-SiO$_2$ FGM-HiDL is the same composition, SiO$_2$, as the lamp envelope. This enables us to adhere them completely and hermeticity sealed without any development of thermal stress during the manufacturing. Nevertheless, Mo-SiO$_2$ FGM-HiDL does not require the cooling system, which results as the stable, compact, and highly reliable HiDLs.

A lamp with a Mo-SiO$_2$ FGM-HiDL and the relative spectral power distribution is shown in Figure 5-a, b. A fiber optics short arc metal halide lamp and a fiber illuminator are shown in Figure 5-c, d. The light from this Mo-SiO$_2$ FGM-HiDL has almost the same special emission characteristics as the light from the sun (Fig.5-b). By placing the lamp inside the reflector mirror, a metal halide lamp with high light efficacy can be fabricated (Fig.5-c). Another manufactured

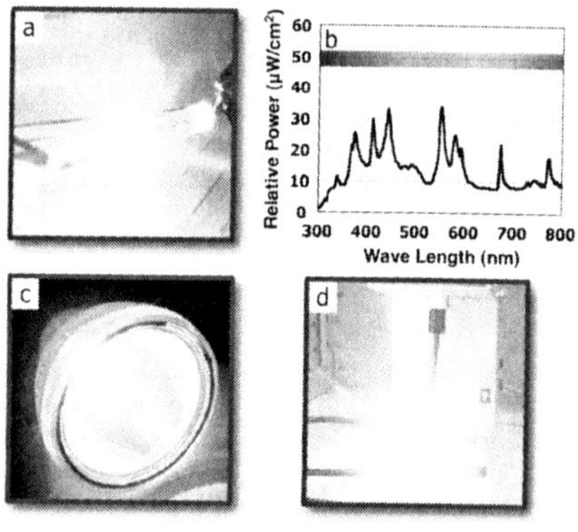

Figure 5. a) Photograph of discharge test of FGM high intensity discharge lamp. b) Relative spectral power distribution. c) Fiber optics short arc metal halide lamp. d) Fiber illuminator (application example).

sample shown is a optical fiber equipment (Fig.5-d), illuminator, where a high luminosity was achieved at low wattage. The newly developed $Mo-SiO_2$ FGM-HiDL has the following characteristics;

1. High luminous efficiency and non-explosive
2. Precise beam control allowing more than 50% of light into a 5mm diameter fiber
3. High luminous stability
4. Super-compact light sources reflector diameter 60mm
5. No cooling fan required
6. More than 4000 hours average life

CONCLUSION

We have described the fabrication of a $Mo-SiO_2$ functionally graded material using a combined sedimentation and pressurized slip casting method with a uniform gradient from the conductive part (Mo-rich phase) to the sealing part (fused silica). We have invented a new type of hermetically-sealed HiDLs with this $Mo-SiO_2$ FGM. Currently, the $Mo-SiO_2$ FGM-HiDLs become practical at 40 to 80 W (dimmable) with less than 50 mm diameter. These lamps will probably be applicable to the field of medicine, science and arts. Furthermore, they are expected to find application as a light source for in the field of automobiles, aircraft, and ships.

ACKNOWLEDGMENT

This work was supported by a Grant-in-aid for "Special Coordination Fund for Promoting Science and Technology" from the Ministry of Education, Culture, Sports, Science and Technology of Japan, which is acknowledged.

REFERENCES

[1] B. H. Rabin and I. Shiota, "Functionally gradient materials", *MRS Bull.*, **20**, 14-18 (1995).

[2] M. Koizumi and M. Niino, "Overview of FGM research in Japan", *MRS Bull.*, **20**, 19–21 (1995).

[3] S. Suresh, A. Mortensen, Fundamentals of functionally graded materials: processing and thermomechanical behaviour of graded metals and metal-ceramic composites, IOM Communications, London, (1998).

[4] A. J. Sanchez-Herencia, R. Moreno, J. R. Jurado, "Electrical transport properties in zirconia/alumina functionally graded materials", *J. Euro. Ceram. Soc.*, **20**, 1611-20 (2000).

[5] L. D. Teng, F. M. Wang and W.C. Li, "Thermodynamics and microstructure of $Ti–ZrO_2$ metal–ceramic functionally graded materials", *Mater. Sci. and Eng.*, **A293**, 130-36 (2000).

[6] C. S. Lee, X. F. Zhang and G. Thmas, "Novel joining of dissimilar ceramics in the $Si_3N_4–Al_2O_3$ system using polytypoid functional gradients", *Acta Mater.*, **49**, 3775-80 (2001).

[7] J. Chu, H. Ishibashi, K. Hayashi, H. Takebe and K. Morinaga, "Slip casting of continuous functionally gradient material", *J. Ceram. Soc. Jpn.*, **101**, 841-44 (1993).

[8] A. P. Tomsia, E. Saiz, H. Ishibashi, M. Diaz, J. Requena and S. Moya, "Powder processing of mullite/Mo functionally graded materials", *J. Euro. Ceram. Soc.* **18**, 1365-71 (1998).

[9] H. Ishibashi, H. Tobimatsu, T. Matsumoto, K. Hayashi, A.P. Tomsia and E. Saiz, "Characterization of $Mo-SiO_2$ functionally graded materials", *Metallurgical and Materials Transactions A*, **31A**, 299-308 (2000).

[10] H. Choe, T. Hsieh and J. Wolfenstine "The effect of powder processing on the coefficient of thermal expansion of $MoSi_2-Si_3N_4$ composites", *Mater. Sci. and Eng.*, **A237**, 250-55 (1997).

11. L. R. Pinckney "Transparent, high strain point spinel glass-ceramics", *J. Non-Crystal. Sol.*, **255**, 171-77 (1999).

DENSIFICATION OF Al_2O_3-SiO_2-TiO_2-BASED CERAMIC COATING FILM BY GRAIN SIZE CONTROL

Y. Muroya,
Interdisciplinary Graduate School of Engineering Sciences, Kyushu University
Kasuga-shi, Fukuoka 816-8580, Japan

K. Shimanoe, Y. Teraoka, N. Yamazoe
Faculty of Engineering Sciences, Kyushu University
Kasuga-shi, Fukuoka 816-8580, Japan

Y. Haruta,
Baham Co.,Ltd, Shibuya-ku, Tokyo 150-6019, Japan

ABSTRACT

The densification of a ceramic coating based on SiO_2-Al_2O_3-TiO_2 by using wet processing was investigated by controlling grain size of materials used. By means of ball-milling method, grain of materials in coating slurry was reduced from initial size of 10μm to less than 5μm in diameter. Ceramic composite films (about 100 μm thick) thus prepared were deposited on a steel plate by slurry coating with a doctor blade, followed by drying and calcinations in air. It was found that the coating prepared was stable to heating up to 800 °C in air, and that, however, the use of too small grains was easy to cause cracks on the surface. By optimizing the amount of alkali silicate, however, such cracks could be eliminated and also the number of through-holes of the coating could decrease drastically. The ceramic coating obtained by both the control of grain size in powders and the optimal addition of the alkali silicate showed excellent densification and thermal stabilities.

INTRODUCTION

Ceramic coatings are very widely used as inorganic functional materials in many fields of application[1-3]. Especially, ceramic coatings on metals have been used for various industrial applications aimed at improving the resistance of metals to corrosion, erosion or abrasion, and at endowing metalwork with a decorative color. The object of thermal barrier coating (TBC) is to provide thermal insulation to metal components at elevated temperature, as auto parts, thermal power parts, gas turbine, and aero-engine applications.

Most of the ceramic coating methods so far developed are based on dry processes, such as CVD[4)-6)] or PVD[7)-10)], which need large-scale equipments with high cost. On the other hand, there have been few studies on ceramic coatings based on wet processes[11, 12], although wet-processes may be more advantageous in some applications such as large surface area coating and complex-shaped body coating. Probably best known in this category is enamel work[13, 14], which has long been used for producing enameled metalware. However, enamel generally used has a weak point to weak thermal shock and physical shock, and preparation of the enamel slurry needs a lot of time and effort. In addition, it is necessary to do many preprocessing before making coating on a metal substrate. Apart from the field of structural materials, ceramic coating has been frequently carried out on metals and ceramics to fabricate functional devices like capaciters[15], memories[16], sensors[17] and so on. In this field, ceramic films are deposited on the substrate from sols or suspensions of ceramic components through dipping, screen printing, spin coating, spray coating, brush coating, etc. These wet-methods are considered to be also applicable in principle for structure materials. This motivated us to explore such a wet-method that is as simple and easy as painting but gives stable ceramic coatings on metals.

Under this situation, we found a promising simple way to deposit a ceramic film (80-100 μm thick) on metals (steel, stainless steel, copper, etc.) by a wet process[18]. The ceramic composite film deposited on steel consists of several inorganic materials (Al_2O_3, SiO_2, TiO_2 etc.). The ceramic composite film exhibited fairly good adhesion strength and thermal stability in air at temperatures below 600 °C. However, the film was spalled in pieces at 700 °C and above, because of the oxidation of the steel plate through through-holes in ceramic film. In order to improve the thermal stability in air, it is imperative to eliminate through-holes in the film, through which oxygen in air penetrates to the plate. Previously we tried to optimize the alkaline silicate component in the ceramic slurry[19]. The optimization gave results that the number of through-holes of the coating decreased drastically with increasing amount of alkali silicates, and that the resistance of the coating to corrosion in 1 mol/l HCl solution was improved. This investigation should be continued to make progress in the properties. In this study, therefore, we aimed to densify the ceramic film by controlling grain size of materials used.

EXPERIMENTAL

By means of ball-milling method, powders of Cristobalite, Kaolinite and Alumina were reduced from initial size of 10 μm to less than 5μm in diameter. The ball-mill container with a 400 ml-capacity loaded with the powder of 50-90 g 450 g with tetragonal zirconia milling balls (diameter of 3 mm) and 70-90 ml of water for a day. The coating slurry was prepared by suspending or dissolving powders of ceramic components, glass-forming-components (Sodium and Potassium silicates involving water 45wt.% and 69wt.%, respectively) and Na_2CO_3 in water under stirring at room temperature. The viscosity of slurry milled was too high to form ceramic coatings. In order decrease the viscosity, therefore, excess of water was added to the slurry. The composition of several coating slurries used is shown in Table 1. The composition ratios of

Tab. 1. Composition of the coating slurries (wt.%).

	Cristobalite	Alumina	Titania	Kaolinite	Sodium carbonate	Sodium silicate	Potassium silicate	water
H-1	15.0	12.0	7.5	18.0	3.0	13.0	13.0	18.5
H-1b	13.2*	10.6*	6.6	15.8*	2.6	11.4	11.4	28.4
H-2b	12.7*	10.4*	6.4	15.3*	2.5	12.5	12.5	27.7
H-3b	12.6*	10.0*	6.3	15.1*	2.5	13.4	13.4	26.7
H-4b	12.1*	9.7*	6.0	14.6*	2.4	15.0	15.0	25.2
H-5b	11.5*	9.2*	5.8	13.9*	2.3	17.5	17.5	22.3

* Milled powder.

Cristobalite, Kaolinite, Titania, Alumina and Sodium carbonate were prepared to be same in all of the samples by controlling the addition of Sodium and Potassium silicates and water. H-1 was original composition ratio, and the others had same amount of water.

Steel plates (JIS G3141 SPCC-SB), cut into a size of 30 mm (width) × 35 mm (length) × 0.6 mm (thick), were used as a substrate for ceramic coating. The steel plates were blasted with alumina grains of 0.1 mm in diameter for surface roughening, and washed with acetone before provided for use. The coating slurries, under continuous agitation by stirring, were applied on the steel plates with a doctor blade at room temperature. The plates with the coating of about 100 μm thick were dried overnight at 120 °C, and pre-calcined at 420 °C for 30 min in air atmosphere. The ceramic coated plates was heated in an electrical furnace to an elevated temperature up to 900 °C at a rate of 5 °C / min, and calcined at temperature for 30 min in air atmosphere and cooled to room temperature.

The prepared coatings were analyzed by means of X-ray diffraction (XRD) and scanning electron microscopy (SEM). In order to visualize through-holes in ceramic coating, electrochemical deposition of copper was performed on each coating in a copper sulfate (0.01 M $CuSO_4$) bath at room temperature under potentiometric condition at -800 mV vs. SCE. The ceramic coating was carried out only on one side of the steel plate. The bare surface of the opposite site was covered with an organic resin to avoid the electrochemical deposition of copper. In addition, the through-holes of ceramic coatings were evaluated when each coating was immersed in 1 M HCl at room temperature by counting the number of the spots of the bubbles generated from the ceramic coating and by measuring the time for the coating to begin exfoliating from the substrate.

RESULTS AND DISCUSSION

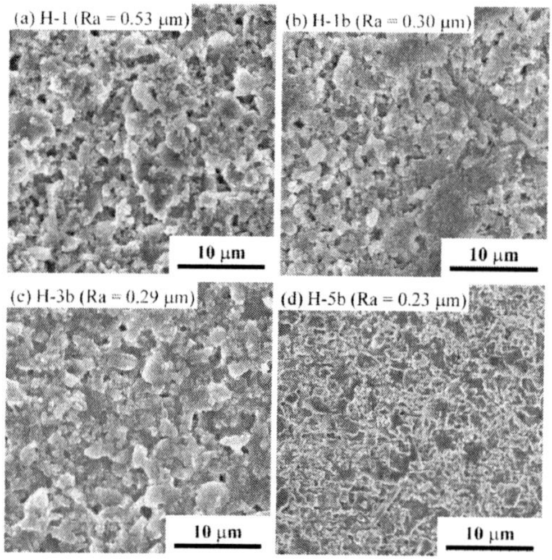

Fig. 1. SEM images of surface for ceramic coatings calcined at 420 °C.

The morphology of the surface of ceramic coatings

Figure 1 shows SEM images of surface and the surface roughness for the several coatings calcined at 420 °C in air. The H-1 and H-1b coatings calcined at 420 °C in air, as shown in Fig. 1(a) and (b), consisted of ceramic grains of various sizes ranging from 0.2 μm to 5 μm and most of the grains aggregated rather porously, although the grain size of ceramic powders decreased. On the other hand, however, the ceramic grains for the H-3b and H-5b, which were prepared by increasing amount of glass-forming-components compared with H-1b, coagulated together more heavily to give smoother, less porous surface morphology, as shown in Fig. 1(c) and (d). The roughness of the H-1b, H-3b and H-5b were half one of H-1. These results indicate that the milled powders decrease the roughness of the coating surfaces, and that the amount of glass-forming-components should be optimized to composition of the small grains ball-milled.

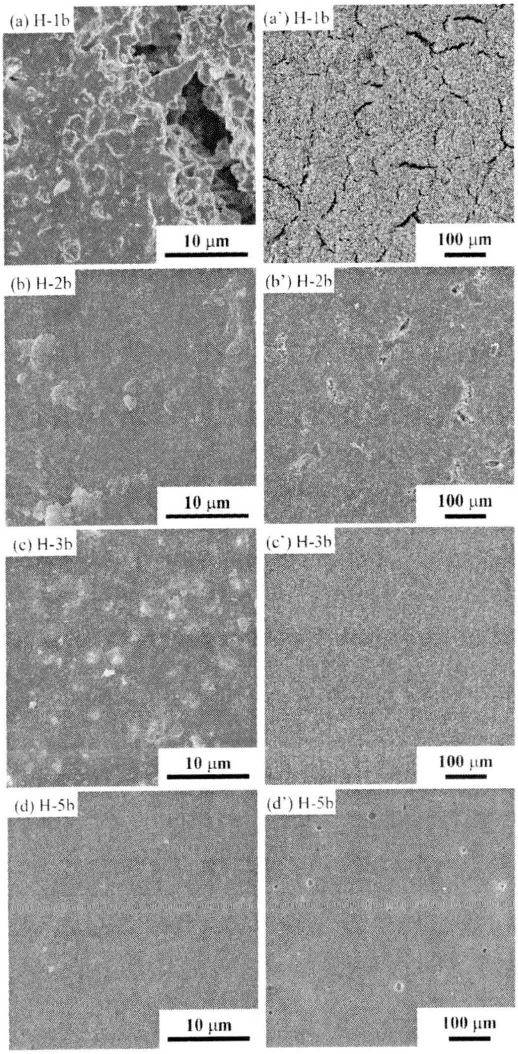

Fig. 2. SEM images of surface for ceramic coatings calcined at 900° C.

After calcination at 900 °C in N_2, the coatings lost granular feature almost completely and gave fairly flat surface morphology, as shown in Fig. 2. However it was found that the addition of small amount of glass-forming-components brought about a lot of small cracks, although large amount of addition gave pinholes as shown in Fig. 2 (d). With increasing amount of glass-forming-components, such cracks remarkably reduced. In order to understand the morphological changes just mentioned, XRD analysis was carried out for the respective coatings. Figure 3 shows XRD patterns of the coatings after calcinations at 420 °C and 900 °C in air and N_2, respectively. Each of the starting ceramic components could be identified in every coatings calcined. However after calcined at higher temperature, XRD peaks for kaolinite and cristobalite became weaker relative to those of titania and alumina, accompanied by appearance of a halo pattern in the 2θ range of 20-30 °. For H-1b and H-2b coatings calcined, XRD peaks were almost the same as those of H-1. In addition, the intensity of halo pattern for the coatings calcined at 900 °C slightly increased with increase in amount of glass-forming-components. From this result, the component of coatings, especially kaolinite and cristobalite, seems to react slightly with glass-forming-components (sodium and potassium silicates). On the basis of above results, the morphological changes of the ceramic coatings

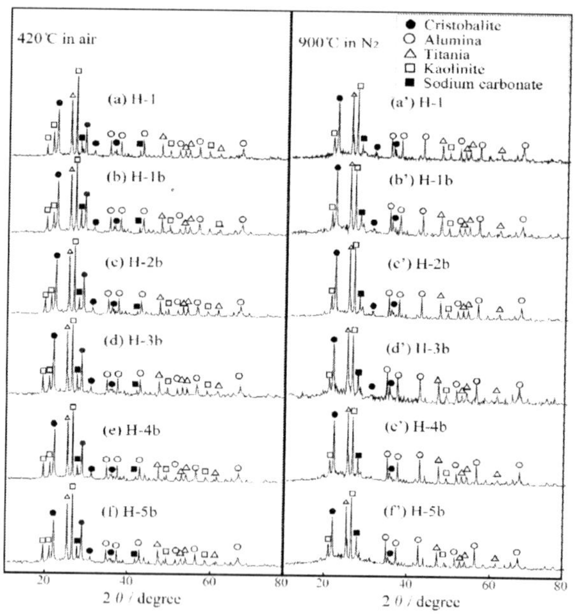

Fig. 3. XRD patterns for ceramic coatings.

observed above may be associated with formation of a glass phase. This indicates that H-4b and H-5b more easily form the glass phase because of the higher content of glass-forming-components and the small grain size of powders. It is thought that the glass phase filled in pores between ceramic grains, eventually eliminating through-holes in the coatings. Furthermore the optimal addition of alkali silicates has a great effect to eliminate cracks from the coatings, because it relaxes the difference of the thermal expansion coefficients of the coating and the substrate.

Thermal stability of ceramic coatings

On calcination in air atmosphere, the ceramic coatings adhered well to the steel plate up to 700 °C. However, the ceramic coatings of H-1 spalled completely when calcined at 800 °C and above. In each spalled piece, the face of coating contacted to the steel substrate was colored black. It is suspected that the precalcined coatings possessed small through-holes, through which oxygen in air penetrates to the surface of the plate. The steel surface was oxidized to FeO and Fe_3O_4 at sufficiently high temperature, resulting in the observed spalling of coatings. The presence of through-holes was experimentally confirmed as stated later. On the other hand, the ceramic coatings of H-1b, H-2b, H-3b H-4b and H-5b did not spall up to 800 °C in air atmosphere. This result seems to be attributed to decreasing cavity in the coating film by using the milled powders and the optimal addition of alkali silicate, which have a great effect to eliminate through-holes from the coatings. This was also confirmed experimentally, as stated later.

The stabilities to thermal shocks were tested for respective ceramic coatings. Each specimen, heated at a designated temperature (Th) for 5 min, was taken out to room temperature. The same procedure was repeated 10 times, while the state of ceramic coating was checked on an optical microscope each time. Th was varied between 700 and 800 °C. The results are summarized in Table 2. All specimens remained intact at all after the thermal shock tests when Th was set to be 700 °C. When Th was set to be 800 °C, the H-1 and H-1b coatings were exfoliated in the second and the fourth runs, respectively. On the other hand, the more addition of alkali silicate could reduce the thermal shock, because no damage was found on the coatings for Th = 700 and 800 °C. These results indicate that all specimens stand the thermal shocks between 700 °C and room. In addition, it is considered that the glass phase formed by adding alkali silicate may absorb strain generated from the difference of the thermal expansion coefficients of the coating and the substrate.

Tab. 2. The stabilities to thermal shocks between Th and room temperature for ceramic coatings.

Temperature of heating (Th)	H-1	H-1b	H-2b	H-3b	H-4b	H-5b
700℃	No damage	No damage	No damage	No damage	No damage	No damage
800℃	Exfoliation after 2 cycles	Exfoliation after 4 cycles	No damage	No damage	No damage	No damage

"No damage" means that no exfoliation and cracks were found after 10 cycles.

Evaluation of though-hole

As mentioned above, the presence of through-holes in the precalcined coating was suspected from the fact that H-1 coatings spalled when calcined at 800 °C and above in air atmosphere. The porous nature of the coating precalcined at 420 °C or calcined at 800 °C was obvious from the SEM images, but it was almost impossible to identify the through-holes of the coatings from such observations. It was thus tried to visualize the through-holes by Cu electrochemical deposition method. If there were through-holes in the coating, copper ions (Cu^{2+}) penetrated in the film and deposited electrochemically on the steel substrate. Each deposit of Cu can be identified clearly from its copper metallic color. As an example, Fig. 4 shows optical microscopic images of the H-1 coating after the Cu electrochemical deposition for 1 h. The through-holes visualized by the Cu-deposition are indicated by arrows in Fig. 4 (a). The cross-sectional image (b) shows that the deposited Cu is found from the surface of the coating to the substrate. This means that the Cu-deposition occurs through pores between particles. Therefore if such pores are very few, visualization by the Cu-deposition seems to be very difficult. In fact, no Cu-deposition was observed in the H-4b and H-5b coatings.

In order to evaluate the through-holes of densified ceramic coatings, a corrosion test was carried out in 1 mol/l HCl aq. The through-holes were evaluated by counting the number of the points of hydrogen bubbles generated from the surface of a ceramic-coated steel plate immersed

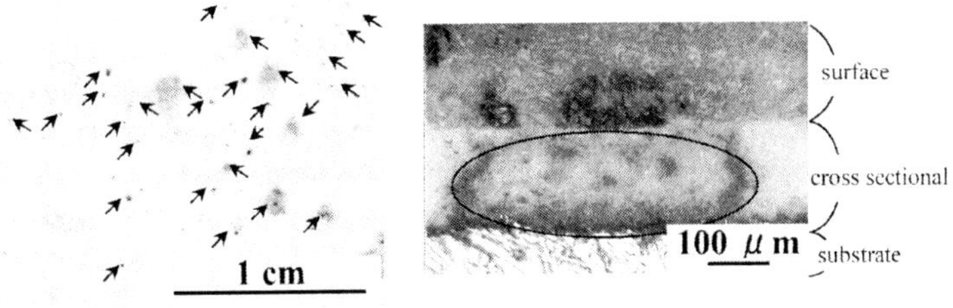

(a) surface
(Arrows indicate copper deposits)

(b) cross sectional
(Circle indicate copper deposit)

Fig. 4. The image of the H-1ceramic coating after electrodeposition of copper.

Tab. 3. The number of the points of bubbles generated from ceramic coating and the time until the ceramic coatings exfoliated when immersed in 1 M HCl at room temperature.

specimen	H-1	H-1b	H-2b	H-3b	H-4b	H-5b
Number of the points of the generated bubbles / cm^{-2}	×	×	×	×	32.1	20.4
Time until exfoliating / h	0.8	1.3	2.0	3.4	4.8	7.2

×: The number of the generated bubbles was too many to count.

in 1 M HCl aq. against time until the ceramic coatings exfoliated. The results of the corrosion test are summarized in Table 3. The bubbles could be found even on the densified ceramic coatings, meaning that this method is useful for evaluating small through-holes. The number of the generated bubbles in H-1, H-1b, H-2b and H-3b coatings were too many to count. On the other hand, it is turned out that the number of the generated bubbles decreased drastically with increasing amount of alkali silicates. The H-5b had especially good densification. Time for Exfoliating also strongly supports the result observed in the number of the bubbles. Acidic solution penetrates the through-holes and reacts with the steel surface, causing the exfoliation of the coating from the steel plate. Therefore it was found that the time for exfoliating was extended. From the above results, it is obvious that both the control of grain size in powders and the optimal addition of the alkali silicate are important to densify the ceramic coating.

CONCLUSION

Densification of ceramic composite coatings in a wet process-based method was investigated by controlling grain size of materials used. The following conclusions are drawn from the present study.

1) The number of through-holes of the coating decreased by both the control of grain size in powders and the optimal addition of the alkali silicate. It is thought that the densification of the ceramic coatings may be associated with formation of a glass phase by the higher content of glass-forming-components and the small grain size of powders, that is, the glass phase filled in pores between ceramic grains, eventually eliminating through-holes in the coatings.
2) It was found that the coating prepared was stable to heating up to 800 °C in air. The stabilities to thermal shocks have been improved with increasing amounts of alkali silicates
3) The corrosion tests in 1 M HCl aq. showed the H-5b had especially good densification.

REFERENCES

1) J. Takadoum, H. H. Bennani, "Influence of substrate roughness and coating thickness on adhesion, friction and wear of TiN films," *Surf. Coat. Technol.*, **96**, 272-82 (1997).

2) Y. Y. Guu, J. F. Lin, "Analysis of wear behaviour of titanium carbonitride coatings," *Wear*, **210**, 245-54 (1997).

3) F. Attar, "Hardness evaluation of thin ceramic coatings on tool steel," *Surf. Coat. Technol.*, **78**, 78-86 (1996).

4) K. L. Choy, "Chemical vapour deposition of coatings," *Progress in Mater. Sci.*, **48**, 57-170 (2003).

5) R. A. Erck, P. S. Maiya, "Fracture behavior of graphite coated with titanium compounds by chemical vapor deposition," *Mater. Sci. and Eng. A*, **251**, 251-4 (1998).

6) C. Li, B. Hua, "Preparation of nanocrystalline SnO_2 thin film coated Al_2O_3 ultrafine pasticles by fluidized chemical vapor deposition," *Thin Solid Films*, **310**, 238-43 (1997).

7) S.Korablov, M.Yoshimura, "Hydrothermal corrosion of TiN PVD films on SUS-304," *Corrosion Sci.*, **45**, 531-43 (2003).

8) Hui-Ping Feng, Cheng-Hsun Hsu, Jung-Kai Lu, Yih-Hsun Shy, "Effects of PVD sputtered coatings on the corrosion resistance of AISI 304 stainless steel," *Mater. Sci. and Eng. A*, **347**, 123-9 (2003).

9) S. Carrera, O. Salas, J. J. Moore, A. Woolverton and E. Sutter, "Performance of CrN/MoS_2(Ti) coatings for high wear low friction applications," *Surf. Coat. Technol.*, **167**, 25-32 (2003).

10) S. Ulrich, T. Theel, J. Schwan, H. Ehrhardt, "Magnetron-sputtered superhard materials," *Surf. Coat. Technol.*, **97**, 45-59 (1997).

11) B.Su, C.B.Ponton, T.W.Button, "Hydrothermal and electrophoretic deposition of lead zirconate titanate (PZT) films," *J. Eur. Ceram .Soc.*, **21**, 1539-42 (2001).

12) Fanglin Chen, Meilin Liu, "Preparation of yttria-stabilized zirconia (YSZ) films on $La_{0.85}Sr_{0.15}MnO_3$ (LSM) and LSM–YSZ substrates using an electrophoretic deposition (EPD) process," *J. Eur. Ceram. Soc.*, **21**, 127-34 (2001).

13) Z.Tang, F.Wang, W. Wu, "Effect of Al_2O_3 and enamel coatings on 900°C oxidation and hot corrosion behaviors of gamma-TiAl," *Mater. Sci. and Eng. A*, **276**, 70-5 (2000).

14) Y. Xiong, S. Zhu, F. Wang, "The oxidation behavior and mechanical performance of Ti60 alloy with enamel coating," *Surf. Coat. Technol.*, **190**, 195-9 (2005).

15) C. J. Hsu, J. H. Jean, "Formulation and dispersion of NiCuZn ferrite paste," *Mater. Chem. and Phys.*, **78**, 323-9 (2002).

16) S. T. Kim, J. W. Kim, S. W. Jung, J. S. Shin, S. T. Ahn, W. J. Lee, "Electrical properties of PZT thin films deposited by electron cyclotron resonance plasma enhanced chemical vapor deposition," *Mater. Chem. and Phys.*, **45**, 155-8 (1996).

17) T. Kida, Y. Miyachi, K. Shimanoe, N. Yamazoe, "NASICON thick film-based CO_2 sensor prepared by a sol–gel method," *Sensors and Actuators B: Chemical*, **80**, 28-32 (2001).

18) Y. Muroya, A. Motoki, G. Sakai, K. Shimanoe, T. Maeda, Y. Haruta, Y. Teraoka, N. Yamazoe, "New type of ceramic composite films coated on steel by wet process," *Proc. of the Electrochem. Soc.* **PV 2003-16**, 443-52, (2003).

19) Y. Muroya, A. Motoki, K. Shimanoe, T. Maeda, Y. Haruta, Y. Teraoka, N. Yamazoe, "Densification of ceramic composite films coated on steel by wet process," *Proc. of the Electrochem. Soc.* In press.

Processing and Fabrication—Non-Oxide Systems

GRAIN GROWTH OF β-SILICON NITRIDE IN RE-Mg-Si-O-N LIQUID

Noritaka SAITO, Daiji NAKATA and Kunihiko NAKASHIMA
Department of Materials Process and Engineering, Graduate School of Engineering, Kyushu University
6-10-1 Hakozaki, Higashi-ku, Fukuoka 812-8581, Japan

Ayumu UMEMOTO
TOTO Ltd., Advanced Ceramic Division
2-1-1 Nakashima, Kokurakita-ku, Kitakyushu 802-8601, Japan

ABSTRACT

Isothermal growth of β-Si_3N_4 crystals dispersed in a series of rare-earth (RE = Y, Gd, Nd and La) oxynitride liquids (RE-Mg-Si-O-N) was studied after heat treatment at 1773 K for 64 to 512 min. The microstructural characterization of α- and β-Si_3N_4 crystals was quantitatively analyzed by electron microscopy. The volume fraction of β-Si_3N_4 crystals in oxynitride glasses increased with heat treatment time. It is determined that the rate of α/β transformation in the oxynitride liquid with Y_2O_3 additive can be faster than that in oxynitride liquids with other additives. Likewise, the radii of rare-earth elements make significant differences in the mean grain size and aspect ratio of β-Si_3N_4 crystals. In all cases mentioned above, a marked change was observed in the order of ionic radius from La to Y. The mean width of the needlelike β-Si_3N_4 crystals heat-treated with La_2O_3 additive was almost constant in the time range of heat treatment in this study. However, the aspect ratio of β-Si_3N_4 crystals with La_2O_3 additive was found to be much larger than that in oxynitride liquids with other additives, indicating negligible growth of β-prism plane.

INTRODUCTION

Since the excellent mechanical properties of Si_3N_4 based ceramics have been attributed to rod-like grain shape due to the highly anisotropic grain growth of β-Si_3N_4 [1], understanding of grain growth kinetics is a key issue in order design high-toughness and high-strength materials. Si_3N_4 based ceramics are generally produced by using α-Si_3N_4 rich raw powder. Densification is achieved by adding oxide additives which form oxynitride melts and promote liquid-phase sintering. Therefore, it is also a fundamental issue for clarification how the grain growth behavior is affected by the physical properties (viscosity, surface tension and wettablity, etc.) of the oxynitride liquid phase.

Rare-earth oxides (RE_2O_3) have been widely investigated as sintering additives for Si_3N_4 ceramics [2-9]. Rare-earth elements including Sc, Y and lanthanides (Ln) have a common ionic valence Z of +3, and are known to be the most electropositive elements. Thus, it has been suggested that they are suitable sintering additives for Si_3N_4 [10]. Because Lanthanide elements continuously decrease their ionic radii r with increasing atomic number, known as lanthanide contraction, their cationic field strength Z/r^2 also continuously change with their atomic number. This causes substantial variations in the various properties of Ln-Si-Al-O-N glasses [11-14] that are the grain boundary phase of Si_3N_4 ceramics sintered with rare-earth oxide additives, and thus, can also affect grain growth of β-Si_3N_4 itself. For example, Hoffmann and co-researchers [15-17] were the first researchers to present the evolution of width-length and width-aspect-ratio distributions during and after the phase transformation of Si_3N_4 in RE-Si-Al-O-N liquid. And, it has been observed that the mean aspect ratio at the end of the phase transformation is proportional to the ionic radius of rare-earth elements [18]. However, the investigated compositions contained Al_2O_3, which would form SiAlON. Kitayama et al. observed that rare-earth oxide elements have an influence not only during phase transformation but also during Ostwald ripening in a loose powder compact of Ln_2O_3-SiO_2-Si_3N_4 [19]. Their data, however, are strongly influenced by the crystal impingement because of the high volume fraction of β-Si_3N_4. Therefore, a direct comparison of their observation with theoretical or with kinetic models is difficult. In this paper, we intend to investigate the grain growth behavior of β-Si_3N_4 in RE-Mg-Si-O-N liquids.

EXPERIMENTAL

Powder Processing and Sintering

Powders of α-Si_3N_4 (SN-E05, β phase content 5% as impurity, UBE Industries Ltd., Japan), RE_2O_3 (RE=Y, Gd, Nd and La), MgO and SiO_2 (Sigma Aldrich Japan Co., Japan) at a molar ratio of α-Si_3N_4:RE_2O_3:MgO:SiO_2=10:8.1:36.9:45 were mixed in a planetary mill for 2 h in a Si_3N_4 jar with Si_3N_4 balls by using methanol as a mixing medium. The resulting slurry was dried and passed through a 60 mesh polyethylene sieve. Portions of 1 g were cold isostatically pressed into pellets at 100 MPa for 3 min. The heat treatment was performed in a graphite-heated gas pressure furnace (Fuji Dempa Kogyo Co., Japan). Pellets of all four samples with rare-earth oxides were placed in a high-purity BN crucible (N1 grade, Denki Kagaku Kogyo Co., Japan) and sintered under 0.9 MPa nitrogen-gas for a specified time. The temperature was controlled to be 1773 ± 2 K with a Pt/Pt-13Rh (mass%) thermocouple and an optical pyrometer. A constant heating rate of 15 K/min from room temperature to 1773 K was used, then after, on reaching the temperature, samples were annealed for 64, 128, 256 and 512 min. The cooling rate was around 80 K/min above 1273 K by the furnace power off. After these heat-treatments the samples have a droplet shapes (Fig. 1), that indicate completed melting.

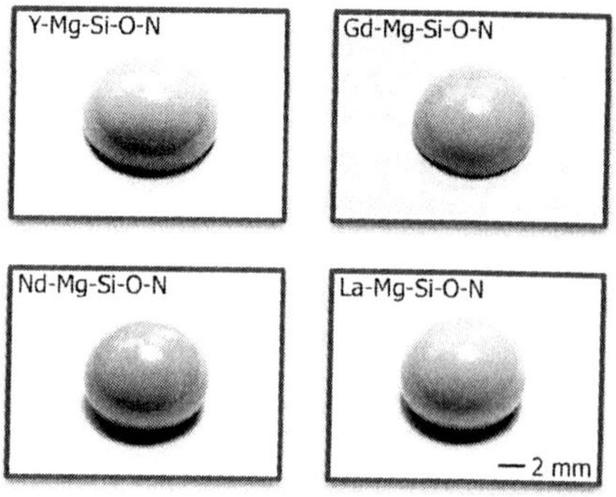

Figure 1. Typical appearance of RE-Mg-Si-O-N compact after heat treatment at 1773 K for 64 min.

Characterization

Phase identification of the sintered bodies was performed by X-ray diffraction (XRD) with CuKα radiation. Coarse-grained silicon powders were used as an internal XRD standard. The α:β ratio was calculated using Gazzara's equation [20]. Scanning electron microscopy (EPMA-1600, Shimadzu, Kyoto, Japan) was used for further phase identification and microstructural characterization. Fig. 2(a) shows a typical image in the BSE (back-scattering electron) mode, revealing the compositional contrast between Si_3N_4 crystals (dark contrast) and RE-Mg-Si-O-N oxynitride glass (bright contrast).

To determine the dimensions of β-Si_3N_4 crystals (widths and lengths) by using the linear intercept method, polished sections of samples were observed in the back-scattered electron mode. For accurate measurement, polished sections of heat treated samples were chemically etched by using molten alkali hydroxides (NaOH and KOH in equimolar) at 573 K [21]. Fig. 2(b) reveals the hexagonal rodlike shapes of β-Si_3N_4 crystals, in contrast with before chemical etching image, cf. Fig. 2(a). For accurate quantitative analysis of β-Si_3N_4 grain dimensions, crystals with only one end visible but with the other end embedded in the oxynitride glass were not considered. Likewise, crystals without at least two side surfaces entirely exposed were excluded. The remaining crystals that were visible on the SEM images were counted and measured. At least 30 images were taken for the determination of the length (L) and width (W) distributions at each samples.

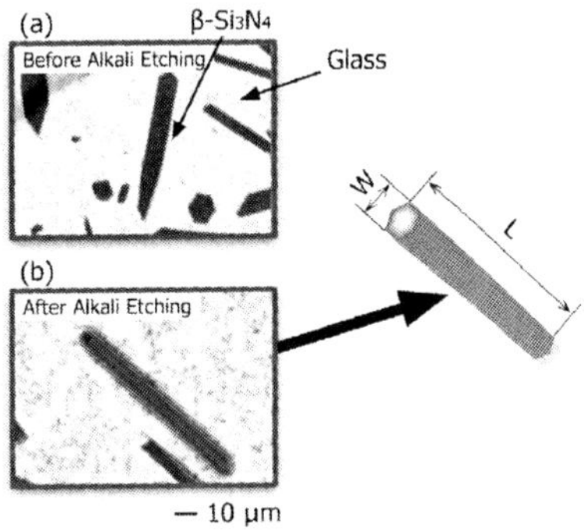

Figure 2. SEM-BSE images of RE-Mg-Si-O-N sinters before and after alkali etching.

RESULTS AND DISCUSSION

A series of XRD patterns taken from Y-Mg-Si-O-N is shown in Fig. 3. Background signal is due to glass matrix. Phase analysis based on XRD clearly reveals the dissolution of α-Si_3N_4 with the attendant formation of β-Si_3N_4. Fig. 3 shows the coexistence of α- and β-Si_3N_4 at 64 and 128 min, but not at 512 min. The sample after 512 min heat treatment, however, contained a negligible amount of α-Si_3N_4; the remainder of the crystalline phase was β-Si_3N_4 only. This development means that the phase transformation is almost completed in Y-Mg-Si-O-N liquid at 1773 K in 512 min, and that grain growth proceeds via Ostwald ripening.

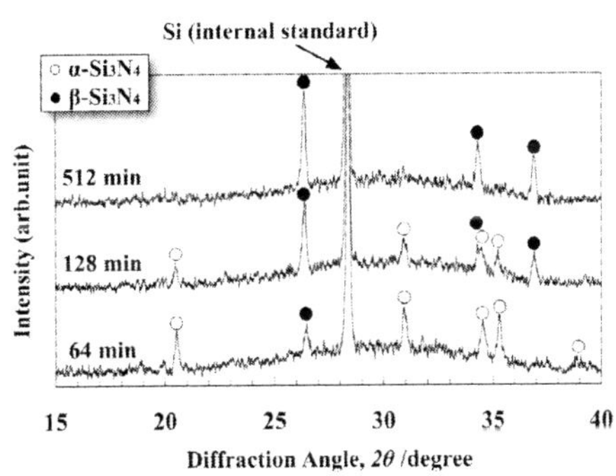

Figure 3. Typical X-ray diffraction pattern of Y-Mg-Si-O-N compacts.

The volume fraction of β-Si_3N_4 crystals derived from XRD intensities is shown in Fig. 4 for all compositions. The initial value stands for β phase content as impurity in α-Si_3N_4 raw powder. The volume fraction of β-Si_3N_4 crystals rapidly increases with heat treatment time, as expected α to β phase transformation and then saturated. The final value (at 512 min) monotonically decreases in the order of ionic radius from Y to La. It is determined that α/β transformation in the oxynitride liquid with Y_2O_3 additive can be faster than that in oxynitride liquid with other additives. Since the RE/Mg/Si and O/N atomic ratios are same for all samples, already mentioned above, this decrease in the volume fraction with ionic radius indicates that the composition range of the liquid is much larger in La_2O_3 additive case and smaller in the Y_2O_3 additive one. It is consistent with our preliminary data that show the liquidus temperature of RE_2O_3-MgO-SiO_2 systems to decrease from Y to La [22].

Fig. 5 shows the typical microstructure of the mechanically polished section of RE-Mg-Si-O-N after 64 (upper row) and 512 min (bottom row), respectively. The round particles (α-Si_3N_4) appear gray and their numbers and sizes decrease with increasing heat treatment time, while the faceted grains (β-Si_3N_4) grow during the heat

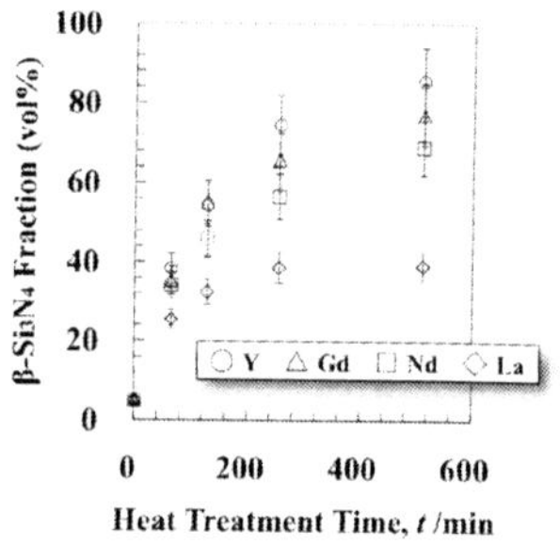

Figure 4. Change in volume fraction of β-silicon nitride in RE-Mg-Si-O-N sinters with heat treatment time.

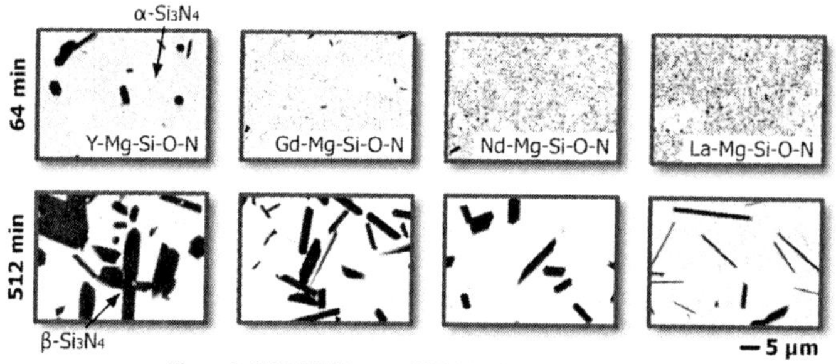

Figure 5. SEM-BSE image of RE-Mg-Si-O-N sinters (polished section) after heat treatment for 64 and 512 min at 1773 K.

treatment by consumption of the dissolving α-Si_3N_4. In La-Mg-Si-O-N liquid heat-treated at 1773 K for 512 min, round particles (α-Si_3N_4) are still observed as shown in Fig. 5 (bottom row right), and this means that the phase transformation is not completed, because of, in our preliminary experiment, lower supersaturation with Si_3N_4 due to higher solubility of Si_3N_4 [23]. After 512 min heat treatment, a fibrous shape of β-Si_3N_4 crystals in the sample with La_2O_3 makes a remarkable contrast with a rather isotropic shape of the sample with Y_2O_3. Krämer et al. performed a periodic bond chain (PBC) analysis of β-Si_3N_4 crystal, and found that its elongated grain shape was originated from a much higher activation energy for nucleation on the (100) facet than on the (001) facet [24]. This strongly suggests that grain growth in the [210] direction is controlled by the interfacial reaction, which results in an atomically flat crystal surface, namely a facet. Thus, the appearance of facet in Fig. 5 is clear evidence of interfacial reaction controlled kinetics.

Fig. 6 shows width-length distribution of β-Si_3N_4 crystals in RE-Mg-Si-O-N liquid heat-treated at 1773 K for various annealing time, in total, at least 200 crystals were used to determine these distributions. After 64 min annealing, rare-earth additives make little difference in width-length distribution. However, they make significant differences as time advances. La_2O_3 yields almost one-dimensional growth in the length direction, while Y_2O_3 gives an appreciable growth in the width direction. As a result, wide distributions are observed in length for La_2O_3 and in width for Y_2O_3. Gd_2O_3 and Nd_2O_3 show an intermediate behavior between both additives.

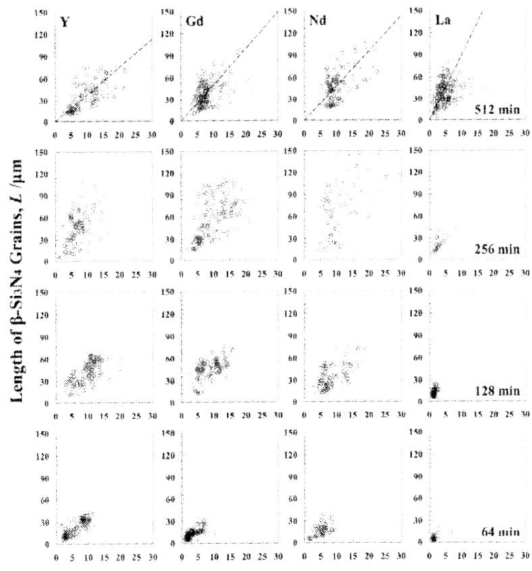

Figure 6. Width-length distribution of β-silicon nitride dispersed in RE-Mg-Si-O-N sinters after heat treatment at 1773 K

Fig. 7 shows time evolution of mean widths and lengths of β-Si_3N_4 crystals dispersed in RE-Mg-Si-O-N liquid heat-treated at 1773 K. The mean lengths are much larger than the mean widths, which reflects the rodlike morphology of β-Si_3N_4. The mean widths of needlelike β-Si_3N_4 crystals increase with heat treatment time. Among them, La_2O_3 causes the lowest width and length growth rates. Also, the mean widths increase much slower than the mean lengths (64~256 min), which suggests interfacial reaction controlled growth in [210] direction. After 256 min annealing, the mean length rapidly decreased expect in La-Mg-Si-O-N melt, which suggests the elongated β-Si_3N_4 crystals re-dissolve from (001) facets into RE-oxynitride liquids, so called anisotropic Ostwald ripening.

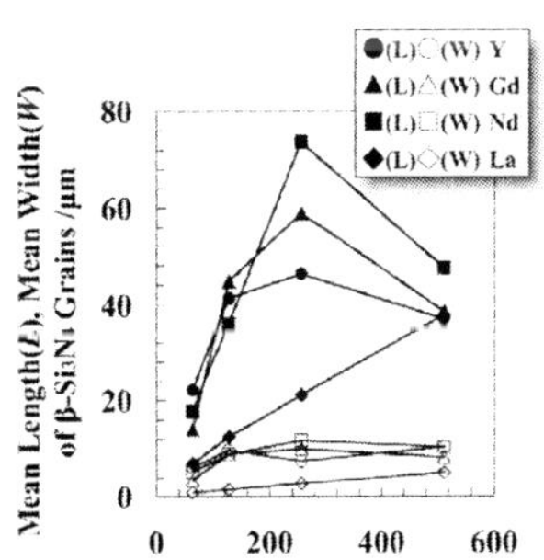

Figure 7. Time evolution of mean length and mean width of β-silicon nitride dispersed in RE-Mg-Si-O-N sinters heat treated at 1773 K

Fig. 8 reveals time evolution of mean aspect ratios of β-Si_3N_4 crystals dispersed in RE-Mg-Si-O-N liquid heat-treated at 1773 K. The evolutions of mean aspect ratios of β-Si_3N_4 strongly reflect the variation of the mean length in each additive, because the increases of the mean widths are much smaller than the variation of the mean length (cf. Fig. 7). In particular, the mean aspect ratios drastically increase (64~256 min), which suggests this period of heat treatment corresponds to α/β phase transformation [15], then after rapidly decrease in Y, Gd and Nd additives, corresponding to anisotropic Ostwald ripening [15]. While, the mean aspect ratio of β-Si_3N_4 crystals gradually increased with heat treatment time, among others in La-Mg-Si-O-N liquid, the aspect ratio was found to be much larger. These developments indicate, in La-Mg-Si-O-N liquid, preferential growth in [001] direction and the α/β phase transformation has not finished after 512 min heat treatment at 1773 K.

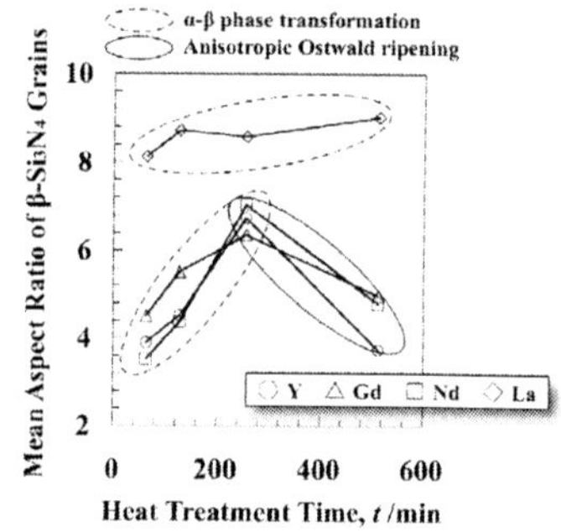

Figure 8. Time evolution of mean aspect ratio of β-silicon nitride dispersed in RE-Mg-Si-O-N sinters heat treated at 1773 K

In Fig. 9, the mean aspect ratio of β-Si_3N_4 crystals after 512 min heat treatment is plotted as a function of the cationic field strength (CFS = Z/r^2) of RE^{3+}, which Z is cation valence and r is cationic radius, under the assumption that all cations present a 6-fold coordination. The mean aspect ratio of β-Si_3N_4 linearly decreases with increasing the cationic field strength, except La additive. Note that La_2O_3 yields much larger aspect ratio than expected. Because the α/β phase transformation has not achieved equilibrium under current experimental conditions. In addition, the coordination numbers (CNs) of rare-earth ions in the RE-Mg-Si-O-N liquid phase are not known exactly, and CNs higher than 6 were reported for rare-earth cations [14]. Thus, the application of cationic radius based on 6-fold coordination would be too simplistic. However, Satet et al. have investigated grain growth anisotropy of β-Si_3N_4 in Me-Mg-Si-O-N (Me = Sc, Lu, Yb, Y, Sm and La) and reported that the near-equilibrium aspect ratio of β-Si_3N_4 increased linearly with increasing radius of the Me^{3+} cation, moreover depending on whether Me^{3+} belong to the 3rd group of the periodic table of elements or to the lanthanides, there were two distinct correlations [25]. As already mentioned in the introduction, it was reported that radii of lanthanide ions showed a good positive correlation with their mean aspect ratios at the end of the phase transformation in Ln-Si-Al-O-N liquids. It was also reported that lanthanide ions have a strong effect on grain growth behavior not during the

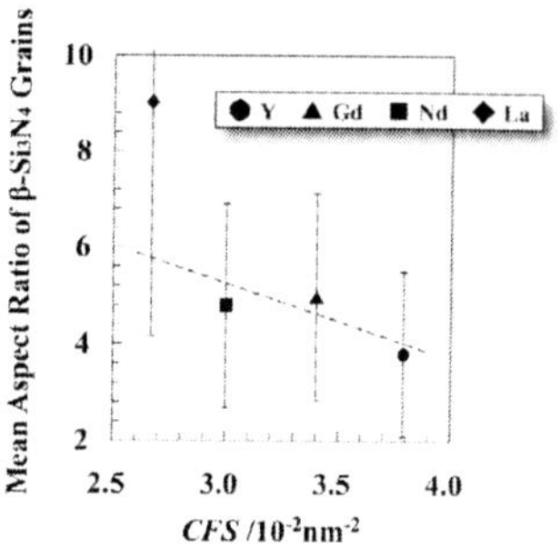

Figure 9. Effect of the cationic field strength of rare-earth elements on mean aspect ratio of β-silicon nitride dispersed in RE-Mg-Si-O-N sinters after heat treatment for 512 min at 1773 K

phase transformation but during Ostwald ripening. In this work, our results basically agree with previous works, however a little change was observed, because of non-equilibrium in grain growth of β-Si_3N_4 under limited annealing time.

CONCLUSION

Isothermal growth of β-Si_3N_4 crystals dispersed in a series of rare-earth (RE = Y, Gd, Nd and La) oxynitride liquids (RE-Mg-Si-O-N) was studied after heat treatment at 1773 K for 1 to 8 h.

1. Phase analysis based on XRD clearly reveals the dissolution of α-Si_3N_4 with the attendant formation of β-Si_3N_4.

2. It is determined that α/β transformation in the oxynitride liquid with Y_2O_3 additive can be faster than that in oxynitride liquid with other additives.

3. After 512 min heat treatment, a fibrous shape of β-Si_3N_4 crystals in the sample with La_2O_3 makes a remarkable contrast with a rather isotropic shape of the sample with Y_2O_3.

4. The mean diameter of the needlelike β-Si_3N_4 crystals in the sample with La_2O_3 additive was almost constant in the time range of heat treatment in this study. However, the aspect ratio of β-Si_3N_4 crystals with La_2O_3 additive was found to be much larger than that in oxynitride melts with other additives, indicating negligible growth of β-prism plane.

ACKNOWLEDGMENT

We thank Prof. Dr. Mikito Kitayama of the Graduate School of Engineering, Fukuoka Institute of Technology and Prof. Dr. Kenji Kaneko of the Graduate School of Engineering, Kyushu University for their helpful discussion and communications with us.

This research was partially supported by the Ministry of Education, Science, Sports and Culture, Grant-in-Aid for Scientific Research (B), 16360376, 2004.

REFERENCES

1. F.F. Lange: J. Am. Ceram. Soc. Vol. 56 (1973), P. 518
2. K.S. Mazdiyasni and C.M. Cooke: J. Am. Ceram. Soc. Vol. 57 (1974), P. 536
3. P.E.D. Morgan, F.F. Lange, D.R. Clarke and B.I. Davis: J. Am. Ceram. Soc. Vol. 60 (1981), P. C-77
4. W.A. Sanders and D.M. Mieskowski: Am. Ceram. Soc. Bull. Vol. 64 (1985), P. 304
5. E. Tani, S. Umebayashi, K. Kishi, K. Kobayashi and M. Nishijima: Am. Ceram. Soc. Bull. Vol. 65 (1986), P. 1311
6. N. Hirosaki, A. Okada and K. Matoba: J. Am. Ceram. Soc. Vol. 71 (1988), P. C-144
7. H. Park, H.-E. Kim and K. Niihara: J. Am. Ceram. Soc. Vol. 80 (1997), P. 750
8. Z.-K. Huang, A. Rosenflanz and I-W. Chen: J. Am. Ceram. Soc. Vol. 80 (1997), P. 1252
9. W.-H. Lee, H.-E. Kim and Niihara: J. Am. Ceram. Soc. Vol. 80 (1997), P. 750
10. K. Negita: J. Mater. Sci. Lett. Vol. 4 (1985), P. 417
11. M. Ohashi, K. Nakamura, K. Hirao, S. Kanzaki and S. Hampshire: J. Am. Ceram. Soc. Vol. 78 (1995), P. 71
12. J.E. Shelby and J.T. Kohli: J. Am. Ceram. Soc. Vol. 73 (1990), P. 39
13. S. Tanabe, K. Hirao and N. Soga: J. Am. Ceram. Soc. Vol. 75 (1992), P. 503

14. R. Ramesh, E. Nestor, M.J. Pomery and S. Hampshire: J. Eur. Ceram. Soc. Vol. 17 (1997), P. 1933-39.
15. M. Krämer, M.J. Hoffmann and G. Petzow: Acta Metall. Mater. Vol. 41 (1993), P. 2939
16. M. Krämer, M.J. Hoffmann and G. Petzow: J. Am. Ceram. Soc. Vol. 76 (1993), P. 2778
17. G. Petzow and M.J. Hoffmann: Mater. Sci. Forum Vol. 113-115 (1993), P. 91
18. M.J. Hoffmann: *Tailoring of Mechanical Properties of Si_3N_4 Ceramics* (Kluwer Academic Publishers, Netherlands 1994)
19. M. Kitayama, K. Hirao, M. Toriyama and S. Kanzaki: J. Ceram. Soc. Jpn. Vol. 107 (1999), P. 995
20. C.P.Gazzara and D.R.Messier: Am. Ceram. Soc. Bull. **56** (1977) 777-780.
21. L. Wang, T.Y. Tien and I.W. Chen: J. Am. Ceram. Soc. **75** (2003) 1578-85.
22. N. Saito, D. Nakata, A. Umemoto and K. Nakashima: J. Jpn. Inst. Metals Vol. 69 (2005) 152-58.
23. N. Saito, S. Shiota, A. Umemoto and K. Nakashima: J. Jpn. Inst. Metals Vol. 68 (2004) 519-22.
24. M. Krämer, D. Wittmüss, H. Küppers, M.J. Hoffmann and G. Petzow: J. Cryst. Growth Vol. 140 (1994), P. 157
25. R.L. Satet and M.J. Hoffmann: J. Eur. Ceram. Soc. Vol. 24 (2004), P. 3437-45.

PREPARATION OF NANOSTRUCTURED SILICON CARBIDE USING A CHEMICAL VAPOR REACTION

H. Tsuda, H. Mabuchi-Nakatani and K. Okamura
Graduate School of Engineering, Osaka Prefecture University
1-1 Gakuen-cho
Sakai, Osaka 599-8531, Japan

ABSTRACT

The reaction between silicon and CO-CO_2 vapor mixtures under an argon atmosphere has been used for the synthesis of nanostructured silicon carbide (SiC). The carbonizing reaction for the synthesis of SiC is the following: $Si(s) + 2CO(g) \rightarrow SiC(s) + CO_2(g)$, and the oxidizing reaction for the formation of pores is the following: $Si(s) + CO_2(g) \rightarrow SiO(g) + CO(g)$. These reactions can be represented by the overall reaction: $2Si(s) + CO(g) \rightarrow SiC(s) + SiO(g)$. By these reactions, porous SiC ceramics with continuously open pores of diameters ranging 2 nm to 200 nm were synthesized by using silicon wafers or silicon powders as a silicon source. The reaction between SiO and CO vapors has been also used for the synthesis of SiC nanofibers. SiO and CO vapors are generated from reaction between silicon and graphite powders and residual oxygen in the chamber. Thus SiC nanofibers with diameters ranging 10 nm to 100 nm and the lengths more than 10 μm have been grown on a graphite substrate. The nanofibers consisted of coaxial two-phase structures: a crystalline β-SiC core and an amorphous silicon oxide (SiO_2) shell. Thus possible reaction for the growth of coaxial SiC/SiO_2 nanofibers was the following: $2SiO(g) + 2CO(g) \rightarrow SiC(s) + SiO_2(s) + CO_2(g)$. These nanofibers were synthesized by a vapor-liquid-solid (VLS) growth mechanism using iron catalysts. The process described appears to be suitable for mass production, and provides a potential technique for synthesizing porous SiC ceramics and SiC nanofibers.

INTRODUCTION

Silicon carbide (SiC) is presently being considered as the high-temperature ceramics for structural applications because of its advantages of excellent high temperature mechanical properties, oxidation resistance and thermal stability. In particular, SiC ceramic fibers are currently being developed and produced primarily as continuous-length structural reinforcement for ceramic matrix composites (CMC) in advanced high-temperature engine components.[1] Furthermore, silicon carbide (SiC) is an important wide-bandgap semiconductor which has a large band gap and other excellent properties such as a high thermal conductivity, and outstanding mechanical and chemical stability at **both room** and high temperatures.[2]

Nanostructured ceramics have been the focus of much attention recently, because they exhibit different electrical, chemical, and mechanical properties at elevated temperatures from those of bulk ceramics. Nanostructured silicon carbide (SiC), particularly with continuously open pores, has high potential applications in a variety of industrial fields. SiC nanofibers are also attracting great interest in the development of electronic nanodevices. We have recently developed porous SiC ceramics with continuously open meso-sized pores and SiC nanofibers via a simple chemical vapor reaction.[3] Objectives of this paper are to describe the new processing of nanostructured silicon carbide (SiC), and the resulting porous SiC ceramics and SiC nanofibers.

PROCESSING OF SILICON CARBIDE

We considered the silicon carbonization for synthesizing silicon carbide (SiC). The SiC can be fabricated by the vapor-solid reaction between CO vapors and silicon under an argon atmosphere. The carbonizing reaction is the following equation:

$$Si(s) + 2CO(g) = SiC(s) + CO_2(g) \quad (1)$$

On the other hand, in the presence of Si, SiC and their equilibrium CO-CO_2 vapor mixtures, two oxidizing reactions for the formation of SiO_2 can be expressed as follows:

$$Si(s) + 2CO_2(g) = SiO_2(s) + 2CO(g) \quad (2)$$

$$SiC(s) + 3CO_2(g) = SiO_2(s) + 4CO(g) \quad (3)$$

In order to search the possibility for formation of the SiC, the standard free energy change (ΔG_0) of the reactions (1), (2) and (3) were calculated. The stability fields in the system Si-SiO_2-SiC at 1573K are shown in Fig. 1 as functions of log P_{CO} and log P_{CO2}. In this study, measured CO and CO_2 vapor pressure were about 10^{-2} atm, and 10^{-7} atm, respectively. This shows that the carbonizing reaction for the formation of SiC is possible with the reaction (1).

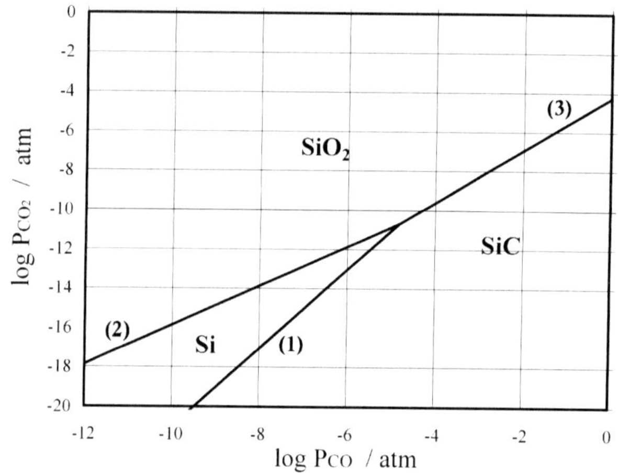

Fig. 1. The stability diagram for the system Si-SiO_2-SiC at 1573 K as functions of log P_{CO} and log P_{CO2}. (1): $Si(s) + 2CO(g) = SiC(s) + CO_2(g)$, (2): $Si(s) + 2CO_2(g) = SiO_2(s) + 2CO(g)$, (3): $SiC(s) + 3CO_2(g) = SiO_2(s) + 4CO(g)$.

EXPERIMENTAL

To investigate the formation of SiC through the carbonizing reaction of CO vapors, silicon wafers as a silicon source were used, and graphite powders were also used as the CO vapor generator. The silicon wafers (thickness 0.3 mm and 0.5 mm; diameter 30 mm) were embedded in graphite powders in a graphite crucible as shown in Fig. 2. The thermal treatments were made in a stainless steel chamber with graphite heating elements. The reaction furnace was heated at 873K for 1 h to degas by the vacuum pump, and then kept at 1573, 1623 and 1653 K for 1 - 48 h under an argon atmosphere at 1 atm. The products formed on the wafer were characterized using X-ray diffraction (XRD), scanning electron microscope (SEM), transmission electron microscope (TEM), and energy dispersive X-ray (EDX) spectrometer.

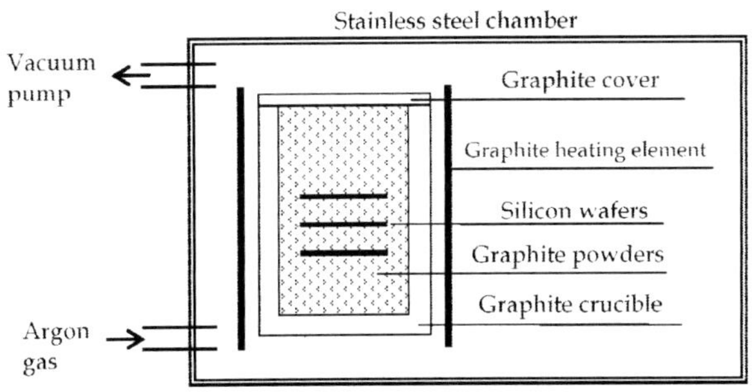

Fig. 2. Schematic diagram of carbonizing apparatus for the synthesis of SiC.

RESULTS AND DISCUSSION

Figure 3 shows an appearance of the silicon wafer (thickness 0.3 mm) reacted at 1623 K for 24 h with a silicon wafer before carbonizing. The result of XRD showed that the reactive product was β-type SiC. A black green-colored, glossy surface was found on the as-reacted wafer, and the shape of the wafer was not changed in comparison with before carbonizing. This suggests that the process is near-net shaping.

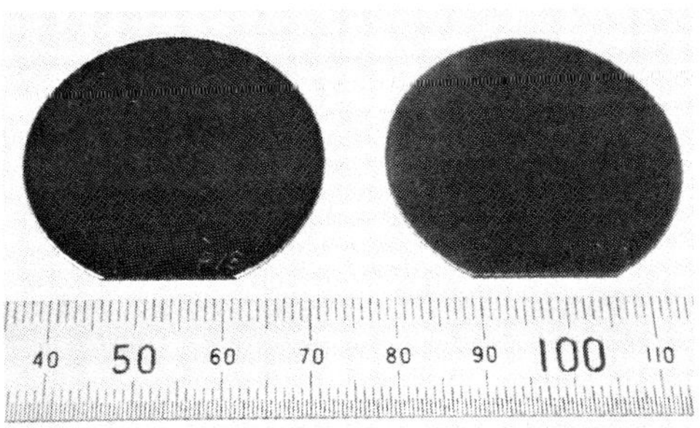

Fig. 3. An appearance of the silicon wafers observed before (left) and after (right) carbonizing.

Figure 4(a) shows the cross section of a wafer reacted at 1653 K for 1 h. The SEM microstructure indicates that the reacted wafer shows the SiC layer of ~10 μm in width from surface. The SiC layer has the porosity of ~50%, and the size of pores is ~100 nm. Figure 4(b) shows the layer width measured as a function of the reaction temperature and duration. The layer width increases rapidly at high temperatures, for example, in the reaction at 1623 K for 24 h the resulting silicon wafer of 0.3 mm is **completely** transformed into SiC wafer (see Fig. 3).

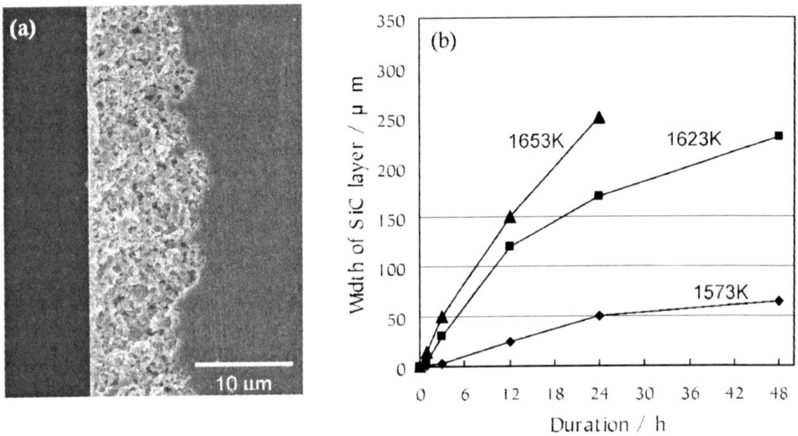

Fig. 4. (a) SEM microstructure of a cross section of the silicon wafer reacted at 1653 K for 1 h, showing the porous SiC layer. (b) The layer width as a function of the reaction temperature and duration.

As mentioned above, porous SiC ceramics with meso-sized pores were synthesized by this process. This indicates that the oxidizing reaction for the formation of pores occurs simultaneously as follows:

$$Si(s) + CO_2(g) = SiO(g) + CO(g) \qquad (4)$$

That is, the reaction (4) shows the active oxidation of silicon with CO_2 vapor, resulting in the formation of pores in the SiC wafer. Therefore, in this process, **these** reactions (1) and (4) are represented by the following overall reaction:

$$2Si(s) + CO(g) = SiC(s) + SiO(g) \qquad (5)$$

Figure 5 shows a typical microstructure of the porous SiC obtained by this process, which was synthesized at different temperatures; (a) 1573K, (b) 1623K, and (c) 1653K. Thus the size of pores increased with increasing the reaction temperature; it was approximately ~25 nm, ~60 nm, and ~110 nm, respectively. Furthermore, as can be seen in Fig. 5, it seems that the porous SiC has pores with continuously connected channels, namely, continuously open pores.

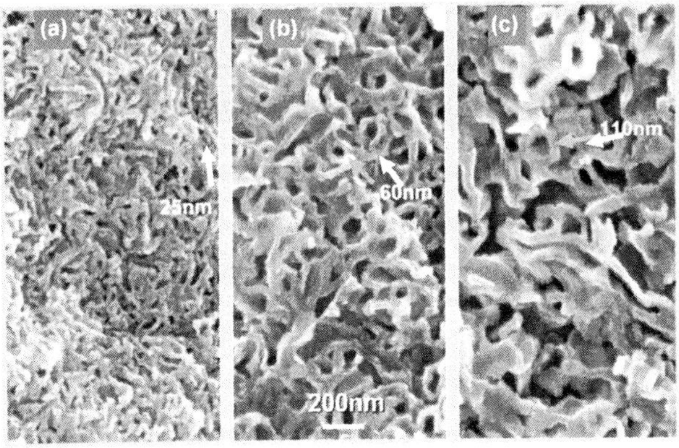

Fig. 5. SEM microstructures of typical porous SiC synthesized by the reactions at (a)1573K, (b)1623K, and (c)1653K.

MESO-/MACROPOROUS SILICON CARBIDE

Porous ceramics are widely used in many applications, such as hot-gas or molten metal filters, catalyst supports, biomedical absorbents, and so on. Porous silicon carbide (SiC) is one of the leading candidates for these applications because of its low thermal-expansion coefficient and good thermal-shock resistance as well as excellent mechanical and chemical stability. Several ways to produce porous SiC ceramics have been reported. One of the recent methods is by an oxidation-bonding process in which the SiC powder compacts are heated in air.[4] We have successfully developed a new processing for the formation of porous SiC ceramics with meso-sized pores, as in the above chapter. The present work further investigates the process, microstructure, pore-size distribution, porosity and the mechanical strength of porous SiC ceramics.

EXPERIMENTAL

Commercial powders of silicon (particle size ~10 μm; purity 99.99 mass %) were used as a silicon source in this work. These powders were weakly cold pressed with small amounts of ethylene-glycol into a cylindrical compact. The silicon powder compacts instead of silicon wafers were embedded in graphite powders (as shown in Fig. 2). The silicon compacts were reacted at 1573 K for 48 h to **ensure complete transformation** into SiC, and then pulverized to **powder** using a mortar and pestle. The obtained SiC powders were uniaxially hot-pressed under 70 MPa at 1873 K. The hot-pressed disks were 20 mm in diameter and 5 mm in thickness.

The porosity was calculated from the density of the hot-pressed samples measured by the Archimedes displacement method. The evolution of porosity was determined by nitrogen gas adsorption; the specific surface area (BET method) and pore-size distribution (BJH method) of samples were evaluated using adsorption data. The microstructure of the cross section was observed by the scanning electron microscopy (SEM).

RESULTS AND DISCUSSION

Figure 6 shows a typical microstructure of the SiC powders prepared by this method using silicon powder compacts. The obtained SiC powder has meso- sized pores. Figure 7(a) shows an appearance of the hot-pressed disk. The SEM microstructure obtained from a cross section of such a disk is shown in Fig. 7(b). The micrograph shows the porous structure and reveals the meso-sized (~30 nm) pores with the macro-sized (~1 μm) pores induced by the sintering of the SiC powders in hot-pressing. Thus the porous SiC ceramics with meso-/ macropore structure are synthesized by a hot pressing using porous SiC powders.

Table 1 shows a summary of main characteristics in the porous SiC ceramics investigated. The porosity calculated is almost **that of** the open porosity. Therefore, it is found that almost all of the pores are continuously open pores. Figure 8 shows the mesopore-sized distribution determined by nitrogen gas adsorption. The distribution has a pore-sized peak at ~ 35 nm with the

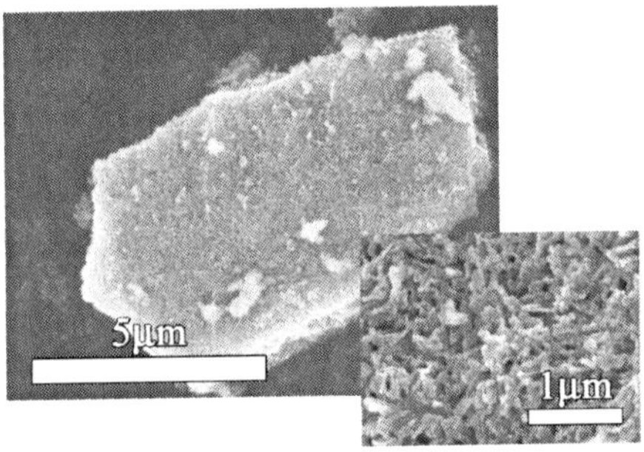

Fig. 6. SEM microstructure of the porous SiC powder prepared, showing meso-sized pores in the powder.

Fig. 7. (a) An appearance of the hot-pressed disk. (b) SEM microstructure of a cross section of hot-pressed disk, showing the meso-/macroporous structure.

Table 1. Summary of main characteristics in the porous SiC ceramics investigated.

Bulk density (g/cm^3)	Open porosity (%)	Closed porosity (%)	Total porosity (%)	Mesopore-size distribution	Macropore-size (average) (μm)	Specific surface area (m^2/g)	Compressive strength (MPa)
1.99	62.0	0.2	62.2	(Fig. 8)	1.5	25	35

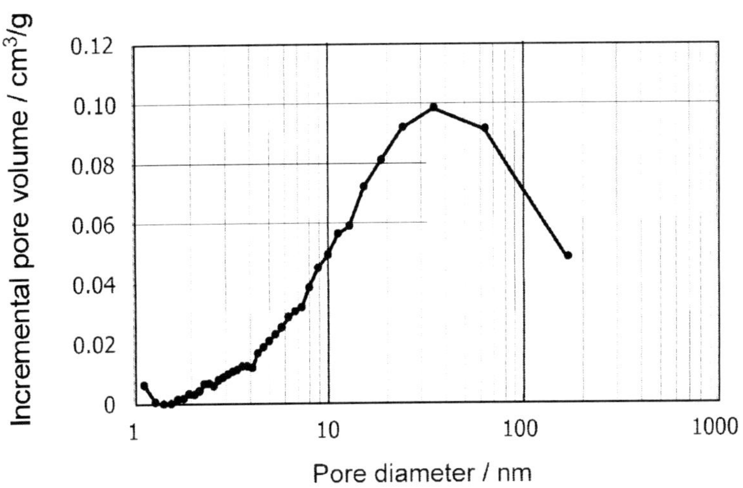

Fig. 8. Mesopore-size distribution of porous SiC ceramics determined by BJH method.

pore diameters ranging **from** ~2 nm to ~200 nm. On the other hand, the porous SiC ceramics synthesized are characterized by the presence of macro-sized (~1.5 μm) pores. Thus, the porous ceramics containing continuously open mesopores and macropores are expected to be used as support materials of catalysts and absorbents and for various other applications. The specific surface area determined by BET method, ~25 m^2/g, and the compressive strength, ~35 MPa, of the porous SiC ceramics are also adequately applicable to the development of porous ceramics.

SILICON CARBIDE NANOFIBERS

Much work has been focused on production of SiC one-dimensional nanostructures, such as nanofibers, nanowires, nanorods, and nanowhiskers, because of their high potential uses in the nano-mesoscopic research and in the development of nanodevices. The elasticity and strength of an individual SiC one-dimensional nanostructure are substantially greater than those of bulk SiC, and they approach theoretical values.[5] Therefore, SiC one-dimensional nanostructures can be considered as an excellent reinforcement for the ceramic-, metal-, and polymer-matrix composites. To date, several methods have been reported for the synthesis of SiC one-dimensional nanostructures, including carbothermal reduction of SiO_2,[6,7] chemical vapor deposition,[8,9] decomposition of organic silicon compounds,[10,11] reaction between silicon halides and CCl_4,[12] conversion of carbon nanotubes to SiC nanorods by reacting with SiO vapor,[13] and laser ablation of metal containing SiC targets.[14] In these methods, in order to grow SiC in one-dimensional morphology it is necessary to supply the reactants in a vapor form in the final step of the reaction. The common vapor species are SiO and CO in an inert atmosphere, or SiO and CH_4 in a hydrogen atmosphere.[15]

In this work, the reaction between SiO and CO vapors in an argon atmosphere has been employed for the synthesis of SiC nanofibers. The raw materials for the vapor-state formation of SiO and CO are silicon and graphite powders, respectively. During heating at high temperatures (e.g., 1523 K), SiO and CO vapors are generated from reaction between silicon and graphite powders and residual oxygen in the chamber. Thus possible reaction for the growth of SiC nanofibers is the following:

$$SiO(g) + 3CO(g) \rightarrow SiC(s) + 2CO_2(g) \tag{6}$$

However, depending on its equilibrium $SiO\text{-}CO\text{-}CO_2$ vapor mixtures at an argon atmosphere, an other reaction may be represented by the following:

$$SiO(g) + CO(g) \rightarrow Si(s) + CO_2(g) \tag{7}$$

The standard free energy change (ΔG_0) of the reactions (1) and (2) can be calculated to search the possibility for formation of the SiC phase from SiO and CO vapors. In the present experiment, measured CO and CO_2 gas pressure were about 10^{-2} atm, and 10^{-7} atm, respectively.

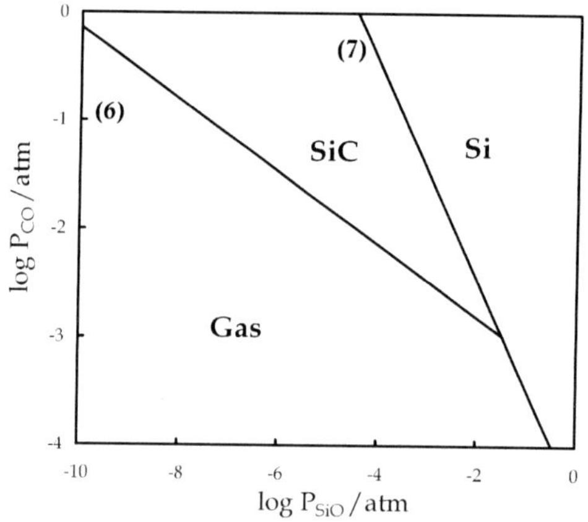

Fig. 9. The stability diagram in the system $SiO\text{-}CO\text{-}CO_2$ vapor phase at 1523 K and $P_{CO2} = 10^{-7}$ atm. (6): $SiO(g) + 3CO(g) \rightarrow SiC(s) + 2CO_2(g)$, (7): $SiO(g) + CO(g) \rightarrow Si(s) + CO_2(g)$.

The calculated stability diagram in the system $SiO-CO-CO_2$ gas phase at 1523 K and $P_{CO2} = 10^{-7}$ atm is shown in Fig. 9 as functions of log P_{SiO} and log P_{CO}. Thus the stability field for SiC certainly exists when an immeasurable SiO gas pressure is the range of about 10^{-4} atm to $10^{-2.5}$ atm. Thereby, this process appears to be suitable for production of SiC nanofibers. It also has the advantages of the availability of low-cost raw materials and the simplicity of the mass production technique.

EXPERIMENTAL

Graphite powders (particle size ~20 μm: purity 99.99 mass%) used as the CO vapor generator were put into a graphite crucible (diameter 50 mm; height 50 mm). On the other hand, silicon powders used as the SiO vapor generator were 99 mass% purity containing 0.2 mass% impurity level of iron. The silicon powders were cold pressed into a cylindrical compact (porosity 50%; diameter 10 mm; height 10 mm) and placed on graphite powders in the graphite crucible. An iron (Fe) film of a 20 nm thickness, as a catalyst, was deposited on a graphite substrate (purity 99.99 mass%: diameter 30 mm; thickness 2 mm) by the electron beam method. The graphite substrate as a lid was mounted on top of the graphite crucible. Figure 10(a) shows the schematic diagram of the graphite crucible used in the nanofiber synthesis. The thermal treatments were made in a stainless chamber with graphite heating elements. The reaction chamber was heated at 873K for 1 h to degas, and then kept at 1523 K for 10 min - 24 h under an argon atmosphere at 1 atm. The products formed on the substrate were characterized using X-ray diffraction (XRD), scanning electron microscope (SEM), transmission electron microscope (TEM), and energy dispersive X-ray (EDX) spectrometer.

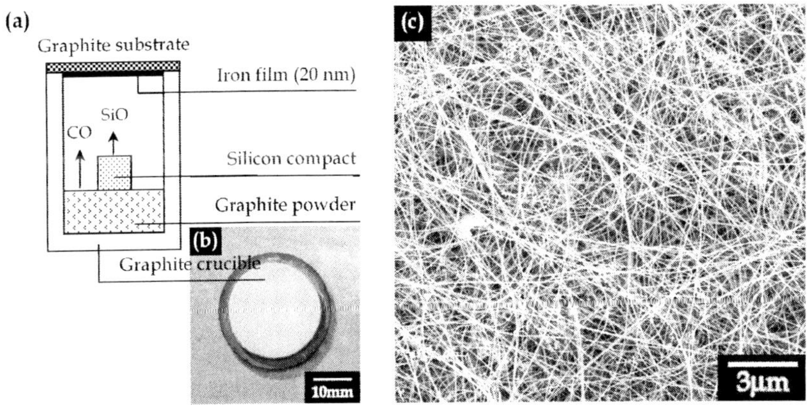

Fig. 10. (a) Schematic diagram of the graphite crucible used. (b) An appearance of the substrate after a thermal treatment. (c) Typical SEM image of nanofibers on the graphite substrate.

RESULTS AND DISCUSSION

Figure 10(b) shows an appearance of the graphite substrate after a thermal treatment at 1523 K for 24 h. A white blue-colored, sponge-like product was found deposited as a sheet with a thickness about 1 mm on the substrate. A typical morphology of the product is shown in the SEM image in Fig. 10(c). Large quantities of straight, curved, and randomly oriented nanofibers are observed in this image. The lengths of the nanofibers are almost up to 10 μm, while the diameters are in the range of 10 - 100 nm.

Figure 11 shows an XRD pattern of the nanofiber product. Main peaks without those of the graphite substrate can be clearly indexed to β-SiC crystal structure. However, two detectable impurity phases are recognized in the pattern. One of the impurity phases can be detected at just left of (111) peak in the β-SiC. It is assumed that a small amount of α-type SiC phase is involved in the β-SiC nanofibers. The other of the impurity phases can be observed for the broad peak at about 22° (**cristobaraite**; (101) peak) of the 2*θ*. Thus the phase may be presumed to be a silicon oxide (SiO_2), and it suggests that the β-SiC nanofibers also contain an amorphous SiO_2 phase.

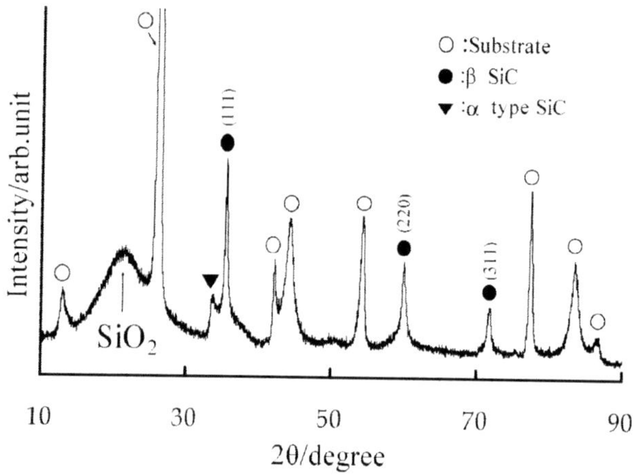

Fig. 11. XRD pattern of the nanofibers grown on a graphite substrate.

The morphologic structure of the individual SiC nanofiber is shown in the TEM image in Fig. 12. Clearly, the nanofibers have a core-shell structure as shown in Fig. 12(a), and a high- resolution TEM image in Fig. 12(b) shows that the nanofiber has a crystalline core and surrounding amorphous-like shell. The selected-area electron diffraction (SAED) pattern (inset in Fig. 12(b)) indicates that the crystalline core is the β-SiC phase with the growth direction of <111>. Diffraction spots have streaks along the <111> direction, representing twinning planes or stackingfaults on {111} planes, which are a typical fault characteristic in one-dimensional SiC structures.[16] Therefore, the image of the crystalline core reveals lattice fringes with the direction perpendicular to the growth direction of nanofibers. Also, the reason why the nanofibers have a small amount of α-type SiC phase, as can be seen in the XRD pattern, may be due to the introduction of the microtwins and stacking faults in the β-SiC crystalline core. The halo ring in the SAED pattern indicates that the shell structure is composed of the amorphous SiO_2 phase (see the broad peak in the XRD pattern). Thus the nanofibers consist of coaxial two-phase structures: a crystalline β-SiC core and an amorphous silicon oxide (SiO_2) shell. We concluded that the reaction between SiO and CO vapors for the growth of coaxial SiC/SiO_2 nanofibers is the following:

$$2SiO(g) + 2CO(g) \rightarrow SiC(s) + SiO_2(s) + CO_2(g) \quad (8)$$

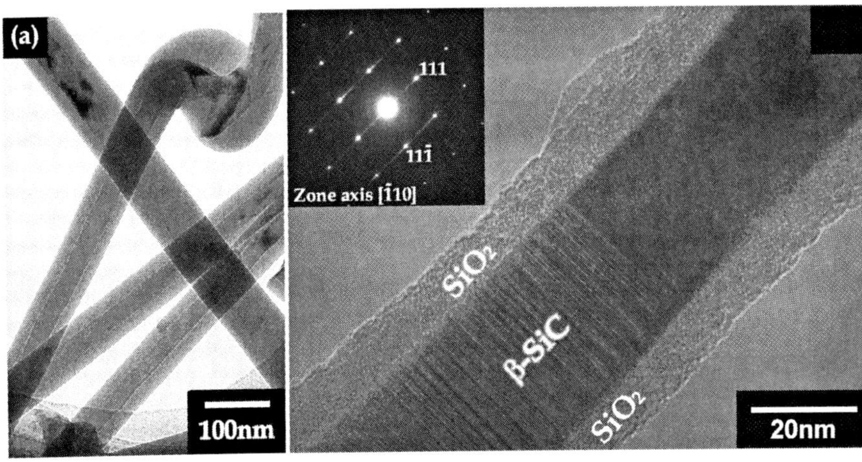

Fig. 12. (a) TEM image of SiC nanofibers. (b) High-resolution TEM image of an individual SiC nanofiber. Inset shows the SAED pattern with the <110> axis parallel to the electron beam.

In order to investigate more about the nanofiber growth, the morphology of the nanofibers were characterized by SEM and EDX. Figure 13 shows the morphology observed on a graphite substrate after a reaction at 1523 K for very short time (10 min). At the initial stage of the nanofiber growth, many drop-like particles (100 - 400 nm) have been observed together with a number of short rod-like nanofibers. On the tip of these short nanofibers, a droplet can be seen whenever as shown by the arrow in Fig. 13. Figure 14 shows the EDX spectrum of the droplet, thus these droplets contain iron impurity with silicon, carbon, and oxygen. Therefore, during heating at high temperatures, the iron film on the graphite substrate seems to condense into the liquid droplets of iron-silicon alloy with a low melting temperature. These droplets are closely connected to the growth mechanism, since the droplet contained iron acts as a catalyst for deposition of SiC nanofibers. This indicates that the SiC nanofibers are synthesized by the vapor- liquid-solid (VLS) growth mechanism.

The typical morphology of the nanofibers grown on a graphite substrate is shown in the SEM image in Fig. 15. Many spherical droplets with the representative SiC nanofibers can be seen in this image. Forethermore, it is noted that many droplets are located at the intermediate of nanofibers, that is, the growing nanofibers are ramified at the droplets. This growth morphology, a repetition of the ramifications, appears to be favorable for mass production of the SiC nanofibers. Therefore, the process in this work provides a potential technique for synthesizing SiC nanofibers.

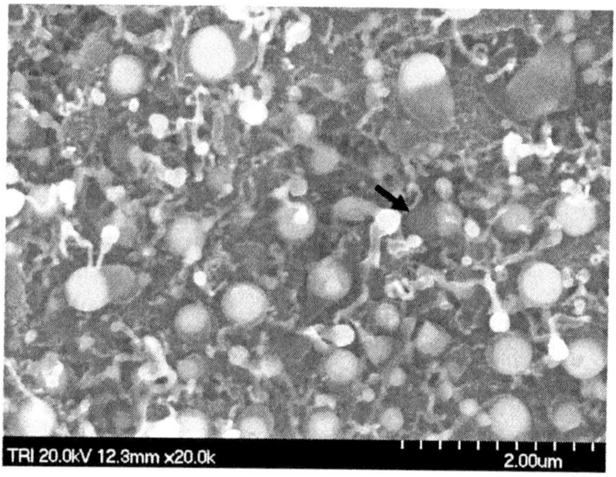

Fig. 13. Initial stage of SiC nanofibers grown on the graphite substrate with 20 nm iron film.

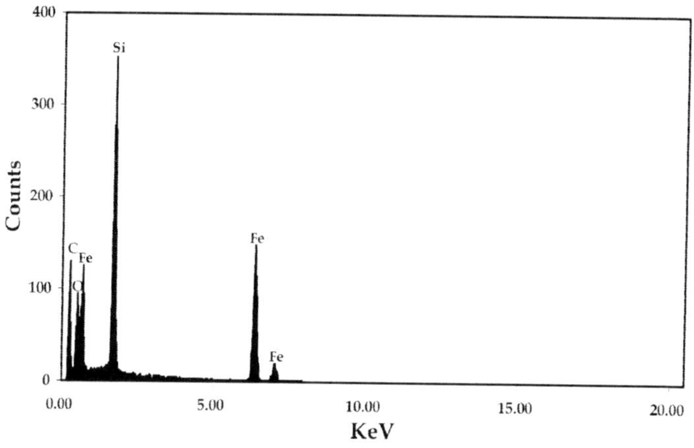

Fig. 14. EDX spectrum of the spherical droplet.

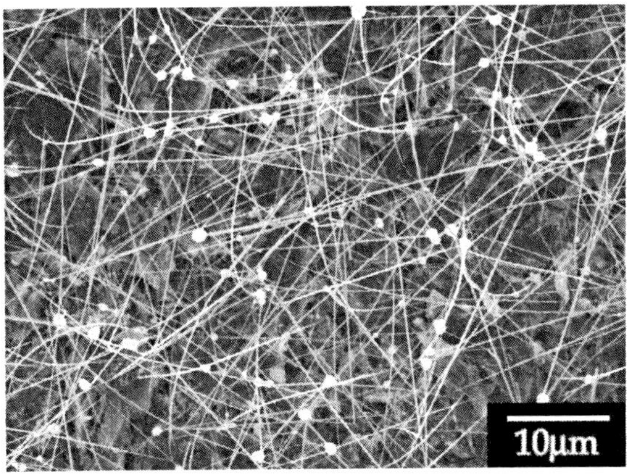

Fig. 15. SEM image of the typical SiC nanofibers on a graphite substrate, showing the spherical droplet as a catalyst.

CONCLUSION

The nanostructured silicon carbide (SiC) was synthesized by the vapor-solid reaction between CO-CO_2 vapor mixtures and silicon under an argon atmosphere. These reactions can be represented by the overall reaction: 2Si(s) + CO(g) → SiC(s) + SiO(g). By these reactions, porous SiC ceramics with continuously open pores were synthesized by using silicon wafers or silicon powders as a silicon source. Consequently, the porous SiC ceramics with meso-/ macropore structure were fabricated by a hot pressing using porous SiC powders. The pore-size distribution has a mesopore-size peak at ~35 nm with the pore diameters ranging ~2 nm to ~200 nm. In addition, the presence of macro-sized (~1.5 μm) pores is characteristic of the porous SiC ceramics synthesized by this method. Thus, the porous ceramics containing continuously open mesopores and macropores are expected to be used as support materials of catalysts and absorbents and for various other applications. Furthermore, the specific surface area, ~25 m^2/g, and the compressive strength, ~35 MPa, of the porous SiC ceramics obtained are adequately applicable to the development of porous ceramics.

The reaction between SiO and CO vapors in an argon atmosphere has been used for the synthesis of SiC nanofibers. The SiO and CO vapors were generated from reaction between silicon and graphite powders and residual oxygen in the chamber. By this reaction, SiC nanofibers with diameters ranging 10 nm to 100 nm and the lengths more than 10 μm have been grown on graphite substrates. The microstructures of SiC nanofibers synthesized were studied in detail by XRD, SEM, TEM and EDX. These nanofibers consisted of coaxial two-phase structure: a crystalline β-SiC core and an amorphous silicon oxide (SiO_2) shell. The growth direction of coaxial SiC/SiO_2 nanofibers was a <111> of crystalline β-SiC core with a high density of microtwins and stacking faults perpendicular to this growth direction. These coaxial SiC/SiO_2 nanofibers were synthesized by a vapor-liquid-solid (VLS) growth mechanism using iron catalyst. The process appears to be suitable for mass production, and provides a potential technique for synthesizing SiC nanofibers.

ACKNOWLEDGMENT

This work was supported by the Analysis Core Center for Department of Materials Science, Osaka Prefecture University. The authors would like to thank Mr. H. Konishi for his experimental assistant and valuable discussions.

REFERENCES

[1]J.A. DiCarlo and H.-M. Yun, "Non-oxide (Silicon crbide) Fibers", *Handbook of Ceramic Composites, Edited by N.P. Bansal, Kluwer Academic Pub.*, 33-52(2005).

[2]W.J. Choyke and G. Pensl, "Physical Properties of SiC", *MRS Bull.*, **22**, 25-29(1997).

[3]H. Tsuda, H. Konishi, H. Mabuchi, T. Matsui and K. Morii, "Synthesis of Silicon Carbide Nanofibers Using Simple Chemical Vapor Reaction", *Transactions of the Indian Ceramic Society*, **63**, 195-98 (2004).

[4]J. She, J-F. Yang, N. Kondo, T. Ohji and S. Kanzaki, "High-strength Porous Silicon Carbide Ceramics by an Oxidation-bonding Technique", *Journal of American Ceramic Society*, **85** [11] 2852-54 (2002).

[5]Z.L. Wang, Z.R. Dai, R.P.Gao and Z.G. Bai, "Side-by-side Silicon Carbide-silica Biaxial Nanowires: Synthesis, Structure, and Mechanical Properties", *Applied Physics Letters*, 77 [21] 3349-3351 (2000).

[6]N.K. Sharma and W.S. Williams, "Formation and Structure of Silicon Carbide Whiskers from Rice Hulls", *Journal of American Ceramic Society*, **67** [11] 715-20 (1984).

[7]H-J. Choi and J-G. Lee, "Continuous Synthesis of Silicon Carbide Whiskers", *Journal of Materials Science*, **30** 1982-86 (1995).

[8]X.T. Zhou, N. Wang, H.L. Lai, H.Y. Peng, I. Bello, N.B. Woug, C.S. Lee and S.T. Lee, "β-SiC Nanorods Synthesized by Hot Filament Chemical Vapor Deposition", *Applied Physics Letters*, **74** [26] 3942-44 (1999).

[9]Y. Zhang, M. Nishitani-Gamo, C. Xiao and T. Ando, "Synthesis of 3C-SiC Nanowhiskers and Emission of Visible Photoluminescence", *Journal of Applied Physics*, **91** [9] 6066-70 (2002).

[10]N. Setaka and Z. Inoue, "Beta Silicon Carbide Whiskers Prepared on a Molybdenum Substrate", *Journal of American Ceramic Society*, **50** [11] 624 (1969).

[11]J. Zheng, M.J. Kramer and M. Akinc, "In Situ Growth of SiC Whisker in Pyrolyzed Monolithic Mixture of AHPCS and SiC", *Journal of American Ceramic Society*, **83** [12] 2961-66 (2000).

[12]S. Motojima and M. Hasegawa, "Chemical Vapor Growth of β-SiC Whiskers From a Gas Mixture of Si_2Cl_6-CH_4-H_2-Ar", *Journal of Crystal Growth*, **87** 311-17 (1988).

[13]Y. Zhang, K. Suenaga, C. Colliex and S. Iijima, "Coaxial Nanocable: Silicon Carbide and Silicon Oxide Sheathed with Boron Nitride and Carbon", *Science*, **281** [14] 973-75 (1998).

[14]W. Shi, Y. Zheng, H. Peng, N. Wang, C.S. Lee and S-T. Lee, "Laser Ablation Synthesis and Optical Characterization of Silicon Carbide Nanowires", *Journal of American Ceramic Society*, **83** [12] 3228-30 (2000).

[15]G. Urretavizcaya and J.M.P. Lopez, "Growth of SiC Whiskers by VLS Process", *Journal of Materials Research*, **9** [11] 2981-86 (1994).

[16]W-S. Seo and K. Koumoto, "Morphology and Stacking Faults of β-Silicon Carbide Whisker Synthesized by Carbothermal Reduction", *Journal of American Ceramic Society*, **83** [10] 2584-92 (2000).

Microstructural and Thermal Characterization

IN-SITU OBSERVATION OF STRUCTURAL TRANSITION IN RARE EARTH MANGANITES BY TEM

Hiromi Nakano[1]
[1]Ryukoku University, Seta Otsu, 520-2194, Japan
Nobuo Ishizawa[2]
[2]Nagoya Institute of Technology, Asahigaoka, Tajimi, 507-0071, Japan
Naoki Kamegashira[3]
[3]Toyohashi University of Technology, Tempaku, Toyohashi, 441-4580, Japan

ABSTRACT

$BaLn_2Mn_2O_7$ (Ln = Eu and Gd) is a layered perovskite that belongs to the Ruddlesden-Popper type series $AO(ABO_3)n$ with n = 2. We successfully observed the multiple phase transition and the intermediate phase by thermal TEM. The orthorhombic phase of this oxide is stable at room temperature when the composition is stoichiometrical, which is caused by a Jahn-Teller effect due to the 3d electron in Mn^{3+} ion. The orthorhombic phase at room temperature ($ortho_1$) was transformed to the tetragonal phase by heating. The tetragonal phase was continuously transformed to the orthorhombic phase ($ortho_2$). In the synthesis procedure of the floating zone (FZ) method, as-grown crystal has a tetragonal phase due to a kind of quenching condition. The tetragonal phase at room temperature ($P4_2/mnm$) was also transformed to the tetragonal phase ($I4/mmm$) and continuously transformed to the orthorhombic phase by heating. The structural transition at relatively high temperature from tetragonal to orthorhombic may be caused by a change in the 3d electron's behavior in Mn^{3+} ion.

INTRODUCTION

Layered perovskite manganites $BaLn_2Mn_2O_7$ (Ln = rare earth) with a double block of oxygen octahedra were synthesized by Deshizeaux-Cheruy and Joubert for Ln = Pr-Gd[1)] and by Kamegashira (one of the present authors) for Ln = Tb[2)]. In recent years, many phases have been synthesized and identified, and the appearance of these phases depended on the synthesis conditions[3)4)]. An orthorhombic phase appeared at room temperature under a low-oxygen-pressure condition[3)]. On the other hand, a tetragonal phase appeared in air or under higher oxygen pressure. The reason for this depends on the oxidation state of the Mn ion, and a

Jahn-Teller effect in the Mn^{3+} ion affects the crystal structure. The phase of $BaLn_2Mn_2O_7$ synthesized under low oxygen pressure was transformed to another structure by heating[5)6)]. Using X-ray diffraction data, Kamegashira et al. suggested that these structures transform the stable phase at high-temperature as an archetype tetragonal (*I4/mmm*)[5)].

In this experiment, the structural transitions in $BaEu_2Mn_2O_7$ and $BaGd_2Mn_2O_7$ are observed *in-situ* by thermal TEM. We compile new transition data by heating and discuss the phase behavior of those oxides from the TEM, XRD, and electrical conductivity data.

EXPERIMENTAL PROCEDURES

$BaLn_2Mn_2O_7$ (Ln = Eu and Gd) was prepared from starting materials $BaCO_3$, Ln_2O_3, and Mn_2O_3 by a ceramics method in an Ar atmosphere. The ceramic was grown by a floating zone method in an Ar atmosphere, and a single crystal was obtained. The detailed procedure was described in the previous paper[5)7)].

The specimen was crushed and scooped onto a copper mesh with carbon-coated microgrids and subjected to high-resolution TEM observations. The TEM (JEM-3000F, JEOL, Japan) was operated at 300 kV with spherical aberration of Cs = 1.0 nm.

In-situ observation was performed using a thermal stage in a range from room temperature to 1073 K in TEM (JEM-2000EX, EM-SHU2 specimen holder, JEOL).

Electric conductivity was measured by a direct current four-probe method[8)].

RESULTS AND DISCUSSION

As-grown crystal of $BaEu_2Mn_2O_7$ has a tetragonal structure (space group of $P4_2/mnm$[7)]) with lattice parameters of a = 0.5520 and c = 2.0150 nm. In the synthesis procedure of the floating zone (FZ) method, as-grown crystal has a tetragonal phase due to a kind of quenching condition. The as-grown crystal was heated by a thermal stage in TEM, and the changes in the selected area diffraction (SAD) pattern by structural transition were recorded at the corresponding temperatures, as shown in Fig. 1.

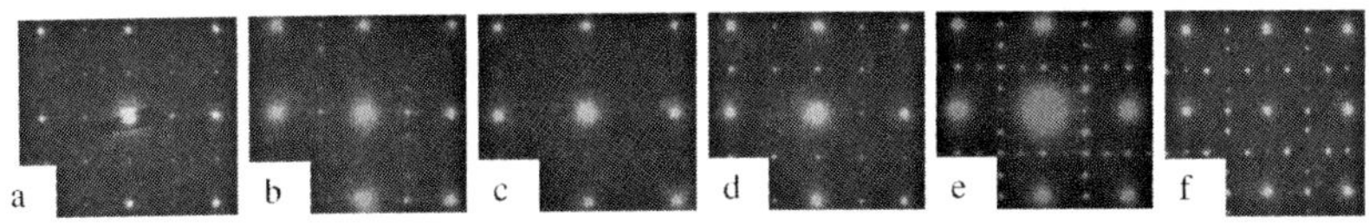

Fig. 1. Selected area diffraction patterns of multiple phases in $BaEu_2Mn_2O_7$ by thermal TEM: (a) at room temperature, (b) 373 K, (c) 433 K, (d) 500 K, (e) 550 K, and (f) 773 K.

We see multiple phases by structural transition in Fig. 1. At first, the tetragonal phase with $P4_2/mnm$ ($tetra_1$) was transformed to another type of tetragonal phase ($tetra_2$) around 433 K. The SAD diffraction pattern of $tetra_2$ shows an archetype tetragonal ($I4/mmm$) in which weak reflections disappeared between the fundamental reflections. We can see the intermediate phase during the transition from $tetra_1$ to $tetra_2$ in Fig. 1(b). Satellite reflections appeared around the weak reflections, and soon these weak reflections disappeared between the fundamental reflections. The phenomenon may be controlled by the tilting octahedra of MnO_6, as seen in classified perovskite phases[9]. We infer that this structure type ($I4/mmm$) may be stable in high-temperature regions. However, at the second stage, the archetypal pattern changed to the original diffraction pattern and continuously transformed to a complex diffraction pattern around 550 K, as shown in Fig. 1(e). The complex diffraction pattern was composed of two diffraction patterns rotated by 90°, as shown in Fig. 2. Therefore, the region was formed by twin domains. The complex SAD pattern shows that the a^*-axis is not equivalent to the b^*-axis. The difference in the lattice distance between the a- and b-axes was slight because no split reflections were observed.

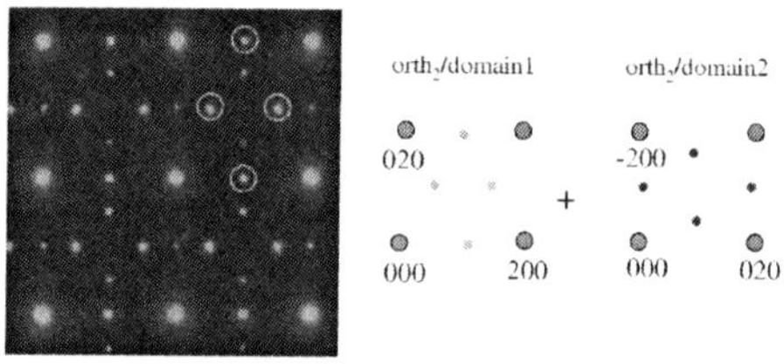

Fig. 2. SAD pattern composed of two patterns above 773 K.

Figure 3 shows electrical conductivity of $BaEu_2Mn_2O_{7+x}$. Transition points were observed at 320 and 520 K in only $BaEu_2Mn_2O_7$ with a stoichiometric composition. The orthorhombic phase of this oxide is stable at room temperature, which was caused by a Jahn-Teller effect due to the 3d electron in Mn^{3+} ion. The orthorhombic phase was stable in a narrow temperature region. In the FZ method, as-grown crystal has a tetragonal phase by the cooling condition. The tetragonal phase may be stable at the temperature region between 320 and 520 K in $BaEu_2Mn_2O_7$. At 520 K, a discontinuous line and slope change were observed. The transition temperature corresponded to the structural change from tetragonal to orthorhombic, as seen in Fig. 1.

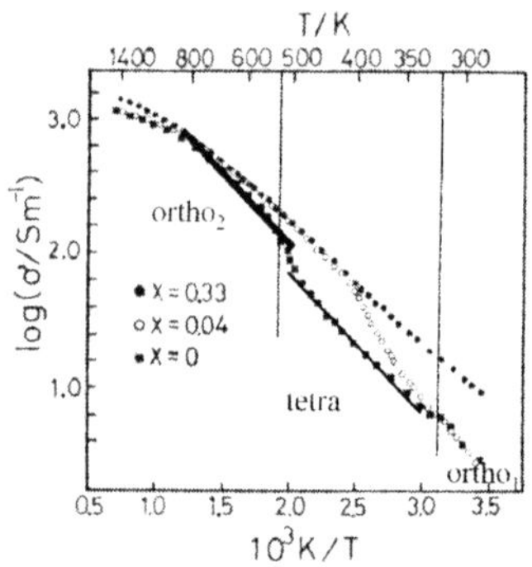

Fig. 3. Electrical conductivity of $BaEu_2Mn_2O_{7+x}$.

We can see a similar structural change in $BaGd_2Mn_2O_7$ by the thermal XRD method. Figure 4 shows two types of orthorhombic phases that Kamegashira et al. called Tb-type-ortho and Sm-type-ortho at room and at high temperatures, respectively[3]. The lattice parameter of the c-axis of $ortho_2$ was longer than that of $ortho_1$ at room temperature.

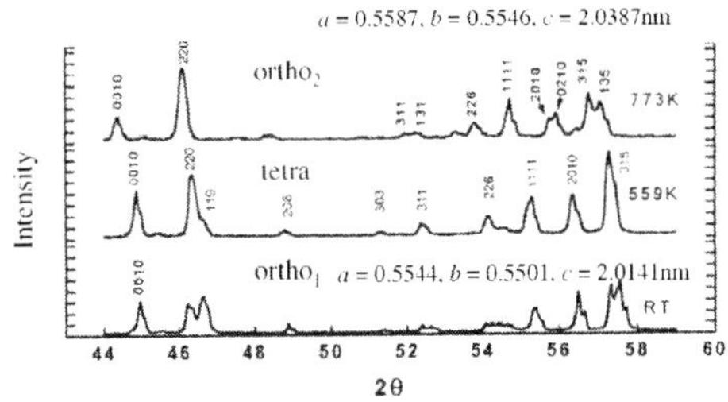

Fig. 4. XRD profiles of $BaGd_2Mn_2O_7$ at various temperatures by thermal XRD.

The electrical conductivity data also show the existence of three types of phases, as shown in Fig. 5. The transition temperatures were higher than that of $BaEu_2Mn_2O_7$, and they were related to the ionic radii of Eu and Gd ions. At 460 K, the slope of the line changed slightly. At 650 K, a discontinuous point was observed, which corresponded to the transition temperature from tetragonal to $ortho_2$.

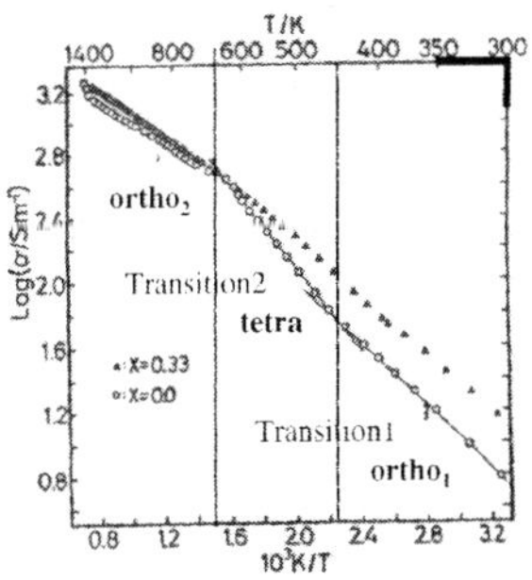

Fig. 5. Electrical conductivity of $BaGd_2Mn_2O_{7+x}$

From this we may infer that the two types of transition were caused by different phenomena. Transition 1 from $ortho_1$ to tetra may be controlled by the tilting of the octahedra of MnO_6, which occurs in the symmetrical subgroup. On the other hand, transition 2 from tetra to $ortho_2$ was related to the 3d electron in the e_g orbit of the Mn^{3+} ion. Therefore, the change in the behavior of the 3d electron is thought to be produced by the structural transition at relatively high temperature. This assumption was supported by magnetic property data, in which slight changes were observed at 640 and 670 K in $BaGd_2Mn_2O_7$, as seen in a previous paper[10].

CONCLUSION

We observed *in-situ* two types of structural transition in layered perovskites $BaLn_2Mn_2O_7$ (Ln = Eu and Gd) with a stoichiometric composition. The transition was examined by thermal TEM, thermal XRD, and electrical conductivity. It was related to the crystal distortion by the Jahn-Teller effect. One type of transition occurred at relatively low temperature, at 320 K in Eu-oxide and at 450 K in Gd-oxide, which was controlled by the tilting octahedra of MnO_6 in a symmetrical subgroup. The other type of transition occurred at relatively high temperature, at 520 K in Eu-oxide and at 650 K in Gd-oxide, which may be caused by a change in the 3d electron's behavior in the Mn^{3+} ion.

REFFERENCES

[1]M. N. Deschizeaux-Cheruy and J. C. Joubert, "Donnees cristallographiques sur une nouvelle serie de manganites mixtes de terre rare et d'alcalino-terreux" *J. Solis State Chem.*, **40**(1), 14-19 (1981)

[2]N. Kamegashira and S. Umeno, "Crystal structure and phase transition of $BaTb_2Mn_2O_7$" *Mater. Chem. Phys.* **16,** 89-98 (1986)

[3]N. Kamegashira, "Phase transition and nonstoichiometry of rare earth manganites" *Shokubai,* **31** (4) 222-226 (1989)

[4]N. Kamegashira, H. Nakano, G. Chen, and J. Meng, "Phase behavior of rare earth manganites" *J. Rare Earths*, **22**(5), 582-585 (2004)

[5]N. Kamegashira et al., "Study on the phase behavior of $BaEu_2Mn_2O_7$ through heat treatment of a single crystal" *J. Alloys and Compods*. (2005) *in press*

[6]N. Kamegashira, A. Shimono, and M. Horikawa, "Synthesis and phase transition of the

$BaGd_2Mn_2O_7$ phase" *Mater. Chem. and Physics*, **24**, 389-397 (1990)

[7]N. Kamegashira et al. "Crystal growth and structure analysis of rare earth manganites with layered perovskite type" Proceedings of 10th Int. Ceramics Congress Part A (2002) CIMTEC, edited by P. Vincenzini, pp. 181-186

[8]K. Naito, N. Kamegashira, and N. Sasaki, "Phase-equilibria in the system between NbO_2 and Nb_2O_5 at high-temperatures" *J. Solid State Chem.* **35(3),** 305-311 (1980)

[9]A. M. Glazer, "The classification of tilting octahedra in perovskites" *Acta, cryst.* **B**28, 3384-3392 (1972)

[10]A. Shimono, K. Hayashi, and N. Kamegashira, "Measurements of magnetic susceptibilities of $BaLn_2Mn_2O_7$ phases (Ln = rare earth elements" *Mater. Chem. and Phys.* **28,** 175-190 (1991)

THERMAL EXPANSION OF NICKEL/SAMARIUM-DOPED CERIA CERMET

Soichiro Sameshima, Yoshihiro Hirata and Keisuke Higashinakagawa
Department of Advanced Nanostructured Materials Science and Technology,
Kagoshima University, 1-21-40 Korimoto, Kagoshima 890-0065, Japan

ABSTRACT

Nickel/samarium-doped ceria (SDC) cermet has attracted an attention as an anode material of the low temperature solid oxide fuel cell. The SDC powder was prepared by heating the oxalate solid solution $(Ce_{0.8}Sm_{0.2})_2(C_2O_4)_3$ at 673 K. The porous SDC powder was infiltrated into a 1.4 M-$Ni(NO_3)_2$ solution and freeze-dried. The powder calcined at 623 K was sintered at 1773 K for 4 h. The NiO-SDC compact was reduced in a 25 vol % H_2- 75 vol % Ar atmosphere at 973 K for 10 h. The thermal expansion coefficient (CTE) of the Ni-SDC cermet was measured at 323 – 1273 K in an Ar atmosphere. The CTE (11.3×10^{-6} K^{-1}) of the porous cermet with 5-25 vol % Ni was close to that of dense SDC. In the range from 30 to 80 vol % Ni, the CTE increased with increasing Ni content.

INTRODUCTION

Solid Oxide Fuel Cell (SOFC) is a multilayer device consisting of electrolyte, anode, cathode and interconnector.[1] This cell has many advantages such as a variety of fuel gases, high energy conversion efficiency, clean product (H_2O), no vibration, no sound, no loss of solid electrolyte and utilization of high temperature exhaust gas. Rare earth-doped ceria is a possible electrolyte for the low temperature SOFC because of its high oxide ion conductivity.[2-4] The decrease of operation temperature of SOFC enhances the thermal stability of electrodes or interconnector and increases the lifetime of SOFC. In our previous papers,[5, 6] Ni/Sm-doped ceria (SDC) cermet was studied as an anode material of SOFC with SDC electrolyte. The cermet was prepared by infiltrating a $Ni(NO_3)_2$ solution into the porous SDC powder. The suspension was freeze-dried and heated at 673 K and sintered in air at 1573 K, followed by the reduction of NiO to Ni in a H_2/Ar atmosphere. The cermet showed a high electronic conductivity at 15 vol % Ni. The coefficients of thermal expansion (CTE) of the solid electrolyte and electrodes should be close to prevent exfoliation or cracking during heat-cycle of SOFC.[7-11] However, little papers have reported the CTE of Ni/SDC cermet. In this paper, sintering behavior, microstructure and CTE of Ni/SDC cermet were reported.

EXPERIMENTAL PROCEDURE

Sample preparation

The fabrication process of SDC with a composition $Ce_{0.8}Sm_{0.2}O_{1.9}$ was reported in our previous paper.[12] The oxalate solid solution $(Ce_{0.8}Sm_{0.2})_2(C_2O_4)_3$ was produced at room temperature by adding the mixed nitrate solution (0.2 M) of Ce and Sm into a 0.4 M oxalic acid solution. The co-precipitated oxalate powder decomposed to a polycrystalline oxide solid solution by heating to 673 K. The SDC solid solution was milled with α-alumina balls (0.3 cm diameter) at a rate of 85 rpm for 24 h. The specific surface area of the SDC powder after the milling was 45.8 m^2/g, which corresponded to 18.4 nm of diameter of spherical particle. The SDC powder was immersed in a 1.4 M $Ni(NO_3)_2$ solution and freeze-dried to make 5-80 vol % Ni/SDC cermet (FRD-50 M Freeze Dryer, Iwaki Glass Co., Ltd., Japan). The dried SDC powder containing $Ni(NO_3)_2$ was heated at 623 K for 1 h in air. An aqueous suspension of 10 vol % NiO/SDC powder at pH 10.5 was stirred for 24 h. The suspension was consolidated by casting in a gypsum mold to a rectangular shape with sizes of 10×10×40 mm. The green compact was sintered at 1773 K for 4 h in air. The sintered NiO/SDC was reduced at 973 K for 10-34 h in a 25 vol % H_2-75 vol % Ar atmosphere. The reduced sample was polished with 1 μm diamond paste.

Microstructure and thermal expansion coefficient

The bulk density and porosity of the Ni/SDC cermet were measured by the Archimedes method using distilled water. The theoretical density (7.135 g/cm^3) of SDC was calculated from the lattice parameter (0.5437 nm).[13] The elemental distribution of Ni, Ce, and Sm in the cermet was evaluated by electron probe X-ray microanalyzer (JXA-8600SX, JEOL, Ltd., Japan). Thermal expansion coefficients of SDC and Ni/SDC cermet were measured by thermomechanical analyzer (TMA; Model 8310 Rigaku Co., Tokyo). The sample was heated to 1273 K in Ar gas (purity:99.99 vol %) at a rate of 10 K/min. The measured data were calibrated by blank test and the reference data of α-alumina.

RESULTS AND DISCUSSION

Sintering behavior of Ni/SDC cermet

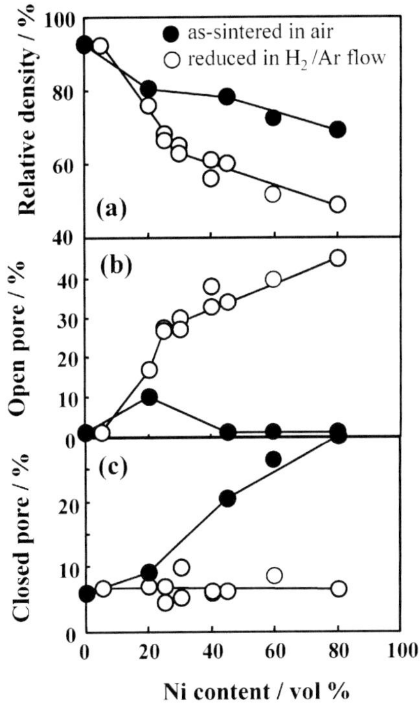

Fig.1 Fractions of solid (a) and pores (b, c) of the Ni/SDC cermet before and after the annealing in a 25 vol% H_2/ 75 vol% Ar atmosphere at 973 K.

Figure 1 shows the relative density and porosity of the NiO/SDC compact sintered at 1773 K in air and the Ni/SDC cermet treated in a 25 vol % H_2/ 75 vol % Ar atmosphere at 973 K for 24 h. The relative density of the as-sintered NiO/SDC compacts with 0, 29, 51 and 82 vol % of NiO (corresponding to 0, 20, 45 and 80 vol % of Ni) were 92.8, 80.6, 78.3 and 69.0 %, respectively. Addition of NiO to SDC inhibited the densification because of the suppression of diffusion of the constituent atoms during the sintering. The produced phases were NiO and SDC and no chemical reaction was suggested from the X-ray diffraction patterns. Figure 1 also shows the fractions of solid and pores of the Ni/SDC cermet annealed in a H_2/Ar atmosphere. The X-ray diffraction patterns of the cermet indicated the reduction of NiO to Ni and no formation of the compound between Ni and SDC. The closed pore of the as-sintered NiO/SDC compact changed to open pores after the reduction of NiO to Ni (See Fig2). As a result, the porosity of closed pores decreased and the porosity of open pores increased after the annealing of the as-sintered compact in a H_2/Ar atmosphere. The above change in the pores

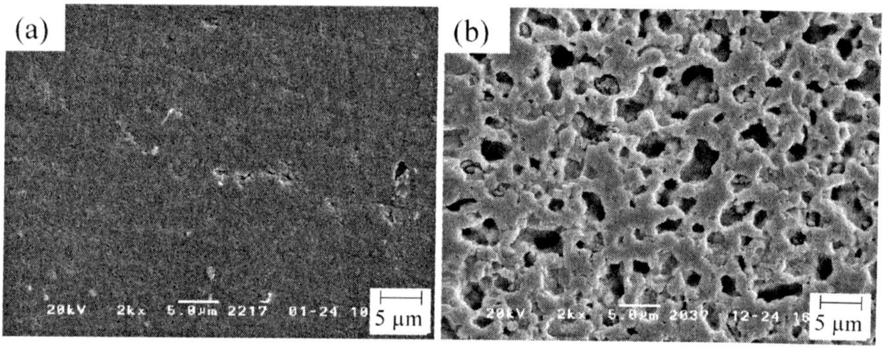

Fig. 2 Microstructures of 68 vol % NiO/SDC compact before (a) and after (b) the annealing in a 25 % H_2 / 75 % Ar atmosphere at 973 K for 10 h.

reduced the bulk density as shown in Fig.1 (a). The measured weight loss of the NiO/SDC compact after the annealing for 10 h in a H_2/Ar atmosphere at 973 K suggested that 0-3 vol % of NiO remained in the Ni-SDC cermet.

Figure 2 shows the microstructures of 68.2 vol % NiO/SDC compact before (a) and after (b) the annealing in a H_2/Ar atmosphere at 973 K for 10 h. The dense structure (a) of the as-sintered compact changed to the porous structure (b) owing to the formation of open pores by the reduction of NiO to Ni. Observation of the annealed Ni-SDC cermet using an electron probe micro analysis indicated that (1) Ni particles were uniformly distributed in the cermet and (2) a network of Ni particles, which act as a path of electronic conduction, was formed.

Coefficient of thermal expansion of Ni/SDC cermet

The thermal expansion ($\Delta L/L_0$, L_0: initial length at room temperature, ΔL: thermal expansion at a higher temperature) of SDC or Ni/SDC cermet was fitted by Eq.(1),

$$\frac{\Delta L}{L_0} = c_1 + c_2T + c_3T^2 + c_4T^3 \qquad (1)$$

where c_1, c_2, c_3, and c_4 are the experimental constants, and T is the measured temperature (K). The coefficient of thermal expansion (α) was determined by differentiating Eq.(1) with respect to T.

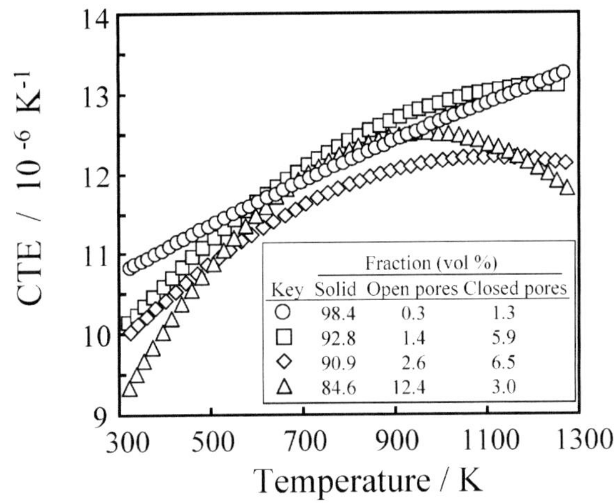

Fig.3 Coefficient of thermal expansion (CTE) of SDC with different porosity.

$$\alpha = \frac{1}{L_0}\frac{dL}{dT} = c_2 + 2c_3T + 3c_4T^2 \qquad (2)$$

Figure 3 shows the α values of SDC with 84.6 – 98.4 % relative density, which were sintered at 1573 -1873 K in air. As seen in Fig.3, the CTE increased at a higher temperature. Decrease in the relative density increased the dependence of CTE on temperature. The SDC samples included two types of pores : open pore and closed pore. The air trapped in the closed pores during the sintering process, increased the pressure of inside of the pores at a high temperature. On the other hand, the open pores have not this type of pressure effect but may give a flexible space for the deformation of grains upon the heating. The above different influences of open and closed pores on the expansion of grains were coupled to increase the nonlinearity of CTE with heating temperature. The rapid change in CTE with temperature results in the origin of cracking during heat-cycle of SOFC.

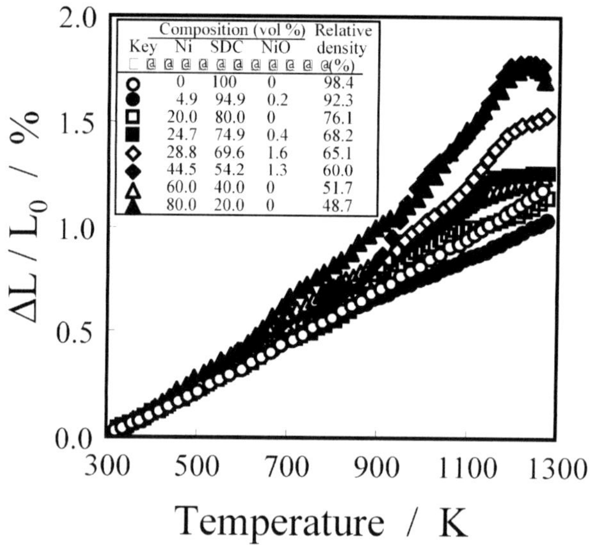

Fig.4 Thermal expansion of the Ni/SDC cermet in an Ar atmosphere.

Figure 4 shows the thermal expansion of 0-80 vol % Ni/SDC cermet in an Ar atmosphere (purity 99.99 %). The difference of thermal expansion with Ni content was small in a low temperature range from 300 to 600 K but became greater at a higher temperature. The cermet with a higher volume fraction of Ni showed a larger thermal expansion at 900- 1300 K. The thermal expansion at 700-1300 K was not essential result of Ni/SDC cermet and was due to the oxidation of Ni by oxygen gas (~10 ppm) in the Ar atmosphere. The thermogravimetric analysis of Ni powder in the same Ar atmosphere, which was prepared by heating a freeze-dried $Ni(NO_3)_2$ powder in a 25 vol % H_2- 75 vol % Ar atmosphere at 973 K for 10 h, showed a weight gain starting at 700 K. When 10 ppm of oxygen gas was introduced in the anode chamber during the operation of SOFC, the Ni/SDC cermet may be exfoliated from the electrolyte because of the rapid thermal expansion. Figure 5 shows the CTE of the Ni/SDC cermet at 323 –673 K where little oxidation of Ni occurred in the Ar atmosphere. The data in Fig.5 depended on the Ni content and temperature. The CTE became larger with increasing Ni content. The temperature dependence of the CTE was also greater for the Ni/SDC cermet with a higher content of Ni.

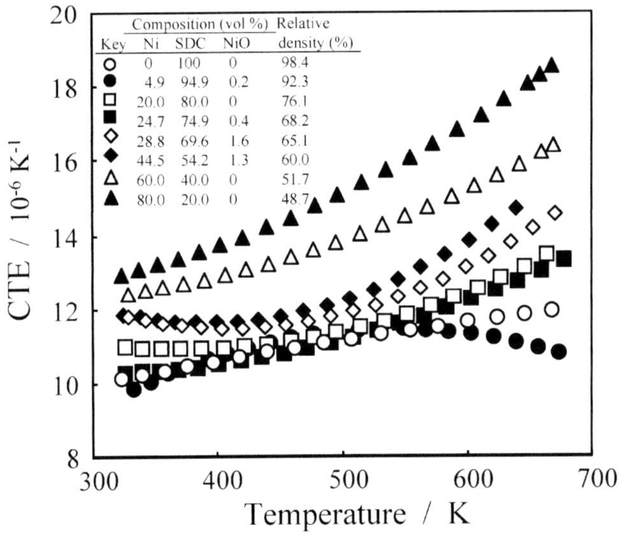

Fig.5 Coefficient of thermal expansion (CTE) of Ni/SDC cermet as a function of heating temperature.

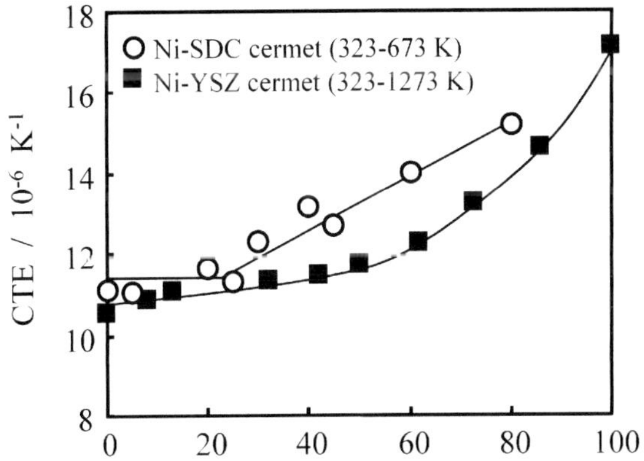

Fig.6 Comparison of averaged CTEs between Ni-SDC cermet and Ni-YSZ cermet.

Figure 6 shows the CTE averaged between 323 and 673 K for the Ni/SDC cermet. The averaged CTE was almost constant (11.3 $\times 10^{-6}$ K^{-1}) at the Ni content below 25 vol % and increased gradually at a higher Ni content. The CTE of 80 vol %

Ni/SDC cermet was close to that increased gradually at a higher Ni content. The above result indicates that Ni content of the cermet should be controlled below 25 vol % to make a good correspondence of CTEs between SDC electrolyte and the Ni/SDC cermet. The electronic conductivity of 25 vol % Ni/SDC cermet at room temperature was 7.2 S•cm^{-1}. In Fig.6, the CTE of Ni/8 mol %-yttria-stabilized zirconia (YSZ) cermet was also shown to compare with that of the Ni/SDC cermet. Both the cermet showed a similar tendency of CTE with Ni content. The Ni/SDC system has a higher CTE than the Ni/YSZ system.[14]

CONCLUSIONS

(1) Addition of NiO to SDC inhibited the densification because of the suppression of diffusion of the constituent atoms during the sintering. As a result, the relative density of the Ni-SDC cermet decreased with increasing Ni content.

(2) The coefficient of thermal expansion (CTE) of SDC increased at a higher temperature. The nonlinearity of CTE of SDC with heating temperature increased with increasing porosity.

(3) The CTE of the porous Ni-SDC cermet averaged at 323-673 K was almost constant (11.3 x 10^{-6} K^{-1}) at the Ni content below 25 vol % and increased gradually at a higher Ni content. The Ni content of the cermet should be controlled below 25 vol % to make a good correspondence of CTEs between SDC electrolyte and the cermet.

REFERENCES

[1]N.Q. Minh, "Ceramic Fuel Cells", *J .Am. Ceram. Soc.*, **76**, 563-588 (1995).

[2]B.C.H. Steel, "Appraisal of $Ce_{1-y}Gd_yO_{2-y/2}$ Electrolytes for IT-SOFC Operation at 500°C", *Soloid State Ionics*, **129**, 95-100 (2000).

[3]K. Eguchi, "Ceramic Materials Containing Rare Earth Oxides For Solid Oxide Fuel Cell" *J.Alloys and Compounds*, **250**, 486-491 (1997).

[4]H.Inaba and H. Tagawa, "Ceria-Based Solid Electrolytes", *Solid State Ionics*, **83**,1-16 (1996).

[5]Y. Okawa and Y. Hirata, "Sinterability, Microstructures and Electrical Properties of Ni/Sm-Doped Ceria Cermet Processed with Nanometer-Sized Particles", *J. Eur. Ceram. Soc.*, **25**, 473-480 (2005).

[6]Y. Okawa, T. Matsumoto, T. Doi and Y. Hirata, "Thermal Stability of Nanometer-Sized NiO and Sm-Doped Ceria Powders", *J.Mater. Res.* **17**, 2266-2274 (2002).

[7]I. Yasuda and M. Hishinuma, "Sintreing Characteristics and Mechanical Properties of CeO_2-Gd_2O_3 Solid Solution", Abstracts of 13th Fall Meeting of The Ceramic Society of Japan, Kanazawa (1996), p.233.

[8]G.A.Tompsett, N.M. Sammes and O.Yamamoto, "Ceria-Yttria-Stabilized Zirconia Composite

Ceramic Systems for Applications as Low-Temperature Electrolytes", *J. Am. Ceram. Soc.*, **80**, 3181-3186 (1997).

[9]P. Shuk and M.Greenblatt, "Hydrothermal Synthesis and Properties of Mixed Conductors Based on $Ce_{1-X}Pr_XO_{1-\delta}$ Solid Solutions", *Solid State Ionics*, **116**, 217-223 (1999).

[10]M.Mogensen, N.M.Sammes and G.A. Tompsett, "Physical, Chemical and Electrochemical Properties of Pure and Doped Ceria", *Solid State Ionics*, **129**, 63-94 (2000).

[11]S. Sameshima, M. Kawaminmi and Y. Hirata, "Thermal Expansion of Rare-Earth-Doped Ceria Ceramics", *J. Ceram. Soc. Jpn.*, **110**, 597-600 (2002).

[12]K. Higashi, K. Sonoda, H. Ono, S. Sameshima and Y. Hirata, "Synthesis and Sintering of Rare-Earth-Doped Ceria Powder by the Oxalate Coprecipitation Method", *J. Mater. Res.*, **14**, 957-967 (1999).

[13]S. Sameshima, H. Ono, K. Higashi, K. Sonoda, Y. Hirata and Y. Ikuma, "Electrical Conductivity and Diffusion of Oxygen Ions in Rare-Earth-Doped Ceria", *J. Ceram. Soc. Jpn.*, **108**,1060-1066 (2000).

[14]H. Tagawa, "SOFC and Environmental of Earth", Agune Shoufu Publishing, Tokyo, Japan (1998), p.184.

THE EFFECT OF CRYSTAL PHASE FORMATION ON LEACHABILITY OF Pb FROM GLASS CERAMICS PREPARED FROM INDUSTRIAL ZINC WASTE

Bussaraporn Patarachao and Sirithan Jiemsirilers
Research Unit of Advanced Ceramics, Department of Materials science, Faculty of Science, Chulalongkorn University, Bangkok 10330 Thailand

Parjaree Thavorniti and Sitthisuntorn Supothina
National Metal and Materials Technology Center, 114 Thailand Science Park, Paholyothin Rd., Klong Luang, Pathumthani 12120 Thailand

ABSTRACT

The correlation between the crystalline phase present in glass-ceramics and the leachability of lead (Pb) was investigated. In this study, glass-ceramics were prepared by the crystallization of glasses obtained by melting raw materials containing 30% of zinc waste from a hydrometallurgical plant mixed with other oxides. The crystalline phases present in the glass-ceramics were identified by x-ray diffractometry (XRD), while the crystal morphologies were investigated by scanning electron microscopy (SEM) and the composition of crystals and residual glasses were confirmed by energy-dispersive x-ray diffractometry (EDS). The leachability of lead (Pb) was examined following the standard procedure of the United States Environmental Protection Agency Toxic Characterization Leaching Procedure (TCLP). The leached out concentration of Pb was then quantified by using atomic absorption spectroscopy (AAS). It was found that the glass-ceramics containing wollastonite-ferroan phases were more stable towards acid leaching, whereas those containing pyroxene phase were less stable.

INTRODUCTION

In recent years, recycling of industrial wastes such as municipal solid waste, incinerator fly ash, blast furnace slag, etc. have been studied by many scientists around the world.[1-5] Glass and glass-ceramic technologies are considered to be the most appropriate technologies to immobilize heavy metals from these wastes in the stable glass matrix with a consistent reduction of volume and the destruction of organic matter associated with the wastes during glass processing.[6,7] In Thailand, the company that produces zinc ingot is located at Tak province. Each year zinc refinery process produces large amounts of waste, which contains significant percentage of Lead (Pb). The possibility of transformation of this zinc waste into more valuable products such as glass-ceramic material has been studied.[8] In this study the effect of crystal phase formation on leached out concentration of Pb from glass-ceramics is the main concerned. The leachability of Pb was examined to determine the quality of obtained glass-ceramics.

EXPERIMENT PROCEDURE

Characterization of zinc waste

Initially, chemical compositions of zinc waste used in this study was analyzed by using x-ray fluorescence (XRF) since the compositions of zinc waste generated from different plants can be varied depending on the type and the source of zinc minerals, as well as the materials adding during zinc recovery process.[9,10]

Preparation of glass

Glass compositions were designed by using Mixsoft Computer software[11] based on eight oxides which were SiO_2, B_2O_3, Na_2O, CaO, MgO, ZnO, Al_2O_3 and Fe_2O_3. Glass compositions are shown in Table I.

Table I Glass compositions in weight percent of oxides

	Composition (%wt)							
	SiO_2	B_2O_3	Na_2O	CaO	MgO	ZnO	Al_2O_3	Fe_2O_3
G1	45.00	15.00	15.00	20.00	0.00	1.00	1.00	3.00
G2	55.00	5.00	13.00	20.00	0.00	1.00	1.00	5.00
G3	55.00	5.00	15.00	18.00	0.00	1.00	1.00	5.00
G4	45.00	15.00	13.00	20.00	0.00	3.00	1.00	3.00
G5	45.00	15.00	11.00	20.00	0.00	1.00	5.00	3.00
G6	63.00	5.00	15.00	10.00	0.00	3.00	1.00	3.00
G7	64.00	5.00	10.00	10.00	0.00	3.00	5.00	3.00
G8	45.00	15.00	5.00	18.00	10.00	3.00	1.00	3.00
G9	45.00	15.00	10.00	10.00	9.00	1.00	5.00	5.00
G10	45.00	15.00	11.00	10.00	10.00	3.00	1.00	5.00
G11	45.00	15.00	11.00	10.00	10.00	1.00	5.00	3.00
G12	56.00	5.00	5.00	15.00	10.00	3.00	1.00	5.00
G13	55.00	5.00	15.00	10.00	8.00	1.00	1.00	5.00
G14	55.00	5.00	15.00	10.00	10.00	1.00	1.00	3.00

In each batch thirty percent of zinc waste was weighted and mixed with other raw materials to produce suitable glass compositions. The mixture was melted at 1450°C for 1 hour to form glasses.

Preparation of glass-ceramics

Preparation of glass-ceramic was performed by heat-treatment process including two main steps. Firstly, the samples were heat-treated up to 750°C and hold at this temperature for 1 hour, and then further heat-treated up to 850°C and hold at this temperature for 2 hours.

Characterization of glass-ceramics

Crystal phase analysis

After heat-treatment, crystalline phases of glass-ceramics were identified by using an x-ray diffractometer (XRD). The size and morphology of the crystals were observed by using a scanning electron microscope (SEM). The chemical compositions of crystal phase and glassy phase contained in glass-ceramics were examined by using an energy-dispersive x-ray spectrometer (EDS).

Chemical durability

Since the zinc waste contained significant amount of lead (Pb), therefore it was very important to examine the leachability characteristic of Pb in glass-ceramics matrix. Noted that the regular standard of heavy metal leached out concentration is different in difference countries.[5-7] In this study the American EPA's toxic characteristic leaching procedure (TCLP) method 1311, which is the acid leaching method, was used to examine the chemical stability of glass-ceramics in terms of leaching of heavy metals ions.[12] The leached out concentration of Pb was then examined by using an atomic absorption spectrometer (AAS).

RESULTS AND DISCUSSION

Chemical compositions of zinc waste

The chemical composition of zinc waste was analyzed quantitatively by using XRF. The compositions in weight percent of oxides are shown in Table II.

Table II Chemical compositions of zinc waste analyzed by XRF

Oxides	%wt
SiO_2	22.45
Al_2O_3	3.96
Fe_2O_3	7.32
CaO	16.43
MgO	0.29
K_2O	0.63
ZnO	3.09
CuO	0.07
PbO	1.07
TiO_2	0.10
MnO	0.10
P_2O_5	0.21
ZrO_2	0.01
SO_3	30.64
L.O.I	13.94

Zinc waste was composed of three main oxides, which were SiO_2, CaO and SO_3, with respect to the zinc refinery process.

Crystal phase formation of glass-ceramics

Five different types of crystalline phase have been identified by XRD as follows: wollastonite-ferroan, anorthite, quartz, cristobalite and pyroxene. Single phase of wollastonite-ferroan was found in G1, G2 and G3. Single phase of pyroxene was found in G8 – G14. Mixed phases of wollastonite-ferroan and anorthite were found in G4 and G 5. Mix phases of wollastonite-ferroan, quartz and cristobalite were found in G6 and G7. The XRD patterns of all five crystalline phases in the selected samples are shown in Fig. 1.

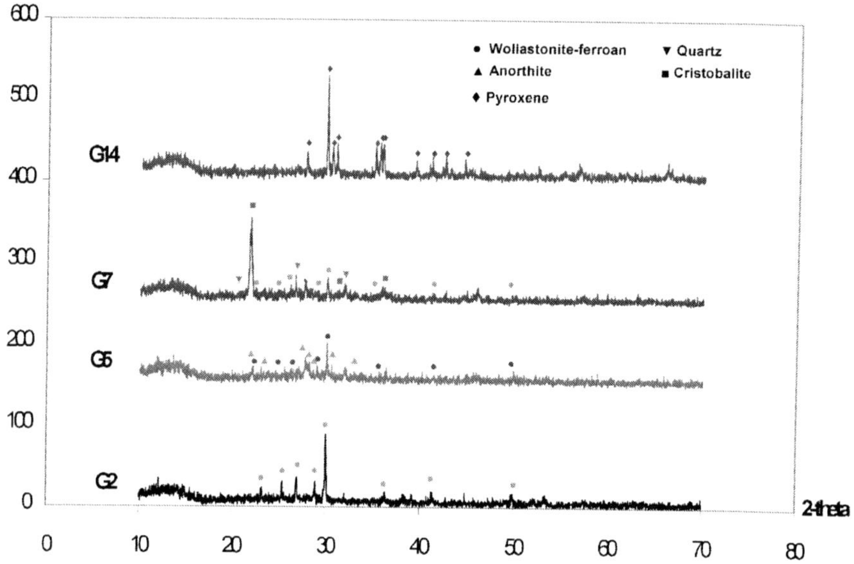

Fig. 1 XRD patterns of glass-ceramics G2, G5, G7 and G14

As described above that single pyroxene phase was found in G8-G14, these glass-ceramics contained 8-10wt%MgO, whereas other phases were found in non-adding MgO glass-ceramics. This is believed to be due to the appropriate of chemical composition of the parent glass with the suitable time and temperature.

The SEM micrograph of G2, G5, G7 and G14 are shown in Fig. 2 (a), (b), (c) and (d), respectively. Acicular crystals dispersed in the glassy matrix were observed in G2. The EDS analysis showed that acicular crystals contain mainly Si, Ca and confirming wollastonite-ferroan ($Ca_{2.87}FeO_{0.13}(SiO_3)_3$) whereas glassy matrix enriched in Si, Al, Na and Fe with small amount of Ca.

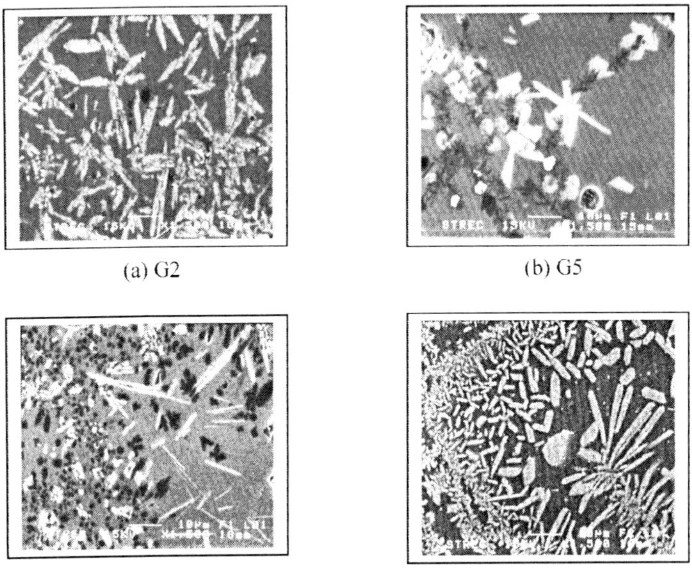

(a) G2 (b) G5

(c) G7 (d) G14

Fig.2 SEM micrograph of the crystalline phases

In fig 2 (b) (G5), two distinct crystal morphologies crystals were observed. The acicular crystals contain mainly Si, Ca and Fe confirming wollastonite-ferroan, the dark dendritic crystals contain mainly Si, Al and Ca confirming anorthite ($Ca(Al_2SiO_8)$). The acicular crystals enriched in Si, Ca and Fe of wollastonite-ferroan.were also observed in G7.

Dendritic crystals were observed in the SEM micrograph of G14. The EDS analysis showed that the dendritic crystals contain mainly Si, Mg, Ca and Fe confirming pyroxene $(Mg_{0.937}Fe_{0.063})(Ca_{0.751}Na_{0.249}Fe_{0.018})(Si_2O_6)$.

Chemical durability

The effect of crystalline phase on leached out concentration of Pb was studied. The leached out concentration of Pb from glass-ceramic matrix are shown in Table III. The concentration of Pb in the leachate of glass-ceramics was compared with the USA regulatory limit (5 ppm). Note that the TCLP test was conducted on the pulverized specimens.

Table III Leached out concentration of Pb in glass-ceramics

	Crystalline phase	Concentration (ppm)
G1	W	2
G2	W	2
G3	W	2
G4	W/A	2
G5	W/A	2
G6	W/C/Q	2
G7	W/C/Q	2
G8	P	22
G9	P	32
G10	P	42
G11	P	26
G12	P	23
G13	P	20
G14	P	18

Note: W is Wollastonite-ferroan, A is Anorthite, C is Cristobalite phase, Q is Quartz

According to Table III, it is very interesting that the leached out concentrations of Pb in glass-ceramics contained pyroxene phase were so much higher than those of glass-ceramics contained wollastonite-ferroan phase. None of glass-ceramics contained pyroxene phase gave the leached out concentration of Pb lower than 5 ppm, whereas glass-ceramics contained wollastonite-ferroan phase gave the leached out concentration of Pb lower than the limit. Not only glass-ceramics contained single wollastonite-ferroan phase, glass-ceramics contained multiple phase of wollastonite-ferroan + anorthite (W/A) and wollastonite-ferroan + cristobalite + quartz (W/C/Q) also gave the leached out concentration of Pb lower than 5 ppm. The leached out concentrations of Pb from glass-ceramics contained pyroxene phase are varied from 18 – 42 ppm. Therefore it is evident that crystal phase formation in glass-ceramics strongly effected on the leachability of Pb.

In this study, it is clear that those glass-ceramics contained pyroxene phase are unstable toward the acid leaching which means they are poor in chemical durability whereas glass-ceramics contained both single and multiple of wollastonite-ferroan phase are much more stable toward the acid leaching.

CONCLUSION

Glass-ceramics obtained in this study are composed of many types of crystalline which are wollastonite-ferroan, cristobalite, quartz, anorthite and pyroxene depending on glass compositions. The crystal phase present in the glass-ceramic strongly effects on a leachability of Pb. The glass-ceramics contained pyroxene phase are very poor in chemical durability and unstable toward acid leaching whereas glass-ceramics contained single and multiple phase of wollastonite-ferroan are stable.

ACKNOWLEDGEMENT

We are very grateful to the National Metal and Materials Technology Center (MTEC), the National Science and Technology Development Agency (NSTDA), Thailand for a financial support of this project. We also would like to acknowledge Padaeng Industry Public Co. Ltd. for supporting zinc waste as a raw material used in this study. Miss Patarachao would like to acknowledge the Thailand Graduate Institute of Science and Technology (TGIST) for financial support during an M.S. study.

REFERENCES

[1]Alexander Karamanov, Mario Pelino and Alessandro Hreglich, "Sintered glass-ceramics from municipal solid waste-incinerator fly ashes-part I: the influence of the heating rate of the heating rate on the sinter-crystallization," *Journal of the European Ceramic ociety* **23,** 827-832 (2003)

[2]Alexander Karamanov, Mario Pelino, Milena Salvo and Ilidko Metekovits, "Sintered glass-
ceramics from incinerator fly ashes. Part II. The influence of the particle size and heat-treatment on the properties," *Journal of the European Ceramic Society* **23,** 1609-1615 (2003).

[3]T.W. Cheng, T.H. Ueng, Y.S. Chen and J.P Chiu, "Production of glass-ceramic from incinerator fly ash," *Ceramics Internationa1* **28**, 779-783 (2002)

[4]Young Jun Park and Jong Heo, "Conversion to glass-ceramics from glasses made by MSW incinerator fly ash for recycling," *Ceramics Internationa1* **28,** 689-694 (2002)

[5]Michihiro Miyake, Chikashi Tamura and Motohide Matsuda, "Resource Recovery of Waste Incineration fly Ash: Synthesis of Zeolites A and P," Journal of the American Ceramic Society 85, 1873 (2002)

[6]P. Pisciella, S. Crisucci, A. Karamanov, M. Pelino, "Chemical durability of glasses obtained by vitrification of industrial wastes," *Waste Management* **21**, 1-9 (2001)

[7]Mazimina Romero and Jesus Ma. Rincon, "Surface and bulk crystallization of glass-ceramic in the Na_2O-CaO-ZnO-PbO-Fe_2O_3-Al_2O_3-SiO_2 system derived from a goethite waste," *Journal of the American Ceramic Society* **82**,1313-1317 (1999)

[8]S. Jiemsirilers, B. Patarachao, P. Thavornniti, and S. Supothina, "Properties of Glass-ceramics synthesized from hydrometallurgical Waste," *Melt Chemistry, Relaxation and Solidification Kinetics of Glasses, Ceramic Transactions* **170**, 237-246 (2005)

[9]M. Pelino, "Recycling of jarosite waste in the production of glass and glass-ceramic materials," *Interceram* **47**, 22-26 (1998)

[10]M. Pelino, "Recycling of zinc hydrometallurgical wastes in glass and glass-ceramic materials," *Waste Management* **20**, 561-568 (2000)

[11]G.F. Piepel, "*Mixsoft-software for the design and analysis of mixture and other constrained region experiments*," User's guide version 2.31, The United States of America (2000)

[12]Method 1311 Toxic characteristic leaching procedure (online). Available from: *http://www.epa.gov/epaoswer/hazwaste/test/pdfs/1311*, July (1992)

Thermomechanical Characterization

EFFECT OF FLUORIDE ADDITIVES ON THE HYDROXYAPATITE LAMINATE COMPOSITES

Sung-Hwan Kim[†] and Sang-Yeup Park[*†]

[*] Department of Ceramic Engineering, Kangnung National University, Kangnung, Kangwondo 210-702, South Korea

[†] Technology Innovation Center for Fine Ceramics, Kangnung National University Kangnung, Kangwondo 210-702, South Korea

ABSTRACT

In this study, the effect of MgF_2 additive on the mechanical properties of Hydroxyapatite/Alumina laminate composites was investigated. When we added MgF_2(5 vol%) to Hydroxyapatite/Alumina laminate composites, the decomposition of Hydroxyapatite was suppressed due to the substitution of F- for OH- in HAp crystal structure. When we compared two additives, such as MgF_2 and CaF_2, MgF_2 showed much more effective for the suppression of decomposition in the Hydroxyapatite/Alumina laminate composites due to more enhanced substitution of F^- for OH^- in HAp. In the case of MgF_2 addition, high flexural strength and Vickers hardness was obtained compared with the MgF_2-free Hydroxyapatite/Alumina composites.

INTRODUCTION

Hydroxyapatite (HAp, $Ca_{10}(PO_4)_6(OH)_2$) is the most promising materials for bone replacement because of its osteoconductivity and bio-compatibility properties. However, HAp has been used only in non-load-bearing areas due to a relatively low mechanical properties and a high dissolution rate [1-3]. Focusing on the crystal structure modification of HAp, it has been studied that these disadvantages can be eliminated by the ion substitution of Ca with Mg, Sr, and Ba [4]. Among those substitution ions, Mg ion is known as the most promising dental caries-promoting element due to a high initial calcification process in the bone formation. The deficiency of Mg ion will result in the degradation of bone structure, the inhibition of bone growth and also decrease of osteoblast adhesion [5]. When we add zirconia as a reinforcing agent, the mechanical properties is reduced due to the phase decomposition from HAp to TCP($Ca_3(PO_4)_2$) [3]. When we added a fluoride compound, however, the phase decomposition is reduced because fluoride ion promote the stabilization of HAp crystal structure. In this study, we attempted to increase the mechanical properties as well as biocompatibility with laminate composite concept. Because the laminate composites is known as to enhance the mechanical properties such as fracture strength and toughness [6], we provide a multilayered laminate composite composed of ZrO_2-rich layer and HAp-rich layer. Also, the effect of fluoride compound such as MgF_2 and CaF_2 was examined focusing on the mechanical properties and the suppression of phase transition in HAp crystal structure.

EXPERIMENTAL PROCEDURE

As a starting powder, non-stoichiometric hydroxyapatite($Ca_{10}(PO_4)_6(OH)_2$, Ca/P ratio: 1.63, Samjo Co., Korea)) and zirconia (TZ3Y, <0.3□, Tosoh, Japan) were used. Fluoride compound was synthesized using the mixing of magnesium nitrate (purity > 99%, Aldrich Chemical, USA) and calcium nitrate ($Ca(NO_3)_2{\cdot}4H_2O$ (purity > 99.99%, Aldrich Chemical, USA) with fluoric acid. After weighing the powder, the ball-milling with ZrO_2 ball was carried out for 24 h, and then dried in an oven for 24h. The composition of powder mixture was indicated in Table 1. Powder mixtures were mold in a die and then cold isostatically pressed at 30,000 psi. Green laminate specimen was fabricated with continuous tape casting method. After ball milling, slurry was ultrasonicated to remove the soft agglomerate. For the good dispersion, polymeric binder and plasticizer were added into the aqueous slip. The green compact was sintered at 1400°C for 2 hours. Densification behavior was measured with dilatometer (DIL 402 C, Netzsch, Germany) up to 1400°C with 10°/min heating rate. Phase change was identified with X-ray diffraction pattern and the microstructure was observed with Scanning Electron Microscopy (S 4700, Hitachi, Japan). Mechanical properties of specimen were evaluated by four-point flexural strength with a cross head speed of 0.5 mm/min. The dimension of mechanical test specimen was 3×4×40mm and the inner span and outer span were 10 and 30 mm, respectively. Vickers hardness was measured with micro vickers hardness tester (HM-124, Akashi, Japan) using 1.96N load on the polished surface of laminate specimen.

RESULTS AND DISCUSSION

Fig. 1 shows the X-ray diffraction analysis of sintered HA series and TZ series laminate composites. HA-0M specimen without fluoride additive was resulted in TCP phase transition from HAp, and also alumina based reaction compounds such as $CaAl_2O_4$ and $CaAl_4O_7$ were formed. There was no phase transition from HAp was observed in HA-5M specimen with 5 vol% MgF_2 additive (Fig. 1(a)). Compared to phase analysis between HA-5M and HA-5C, the addition of MgF_2 is more effective for the inhibition of phase transition. With increasing the MgF_2 content, the phase transition form HAP was suppressed in TZ-series composites, and the tetragonal ZrO_2 transformed into cubic ZrO_2 (Fig. 1(b)).

Table 1. Specimen notation of HAp with addition of CaF_2 and MgF_2.

Notation	Composition
P-HAp	Pure HAp
HA	HAp + 20 vol% Al_2O_3
HA-xC	HAp + 20 vol% Al_2O_3 + x vol%CaF_2 (x: 2.5, 5)
HA-xM	HAp + 20 vol% Al_2O_3 + x vol%MgF_2 (x: 2.5, 5)
TZ-yM	TZ3Y + 40 vol% Al2O3 + y vol%MgF_2 (y: 0, 1, 2, 4)
THL-yM	TZ3Y + 40 vol% Al2O3 + y vol%MgF_2 / HA-5M (Laminate Composite) (y: 0, 1, 2, 4)

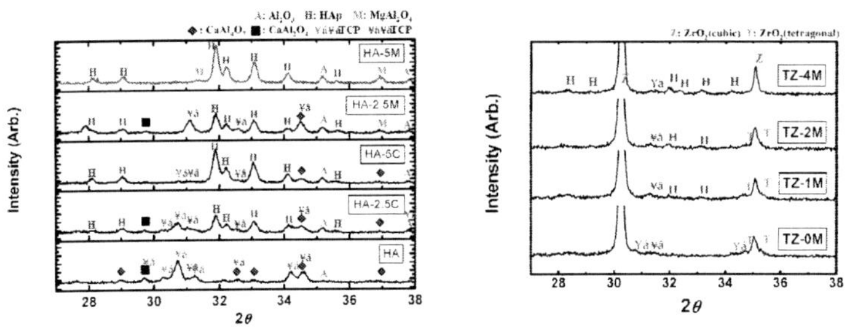

(a) (b)

Fig. 1. . X-ray diffraction analysis of HAp composites sintered at 1400°C for 2h.

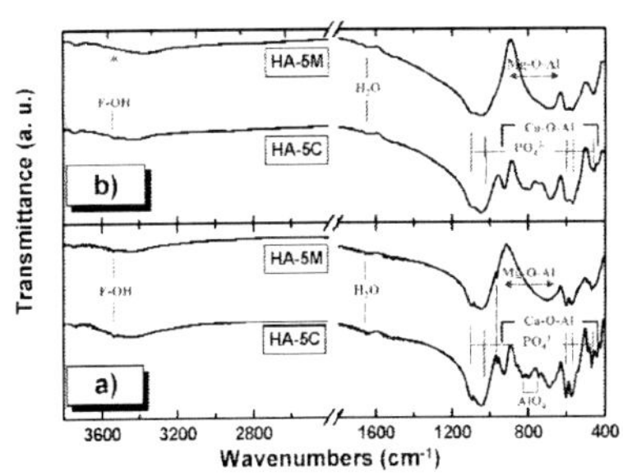

Fig. 2. FTIR analysis of HAp composites with sintered at (a)1300 and (b)1400°C.

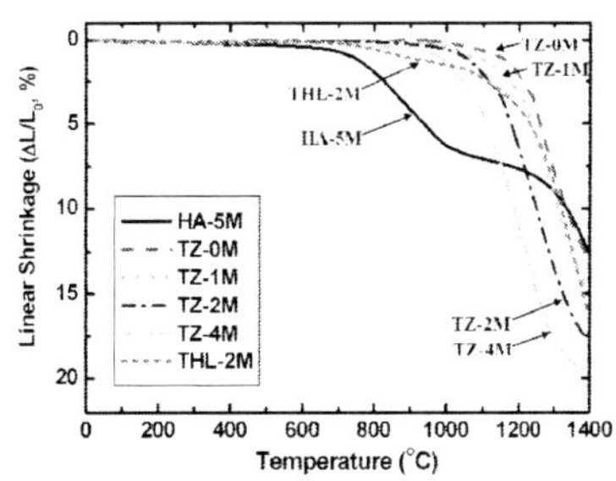

Fig. 3. Variation of linear shrinkage with sintering temperature.

Fig.2 shows the analysis of FT-IR result, which was carried out for the identification of ion substitution between OH and F. In the HA-5M, Mg-O-Al peak was observed in the wide vibration range of 600-900 cm^{-1}. In the HA-5C, Ca-O-Al peak was observed at the vibration range of 437 cm^{-1} and 930 cm^{-1}. Both of specimen as shown in Fig. 2(a), F-OH bond (3535 cm^{-1}) was observed due to the ion substitution between OH and F. When the temperature was increased up to 1400 °C, however, F-OH bond peak was disappeared in HA-5M specimen, as shown in Fig. 2(b). These results indicate that the F ion substitution content in HA-5M is much higher than HA-5C due to the lower dissociation temperature of MgF_2.

Fig. 3 shows the variation of linear shrinkage of composites with different composition. Shrinkage rate was increased with increasing the MgF_2 addition. HA-5M composite shows the first shrinkage at 700°C due to the crystallization of HAp and the phase decomposition of MgF_2. The second shrinkage around 1,200°C was resulted from the melting behavior of MgF_2. Shrinkage rate was similar between HA-5M and TZ-0M specimen. Laminate composites composed of TZ-2M and HA-5M showed a medium shrinkage behavior compared to other specimens.

Fig. 4 shows the microstructure of laminate composites sintered at 1,400°C. With increasing the MgF_2 contents, the densification was enhanced due to the porosity decrease. Thermal mismatch crack was formed in THL-4M laminate composite due to the different thermal expansion coefficient between HA-5M layer and TZ-4M layer (Fig. 4(d)).

Fig. 5 shows the microstructure of ZrO_2-rich layers in TZ-yM/HA-5M laminate composites. With increasing the MgF_2 contents, porosity was reduced. In the case of THL-4M specimen (Fig. 5(d)), particle size of ZrO_2 was increased compared to the addition of 2 vol% MgF_2 specimen (Fig. 5(c)).

Fig. 6 shows the mechanical properties of laminate composites sintered at 1400°C. HA-0M composites without MgF_2 showed relatively low flexural strength (~70 MPa) and vickers hardness (~2 GPa) compared to pure HAp specimen. However, HA-5M composites showed more enhanced flexural strength (~160 MPa) and vickers hardness (~6.5 GPa). THL-2M laminate composite composed of TZ-2M layer and HA-5M layer showed a higher fracture strength (~240 MPa) and vickers hardness (~9.5 GPa). However, THL-4M laminate composite showed low fracture strength due to the high grain growth of ZrO_2 and thermal mismatch crack formed in ZrO_2-rich layer.

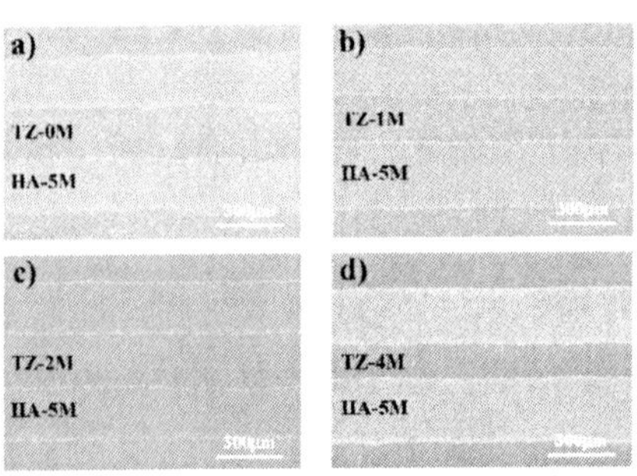

Fig. 4. SEM microstructures of HAp laminate composites sintered at 1400°C for 2h (a) THL-0M, (b) THL-1M, (c) THL-2M, and d) THL-4M

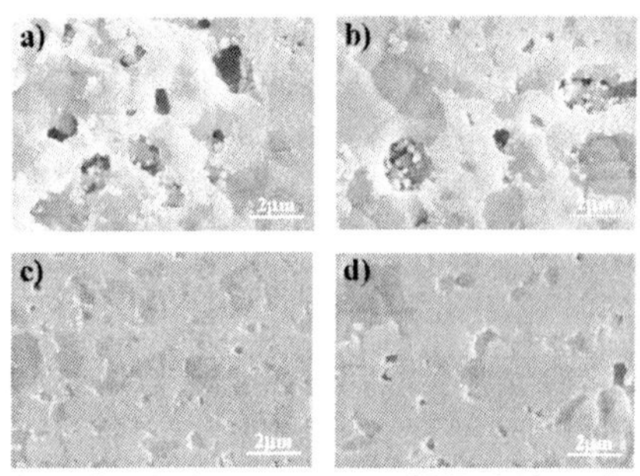

Fig. 5. SEM microstructures of TZ-yM layer in HAp laminate composites sintered at for 2h: (a) THL-0M, (b) THL-1M, (c) THL-2M, and (d) THL-4M.

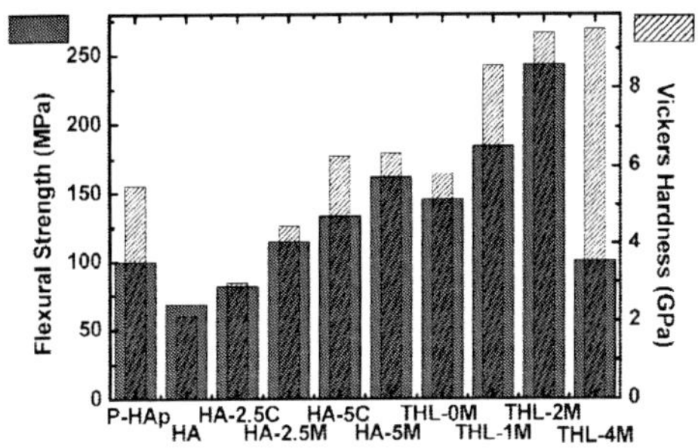

Fig. 6. Variation of mechanical properties of specimens sintered at 1400°C for 2h.

SUMMARY

When MgF_2 (5 vol%) was added into HAp-20 vol% Al_2O_3, the phase transition of HAp was completely inhibited below 1400°C. These results indicate that the F ion substitution content in HA-5M is much higher than HA-5C due to the lower dissociation temperature of MgF_2. With increasing MgF_2 contents in TZ series composites, the phase transition from HAp into TCP ($Ca_3(PO_4)_2$) was inhibited and the ZrO_2 was transformed from tetragonal phase to cubic phase. THL-2M laminate composites composed of TZ-2M layer and HA-5M layer showed more higher fracture strength (~240 MPa) and Vickers hardness (~9.5 GPa) compared to fluoride-free composites.

REFERENCES

1. L.M. Rodriguez-Lorenzo, J.N. Hart, K.A. Gross, "Influence of fluorine in the synthesis of apatites. Synthesis of solid solutions of hydroxy-fluorapatite," Biomaterials 24: 3777-85(2003).
2. Thomas J. Webster, Elizabeth A. Massa-Schlueter, Jennifer L. Smith, Elliot B. Slamovich, "Osteoblast response to hydroxyapatite doped with divalent and trivalent cations," Biomaterials 25: 2111-21(2004).
3. Hae-Won Kim, Young-Hag Koh, Seung-Beom Seo, Hyoun-Ee Kim, "Properties of fluridated hydroxyapatite-alumina biological composites densified with addition of CaF_2," Mater Sci Eng C23: 515-21(2003).
4. Soo-Ryong Kim, Young-Hee Kim, Sang-Jin Jung and Do-Hyung Riu, "Synthesis and Characterization of Silicon Substituted Hydroxyapatite," J Kor Ceram Soc Vol. 38, 1132-6(2001).
5. I. Mayer, R. Scblam, "Magnesium-Containing Carbonate Apatites," Elsevier Sci Inc (1997).
6. Jun-Ho Song, Sang-Yeup Park, "Microstructure Orientation of Alumina Laminate Composites," J. Kor. Ceram. Soc., 38[4] 351-359 (2001).

EFFECT OF H-BN ON THE MICROSTRUCTURE, MECHANICAL PROPERTIES, AND MACHINABILITY OF Si_3N_4-BASED CERAMIC COMPOSITES

Z. H. Piao[1], and W. S. Cho[1]

[1]School of Materials science and Engineering, Inha University, Incheon, 402-751, Korea

J. H. Lee[2]

[2] School of Metallurgical and Materials Engineering, Yeungnam University, Gyonsan 712-749, Korea

M. W. Cho[3], and C. Cho[3]
[3]School of Mechanical Engineering, Inha University, Incheon, 402-751, Korea

Z. A. Munir[4]

[4]Department of Chemical Engineering and Materials Science, University of California, Davis, CA, 95616, USA

ABSTRACT

Si_3N_4/h-BN composites were prepared by hot-pressing and pressureless-sintering at 1800°C for 2 h. The effects of h-BN content on microstructure, mechanical properties, and machinability of the Si_3N_4/h-BN composite were investigated. The results of the pressureless-sintered composites were compared with those of the hot-pressed composites. Relative density of the sintered compacts decreased with increasing h-BN content. Also, flexural strength decreased with increasing h-BN content, mainly due to lower Young's modulus of h-BN than that of Si_3N_4. Strength drop was also predicted from the residual stress analysis in the vicinity of h-BN within the matrix. With increasing h-BN content, Si_3N_4/hBN based ceramic composites revealed enhanced crack resistance (R-curve) behavior. Both hot-pressed and pressureless-sintered Si_3N_4/h-BN composites revealed a textured microstructure, in which c axes of h-BN tended to be oriented toward hot-pressing and cold-pressing directions. The Vickers crack interactions with h-BN on a plane parallel to hot-pressing or cold-pressing directions showed anisotropic behavior. The Vickers indentation crack paths in specimens are sinusoidal due to pull out grains during crack propagation. Si_3N_4 monolith could not be machined due to its brittle fracture, but Si_3N_4/h-BN based ceramics could be machined without fatal fracture. With increasing h-BN content, the cutting resistance decreased.

1. INTRODUCTION

Silicon nitride (Si_3N_4) gives great promises for engineering applications due to its excellent thermomechanical properties, such as high bending strength, high thermal shock resistance, and high fracture toughness. However, it is difficult to machine due to its brittle fracture behavior. This disadvantage leads to high machining cost and limits the application Si_3N_4 greatly. Thus investigating the feasibility of producing machinable silicon nitride would significantly contribute to the utilization of the composites for wider applications. It was found that the addition of hexagonal boron nitride (h-BN), which has excellent machinability due to its structure similar to that of graphite, improves the machinability of ceramics. Si_3N_4-BN based ceramics can be easily machined since the cleavage plane (basal plane) of h-BN can facilitate crack propagation during machining, thereby decreasing the cutting resistance.[1-3]

Most of Si_3N_4-BN composites are being prepared by hot-pressing due to its poor sinterability.[1,2] However, such hot-pressing is an expensive sintering method compared with pressureless sintering, furthermore, it is impossible to manufacture precision parts with complicated shapes.

In this paper we investigated both hot-pressing and pressureless sintering conditions for Si_3N_4-BN ceramics, and also present investigation results of the effect of h-BN content on microstructure, mechanical properties, R-curve behavior, and machinability of silicon nitride.

2. EXPERIMENTAL PROCEDURE

The starting powders were Si_3N_4 (α-crystal phase>95%, average particle size, 0.17□, Ube Industries., Ltd, Japan), and h-BN (>99% pure, 10 μm, Kojundo Chemical Laboratory Co. Ltd, Japan). For the hot-pressed compacts, added amounts of h-BN varied from 0 to 30 vol%. For the pressureless-sintered compacts, particle size of h-BN was reduced to 6.4 □ using planetary ball mill, and it is added up to 20 vol%. To all specimens, 8 mol% of Y_2O_3 (>99.9% pure, < 0.5 μm, Shinestu Chemical Co. Ltd. Japan) and 6 mol% Al_2O_3 (>99.99% pure, 0.39□, Sumitomo Chemical Co. LTD, Japan), were added as sintering aids. The powders were mixed by wet ball-milling using Al_2O_3 jar, Si_3N_4 balls and ethanol for 72 h. The mixed wet powders were dried on a hot plate using a rotating stirrer to avoid gravity-induced segregation. To obtain fully dense compacts, the dried and sieved (to 16 mesh) powders were hot-pressed at 1800°C for 2 h under a pressure of 25 MPa in N_2 atmosphere. For the fabrication of pressureless-sintered compacts, the dried and sieved (to 16 mesh) powders were cold-pressed at 160 MPa using WC-Co die into rectangular green compacts (6×15×40 mm) until the relative density reached to approximate 50%. Green compacts were pressureless-sintered at 1800°C for 2 h under the N_2 atmosphere. Rectangular specimens (3x4x36 mm) were prepared by cutting, grinding, and polishing the sintered compact. The relative density, four-point flexural strength, Vickers hardness (load; 98 N, dwell time; 30 s), and Young's modulus were measured.

For the R-curve characterization, indentation under loads ranging from 4.9 to 196 N was employed on the polished surface of the hot-pressed specimens, and the 4-point indentation strength was determined. The inner and outer span were 10 and 20 mm, respectively. R-curves were obtained from the relationship between the indentation load, indentation crack length and the strength after indentation following Krause's procedure.[4] At least five specimens were tested for each experimental condition for the R-curves.

The fracture surfaces of specimens were investigated by SEM. The interaction between the Vickers crack and h-BN was also investigated by SEM.

Machinability was evaluated by end-milling test. During milling test (Cutting tool; WC-Co, feed rate; 30, 50, 100 mm/min, cutting speed; 398 rpm, depth of cut; 0.1, and 0.2 mm), thrust force was detected by piezoelectric sensor and tool dynometer (Kistler 9257A).

3. RESULTS AND DISCUSSION

Fig. 1 shows the relative density of specimens with varying content of h-BN. The relative density

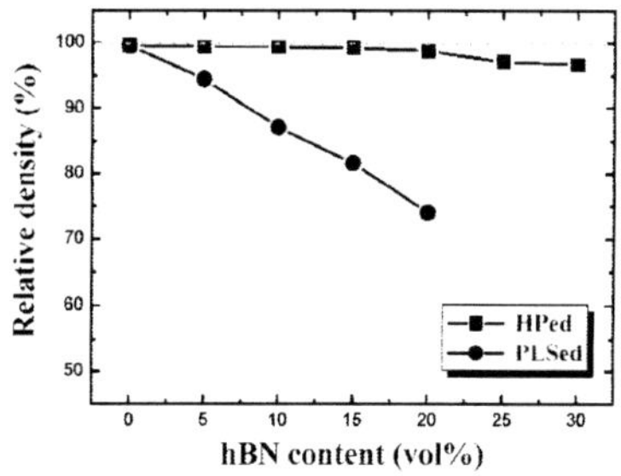

Fig. 1. Effect of h-BN content on the relative density of Si_3N_4-BN based ceramics.

of hot-pressed specimens decreased slightly with increasing h-BN content; while that of pressureless sintered specimens decreased very rapidly with increasing h-BN content. This is the consequence of the role of the plate-like h-BN particles, which restrict the rearrangement of Si_3N_4 particles during liquid phase sintering.

Fig. 2 shows the effect of h-BN content on the four-point flexural strength of Si_3N_4. Flexural strength decreased with an increase of the additive levels. The flexural strength of hot-pressed specimen decreased about 54%, from 740 MPa for monolithic Si_3N_4 to 340 MPa for Si_3N_4 containing 30 vol% h-BN. As anticipated, the flexural strength drop by the addition of h-BN was large for pressureless-sintered specimens. According to Griffith and Orowan,[5] the fracture strength of brittle materials depends on Young's modulus and fracture energy, and inversely depends on the size of fracture source.

In order to investigate the cause for the decrease in strength, resulting from the addition of h-BN, the Young's modulus was measured as a function of h-BN content. The results are shown in Fig. 3.

The modulus of hot-pressed specimens decreased by about 36% with the addition of 30 vol% h-BN, from 300 GPa for monolithic Si_3N_4 to 192 GPa for samples with 30 vol% h-BN. The modulus of pressureless-sintered specimens showed much lower values than that of hot-pressed

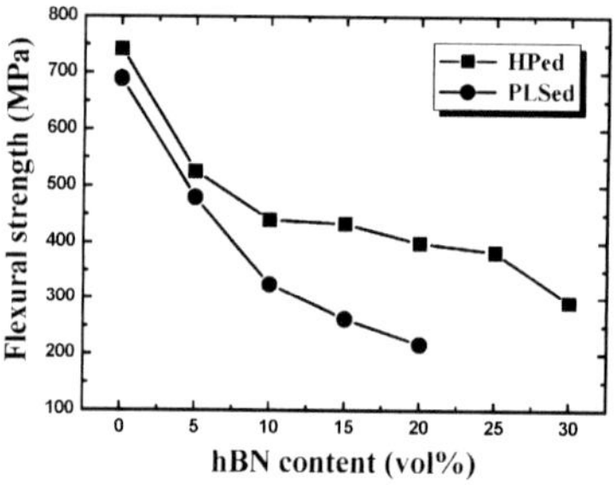

Fig. 2. Effect of h-BN content on the four-point flexural strength of Si_3N_4-BN based ceramics.

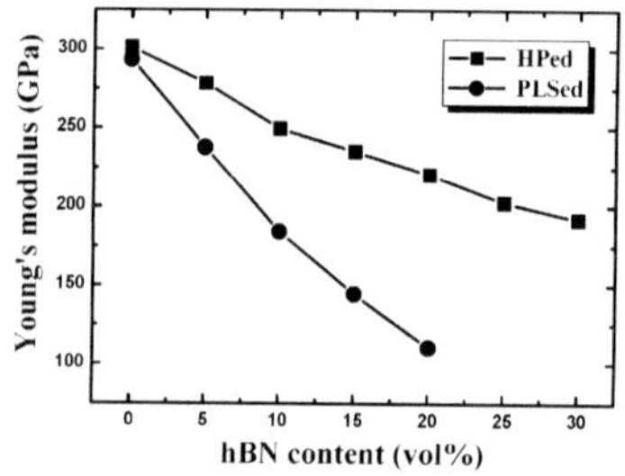

Fig. 3. Effect of h-BN content on the Young's modulus of Si_3N_4-BN based ceramics.

specimens. In general, the Young's modulus is inversely proportional to the porosity. Thus, it can be seen that the modulus of pressureless-sintered specimens greatly decreases due to increased porosity. Since the Young's modulus (86 GPa along to a-axis, 34 GPa along to c-axis)[6] of h-BN is lower than that (287.5 GPa) of Si_3N_4,[6] the Young's modulus of the composite decreased with h-BN content. Thus, the addition of BN resulted in a decrease in the modulus as well as an increase in the size of fracture source, both contribute to the observed decrease in flexural strength. The decrease of strength may also be caused by the residual tensile stress developed by the thermal expansion coefficient (TEC) difference between Si_3N_4 and h-BN. The h-BN has an anisotropic TEC ($7.15\times10^{-6}\ C^{-1}$ along the c-axis and $0.77\times10^{-6}C^{-1}$ along the a-axis), but the TEC of Si_3N_4 is isotropic ($2.87\times10^{-6}\ C^{-1}$).[6] The TEC mismatch between Si_3N_4 and h-BN can lead to a residual tensile stress along to plane vertical to c-axis of h-BN. In order to predict residual stresses formed in the hot-pressed composite, we develop a simple finite element model of a quarter of representative volume. The model is consisted of 1600 axisymmetric linear elements, which are cooled to simulate the fabrication process from initial 1000℃ to 20℃. The Si_3N_4/10 vol% composite is analyzed to observe residual stresses after fabrication. Fig. 4 shows the simulation results shows

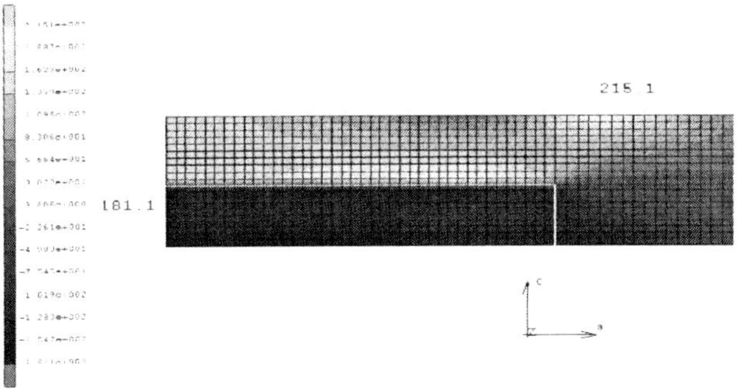

Fig. 4. Residual stress distribution of Si_3N_4/10% hBN composite in a-axis direction of h-BN.

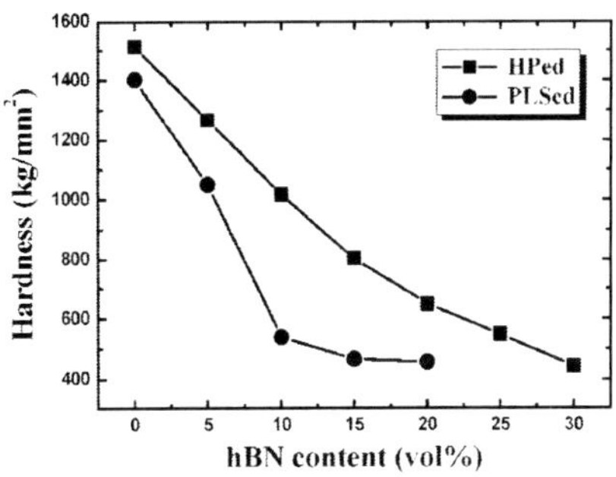

Fig. 5. Effect of h-BN content on Vickers hardness of Si_3N_4-BN based ceramics.

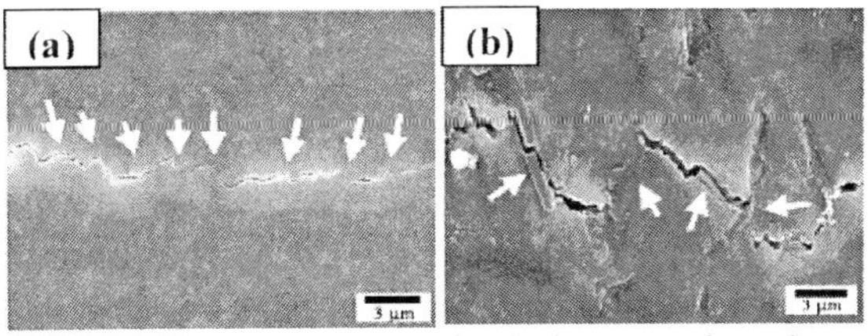

Fig. 6. SEM micrographs of Vickers crack paths for the hot-pressed specimens; (a) monolithic Si_3N_4 and (b) Si_3N_4-30 vol% BN composite. Arrows in (a) indicate grain bridging while those in (b) indicate BN platelets.

expected residual stress (in MPa) distributions. The maximum residual tensile stress of 215.1 MPa occurs in the Si_3N_4 ceramic matrix.

Fig. 5 shows the effect of h-BN content on the hardness of Si_3N_4. Hardness also decreased with h-BN content, mainly due to the low Young's modulus of h-BN. For example, the hardness for the hot-pressed specimens decreased by about 70%, from 1,510 kg/mm^2 for monolithic Si_3N_4 to 442 kg/mm^2 for Si_3N_4 containing 30 vol% h-BN.

From the XRD analysis for Si_3N_4-BN composites, the phase transition of α- to β-Si_3N_4 was completed in all specimens. $Y_2Si_2O_7$ and YSi_2N phases, formed by a reaction with sintering aid (Y_2O_3) and with SiO_2 which existed on the surface of Si_3N_4 powders, were detected in addition to Si_3N_4 and h-BN as the major phases.

Fig. 6 shows the crack path for the hot-pressed Si_3N_4-BN composites containing 30 vol% BN as compared to that for the monolithic Si_3N_4. In the monolithic Si_3N_4, the traces for crack bridging by elongated rod-shape grains were observed frequently as indicated with arrows in the micrograph. The grain bridging behind the crack tip should increase fracture toughness of the monolithic Si_3N_4 as the crack grows and exhibits the R-curve behavior.[7] On the other hand, the crack for the

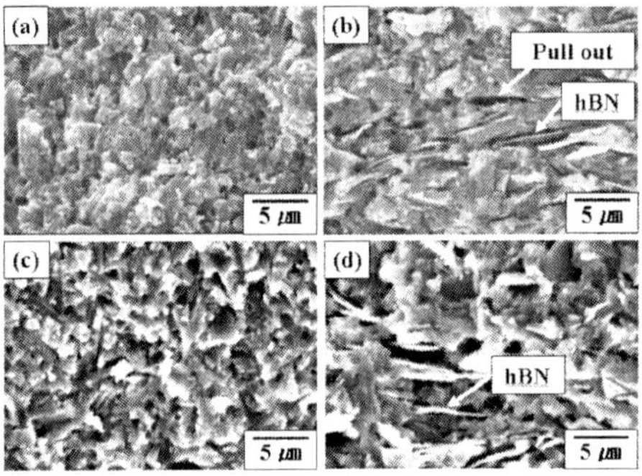

Fig. 7. Fracture surfaces for Si_3N_4-BN based ceramics; (a) hot-pressed Si_3N_4, and (b) hot-pressed Si_3N_4-15 vol% BN, (c) pressureless-sintered Si_3N_4, and (d) pressureless-sintered Si_3N_4-15 vol% BN.

composite containing 30 vol% BN seemed to propagate preferentially along the weak BN platelets or their interfaces, which are indicated with arrows in the figure. Crack path was sinusoidal in a much larger scale because of the lager BN platelets. As a result, while the scale of crack bridging in the monolithic Si_3N_4 was on the order of its grain size, it was much bigger for the BN composites.

Fig. 7 shows examples of SEM images of fracture surfaces after bending tests. In the case of Si_3N_4-BN composites, evidence for pull-outs of Si_3N_4 grains and h-BN platelets was observed. Based on these microstructural observations, bridging, and pull-out will be possible toughening mechanisms for Si_3N_4-BN composite.

In general, hot-pressed Si_3N_4-BN composites exhibit strong texture of BN platelets oriented with the c-axis parallel to hot-pressing direction. It is noteworthy that pressureless-sintered

Si_3N_4-BN composite also exhibit strong texture of h-BN grains due to the orientation of h-BN in powder mixtures during cold-pressing. This anisotropic texture might enhance rising crack resistance curve (R-curve) by more effective bridging and pull-out of grains oriented vertically to the crack propagating direction. In this study, R-curve behaviors for the pressureless-sintered sample could not be characterized, since the effect of residual pores on the R-curve behavior is not simple to evaluate. Thus, based on our previous study[8] on the Si_3N_4-BN composites, we introduce the R-curve behaviors of hot-pressed AlN-BN composites.

R-curves, obtained from the indentation strength data for the Si_3N_4-BN composites as compared with that of monolithic Si_3N_4, are presented in Fig. 8. For the monolithic Si_3N_4, fracture toughness rose rapidly initially (at the relatively short crack lengths) but leveled off at higher crack lengths. The rapid rise of the fracture toughness is caused by the relatively small grain size in Si_3N_4.

For the Si_3N_4-BN composites, the R-curve rises slowly with the initial fracture toughness being relatively low. However, as the crack grew, fracture toughness increased and became higher than that of the monolithic Si_3N_4 when the crack was longer than approximately 700 μm. This R-curve behavior was expected from the earlier observation of the fracture surfaces and crack paths as the crack bridging scale in the composites was significantly lager than that in monolithic Si_3N_4.

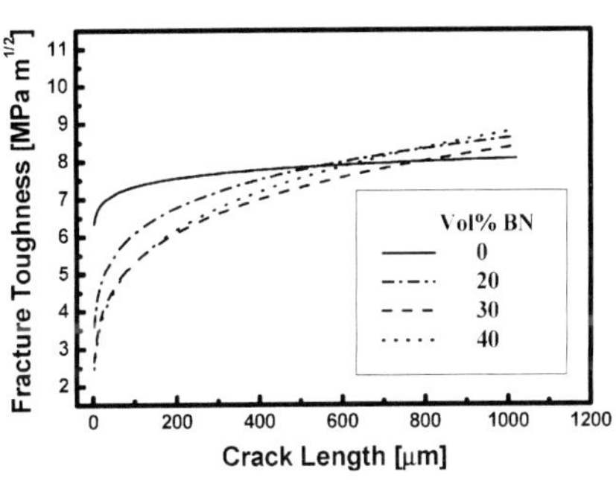

Fig. 8. R-curves for the monolithic Si_3N_4 and Si_3N_4-20, 30, and 40 vol % composites. Specimens are prepared by hot-pressing.

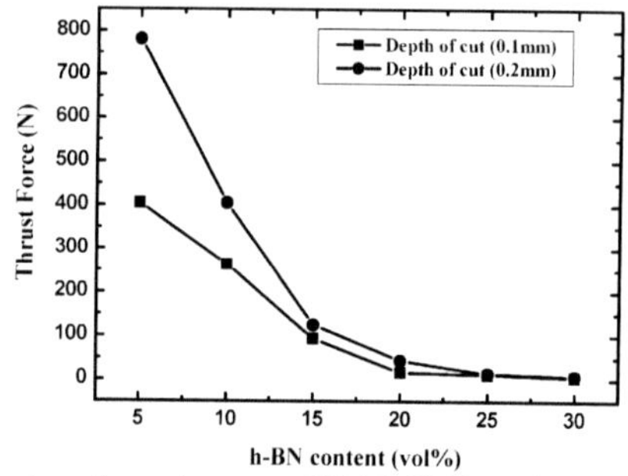

Fig. 9. Effect of h-BN content on the resistance of Si_3N_4-BN based machinable ceramics. Thrust forces are measured under the feed rate of 30 mm/min for the hot-pressed specimens.

A slowly rising R-curve behavior is very important for the machinability of these composites. The toughness is low with small cracks, which enable easy machining with a relatively low cutting force. However, the high toughness at long cracks prevents the cracks from propagating deeply to the inside of the composites, thereby minimizing machining damage.

Fig. 9 shows the cutting resistance (thrust force) as a function of h-BN content for the hot-pressed Si_3N_4-BN composites. The cutting resistance of Si_3N_4 monolith could not be measured due to its brittle fracture. The thrust forces decreased with increasing h-BN content. It is seen that the cutting resistance for the thrust force becomes identical when the BN content is 20 vol% and higher, showing excellent machinability.

Fig. 10 shows the cutting resistance (thrust and feed force) as a function of h-BN content for the pressureless-sintered Si_3N_4-BN composites. The cutting resistance decreased with increasing h-BN content, and increased with increasing feed rate. In general, ease in machining is inversely proportional to the strength. Thus, It was estimated that the optimum content of h-BN in

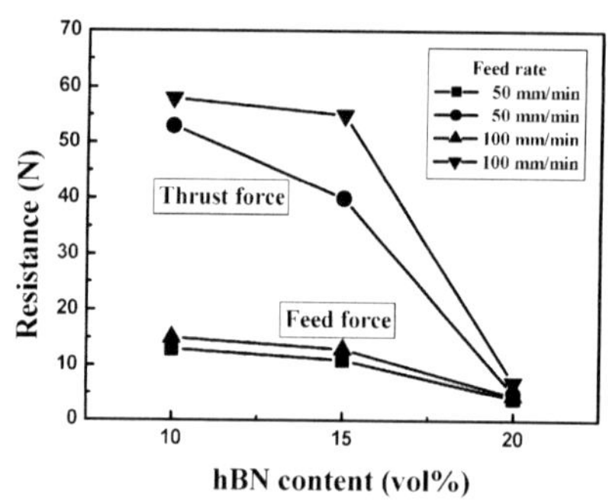

Fig. 10. Effect of hBN content on the resistance of Si_3N_4-BN based ceramics. Cutting resistances are measured under the depth of cut of 0.2 mm for the pressureless-sintered specimens.

pressureless-sintered specimen was about 10 vol%. Although increased residual pores deteriorated the strength of the composite, it is regarded to improve the resistance to thermal shock.[5] Also, crack blunting effect by micro-pores existing in the advancing crack front can be expected. It would relieve the stress concentration and triaxial stress field developed around the crack tip.[5] The residual pores can be attributed to another reason for the improvement of machinability because of preventing growth of cracks associated with machining.[2] Moreover, it appears that the more porous ceramic, the weaker and more machinable it becomes.

4. CONCLUSION

The effect of h-BN on the microstructure and mechanical properties of Si_3N_4 ceramics were investigated. The results can be summarized as follows.

1. Flexural strength, Young's modulus, and hardness decreased with increasing h-BN content. The flexural strength decrease is mostly due to the increase in the size of fracture source.

2. Si_3N_4-BN composites showed a slowly rising R-curve behavior owing to the BN platelets, which increased the crack bridging scale. The low fracture toughness at the small crack length makes possible machining with a relatively low cutting force, while machining damage as well as the surface roughness are minimized due to the high fracture toughness for the long cracks.

3. Si_3N_4 monolith couldn't be machined due to its brittle fracture, but Si_3N_4-BN based ceramics could be machined without fracture, showing excellent machinability. With increasing h-BN content, the cutting resistance decreased. The residual pores in pessureless-sintered Si_3N_4-BN composite can be attributed to another reason for the improvement of machinability because of preventing growth of cracks associated with machining.

ACKNOWLEDGEMENT

This work was supported by grant No. R01-2001-000-00257-0 from the Basic Research Program of the Korea Science & Engineering Foundation.

REFERENCES

[1]T. Kusunose, T. Sekino, Y. H. Choa, and K. Niihara, "Machinability of Silicon Nitride/Boron Nitride Nanocomposites," *J. Am. Ceram. Soc.*, **85**, 2689- 95(2002).

[2]W. Ruigang, P. Wei, C. Jian, and L. Yongming, "Investigation of the Physical and Mechanical Properties of Hot-pressed Machinable Si_3N_4/h-BN Composites and FGM," *Mater. Sci. Eng. B*, **29**, 261-68 (2002).

[3]T. Kusunose, T. Sekino, Y. H. Choa, and K. Niihara, "Fabrication and Microstructure of Silicon Nitride/Boron Nitride Nanocomposites," *J. Am. Ceram. Soc.*, **85**, 2678-88 (2002).

[4]Krause, Jr. R. F., "Rising fracture toughness from the bending strength of indented alumina beams," *J. Am. Ceram. Soc.*, **71**, 338-42. (1988).

[5]Hertzberg R. W., Deformation and Fracture Mechanics of Engineering Materials, John Wiley & Sons, New York, 1996, pp. 261-374.

[6]J. F. Shackerfold and W. Alexander: *Materials Science and Engineering Handbook, 3rd ed.* CRC Press, New York, 2001.

[7]S. R. Choi, and J. A. Salem, "Crack-Growth Resistance of in Situ-Toughened Silicon Nitride," *J. Am. Ceram. Soc.*, **77**, 1042-46 (1994).

[8]Y. S. Yoon, S. W. Na, J. Lee, M. -W. Cho, E. S. Lee, and W. S. Cho, "R-curve Behavior of Si_3N_4–BN Composites," *J. Am. Ceram. Soc.*, **87**, 1374-77 (2004).

WEIGHT FRACTION EFFECTS OF NANOPARTICLE INCLUSION ON MECHANICAL PERFORMANCE OF CONTINUOUS FIBER CERAMIC NANOCOMPOSITES

Vamshi M.Gudapati, Vinod P. Veedu, Anyuan Cao, and Mehrdad N. Ghasemi-Nejhad
Advanced Materials Manufacturing and Hawaii Nanotechnology Laboratories, Department of Mechanical Engineering, University of Hawaii at Manoa
2540 Dole Street, Holmes 302
Honolulu, HI 96822, USA

ABSTRACT

Continuous Fiber Ceramic Composites (CFCCs) have attracted immense attention for use as high-temperature structural materials. However, efforts are still under way to solve different problems associated with advanced ceramics, such as relatively low fracture toughness and strength, degradation of mechanical properties at high temperatures, and poor resistance to creep, fatigue, and thermal shock. Addressing these problems may involve incorporating second phases such as micron size range particulates at the matrix grain boundaries. But the results were generally disappointing when micron-size fillers were used to achieve these goals. To circumvent this problem, nanoparticle fillers were used in CFCCs. The smaller particle size and the fact that their distribution can be stabilized are some of the advantages of using nanoparticles. In this experimental study, effect of using nanoparticle fillers on mechanical performance of CFCCs by preceramic polymer pyrolysis processing method is being examined. A wet-lay-up process, involving ceramic fiber reinforcements (Nicalon™) and preceramic polymer (KiON CERASET®) readily mixed with nano size fillers (Yttrium Oxide - 29 nm) in the presence of a surfactant agent, is used for manufacturing followed by preceramic polymer reinfiltration/pyrolysis. To study the effects of weight percentage of nanoparticle on mechanical performance of CFCC, test specimens varying in nanoparticle weight percentage of 0, 5, 10, 15, and 20 were manufactured and tested. Four-point flexure test was performed in order to evaluate the mechanical performance of the nanoparticle filled CFCC samples at room temperature. To analyze the part quality, a characterization study of the samples using scanning electron microscopy is also reported.

INTRODUCTION

Recently researchers have investigated nanocomposites containing polymer, nanoparticle to improve their physical, mechanical, and chemical properties. Nanoparticles embedded in polymer matrix have attracted increasing interest because of the unique properties displayed by nanoparticles. Due to the nanometer size of these particles, their physicochemical characteristics differ significantly from those of molecular and bulk materials. Nanoparticle polymer nanocomposites synergistically combine the properties of both the host polymer matrix and the discrete nanoparticles, thereby tailoring a wide variety of magnetic, electric, thermal, and optical properties[1,2].

The preceramic polymer pyrolysis route to ceramic chemistry has attracted much attention as it offers a promising and cost-effective way to manufacture CMCs/CFCCs with little to no final machining[3-8] and parts with greater compositional homogeneity[9-11]. Preceramic polymers are organo-metallic polymers which, after curing, are chemically transformed during their pyrolysis step to yield a contributing ceramic phase(s) in the final ceramic part. It has been reported that the mechanical properties of the composites could be improved by adding an optimal amount of nanoparticles into polymer-based composites. The optimum amount of nanoparticles in the composites depends on the filler size, shape, homogeneity of particle distribution and the interfacial bonding properties between the fillers and matrix[12-14]. In this study, we integrate the concept of CFCC with that of nanocomposites, which is based on passive control of the microstructures by incorporating nanometer-size second-phase dispersions into the ceramic matrix of the CFCC. Continuous Fiber Ceramic Composites were manufactured using CERASET preceramic polymer with increasing weight percentage to find the optimum amount of nanoparticles. Five different types of CFCCs are manufactured with varying weight percentage of nanoparticles 0, 5, 10, 15, and 20. The nanoparticles used is yttrium oxide with an average size of 29 nm. The effects of varying weight percentage of nanoparticle reinforcements on processing and flexural mechanical performance of CFCCs using Polymer Infiltration Processing (PIP) were investigated.

COMPOSITE MATERIAL SYSTEMS

The inorganic polymer used in this study was KiON CERASET® preceramic polymer[15]. KiON CERASET® materials are novel patented liquid compositions which enable the fabrication of silicon carbide, silicon nitride, and silicon carboxide monolithic ceramics or ceramic composites. The final composition of the composite depends on the pyrolysis cycle and inert gas used[2]. In this work, nitrogen gas was used to produce silicon carbide matrix. Table 1 shows physical properties of KiON CERASET®.

Table I. *Properties of KiON CERASET®*[15].

Melting point	-78 °C
Boiling point	150 °C
Viscosity, 25°C / 50°C / 75°C / 100°C	50 cp / 30 cp / 20 cp / 15 cp
Density of liquid	0.96 g/cm^3
Density of ceramic from Pyrolysis of cured polymer	2.4 g/cm^3
Incompatible Reagents	acids, bases, water, alcohols

Liquid CERASET Polysilazanes are readily thermoset to a solid, through a vinyl cross-linking mechanism, by heating with a small quantity of a free radical generator, typically peroxide. Around 0.1 to 5 wt% of peroxide is sufficient to initiate cure. Dicumyl peroxide obtained from Aldrich Chemical is used as the free radical initiator. Nicalon™ ceramic fiber was chosen as the fiber system reinforcement. Nicalon™ is a silicon carbide (SiC) type fiber and has a desirable combination of modulus, strength, density, and electrical properties with retention of these properties at elevated temperatures up to 1400°C. Nicalon™ is black in color and has a composition of 58% silicon (Si), 31% carbon (C), and 11% oxygen (O)[16]. Table 2 gives the physical properties of the Nicalon™ fibers. The architecture of the fiber is plain weave cloth with 16 tread count/inch in both warp and weft directions.

Table II. *Properties of Nicalon™ Fibers*[16].

Thickness (m)	0.000318
Cloth Weight (g/m^2)	271
Density (g/cc)	2.55
Tensile Strength (MN/m^2)	2944
Tensile Modulus (MN/m^2)	209600

Yttrium oxide (Y_2O_3) nanoparticles were used in varying percentages in this work. These particles were white in color, insoluble in water, and stable in air. The average particle size of the nano-sized inclusion was 29 nm, with density of 1gm/cc and a surface area of 42 sq.m/gm[17]. Glycerol monooleate obtained from BASF Corporation[18] was used as surfactant agent. Glycerol monooleate was mixed with preceramic polymer to provide good dispersion of nanoparticles in KiON CERASET®. In the following, Ni/CE refers to a CFCC that is not reinforced with nanoparticles and will be considered as a base material. Ni/CE-n refers to a base composite reinforced with yttrium oxide nanoparticles, with n referring to the percentage of nanoparticles in the polymer, manufactured with preceramic polymer pyrolysis/reinfiltration route, where the initial matrix has nanoparticles but reinfiltration matrix does not.

MANUFACTURING PROCEDURE

Manufacture of Ni/CE CFCC's

The general process of manufacturing continuous fiber ceramic composites encompasses the following basic steps: wet lay-up, curing using autoclave, and polymer pyrolysis/reinfiltration cycles. In wet lay-up process, preceramic polymer KiON CERASET® is mixed with 2 wt. % of Dicumyl peroxide for initiating the cure through vinyl cross-linking mechanism at 200^{0}C. The fiber cloth is now impregnated with the as prepared polymer mixture and are stacked on top of each other until a desired thickness is obtained. This is followed by the vacuum bagging process and the cure in an autoclave to complete the B-staging process[2]. The sample, after this step, is at the green stage. For the final step of pyrolysis/reinfiltration cycle, the edges of the laminated composite plates are trimmed and samples of required dimensions (ASTM C 1341-97) are cut using a diamond blade cutter.

The cut samples are then pyrolyzed in a high temperature tube furnace in an inert nitrogen gas environment to convert the preceramic polymer into ceramic[2]. At the end of the pyrolysis cycle, the test samples are removed from the high-temperature tube furnace, weighed, and measured. A sample at this stage, pyrolyzed for the first time, is classified as CFCC-0 state. Pyrolyzed parts were immersed in preceramic polymer mixed with 2 wt% Dicumyl Peroxide for reinfiltration. Reinfiltration/pyrolysis cycles are necessary to increase the part density and fill the voids and microcracks inside the samples created by matrix shrinking as well as the conversion of polymer to ceramic during the pyrolysis step. The samples were allowed to stay in the neat polymer until all the bubbling subsides in the reinfiltration stage, which took 6-8 hours. Once the bubbling subsided, the samples were wiped free of excess resin, weighed, and subjected to cure according to the cure cycle, without pressure. After curing, the samples go through the next pyrolysis stage in an inert environment. After this stage, the samples received their first densification, and were classified in the CFCC-1 state. This densification cycle was repeated five more times until all plates were in the CFCC-6 state. Figure 1 shows the CFCC manufacturing methodology we have used in this work.

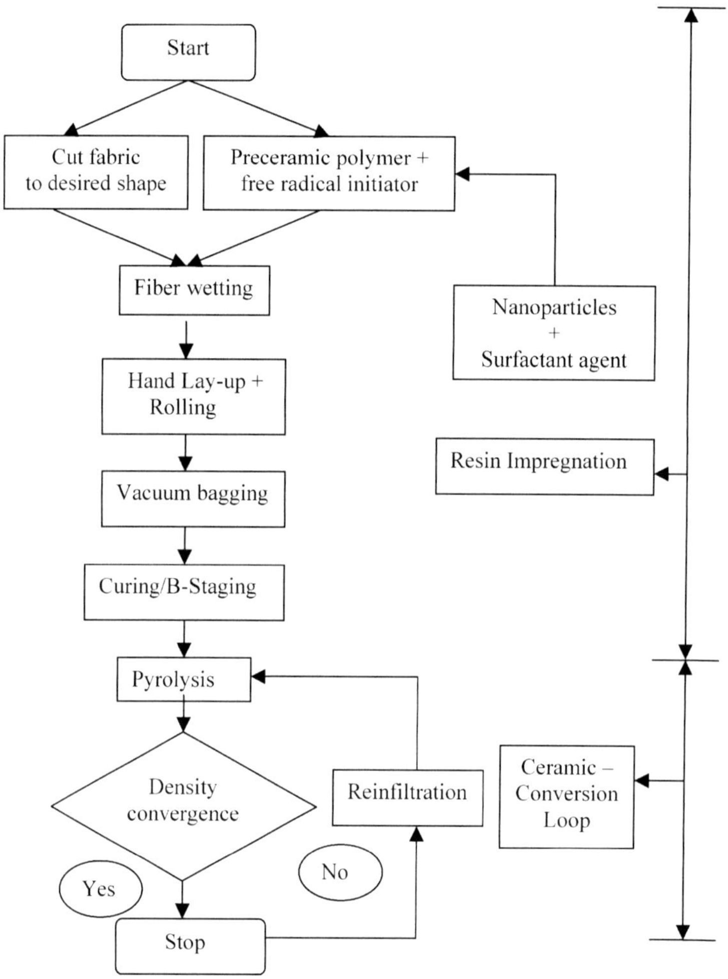

Figure 1. *Nanoparticle filled CFCC manufacturing methodology.*

Manufacture of Ni/CE-n CFCCs

To evaluate the effects of nanoparticles on processing and mechanical properties, CFCCs were reinforced with different weight percentages of yttrium oxide nanoparticles. The manufacturing process remains the same as mentioned in the section above, except for the use of nanoparticle reinforced preceramic polymer (see Figure 1). For the present study, the KiON CERASET® polymer was first reinforced with 5 wt. % of yttrium oxide nanoparticle and 5 wt. % of surfactant glycerol monooleate. This was followed by samples with 10 wt. %, 15 wt. %, and 20 wt. % of nanoparticles. Same amount of surfactant agent was used for all the samples manufactured. The nanoparticle filled polymer was mixed with the surfactant agent using a high speed stirrer for 10-15 minutes to achieve good dispersion. The addition of surfactant agent was used to aid in proper dispersion of nanoparticles. It is important to note that the reinfiltration process was carried out using neat preceramic polymer without any nanoparticle reinforcement. The authors have reported a detailed description of the steps involved in manufacturing process, in their previous work[2].

MECHANICAL PERFORMANCE EVALUATIONS

The main objective of this part of the study is to experimentally evaluate the performance of CFCC specimens with varying weight percentage of nanoparticles in flexure. A four-point bend fixture was used to measure the flexural load-deflection response of CFCC specimens loaded in four-point bending. The test specimens were selected so that they have a span to depth ratio (*L/d*) that produces tensile and compressive failure at the outer surfaces of the sample under the bending loading. In this study, the *L/d* value is 43.5. All specimens were nominally 180 mm long, 12.7 mm wide, and on an average 3.67 mm thick. Five samples were used for each type of CFCC flexure test. Test method for flexural properties of continuous fiber reinforced advanced ceramic composites, ASTM C 1341[19], is used to determine the flexural properties of the CFCC specimens in the form of rectangular bars. Figure 2 shows a schematic of the four-point bending test. The support span length was 160 mm and load span length was 53.3 mm.

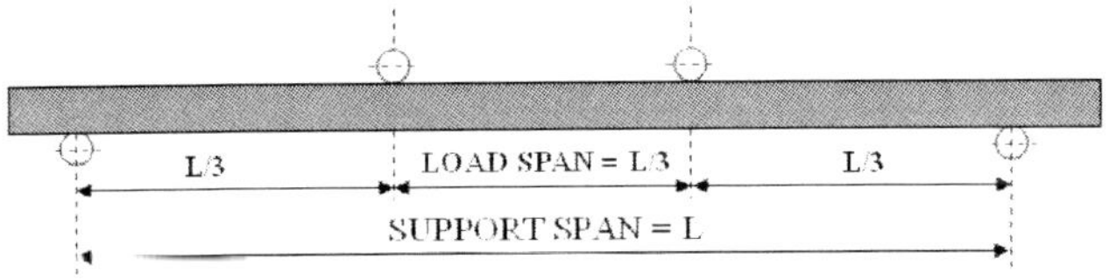

Figure 2. *Schematic of four-point bending test, ASTM C1341[19].*

Flexure test provides information on the strength and deflection of materials under complex flexural stress conditions. In CFCCs, nonlinear stress-strain behavior may develop as a result of cumulative damage processes, for example, matrix cracking, matrix/fiber debonding, fiber pull-out, fiber fracture, and delamination. This may be influenced by testing mode, testing rate, processing effects, or environmental influences. Some of these effects may be consequences of stress corrosion or sub-critical (slow) crack growth, which can be minimized by testing at sufficiently rapid rates. In this study, the test rate was 1.18 mm/sec[19]. The maximum stress in the outer surface at the point of maximum stress, flexural strength (S_U), can be calculated using the maximum load and deflection obtained from the four-point bending test, and is given in Equation (1)[19].

$$S_U = \frac{P_U L}{b d^2} \tag{1}$$

where, P_U is the maximum load in the flexure test in N, b is the specimen width in mm, d is the specimen thickness at the point of break in mm, and L is the length of the specimen. Equation (2) gives the flexural modulus of elasticity, E, for samples tested in four-point bending[19].

$$E = \frac{0.21 L^3 m}{b d^3} \tag{2}$$

where, m is the slope of tangent to the initial straight-line portion of the load-deflection curve (N/mm).

PROCESSING RESULTS AND DISCUSSION

Quality assessment of the CFCC test samples

To prepare the CFCC samples for SEM analysis, first a four step polishing procedure, using diamond abrasives varying in size from 45 μm to 1 μm, was performed. Figure 3 shows images obtained using SEM for a CFCC (Ni/CE type) sample. From such SEM images, fiber volume fraction and void volume fraction were calculated using a PLACOM® planimeter. The void content for samples with and without nanoparticles was found to be well below 4% indicating a good part quality (part with void volume fraction below 5% is well accepted for ceramics). Fiber volume fraction for samples with and without nanoparticles remained the same as well.

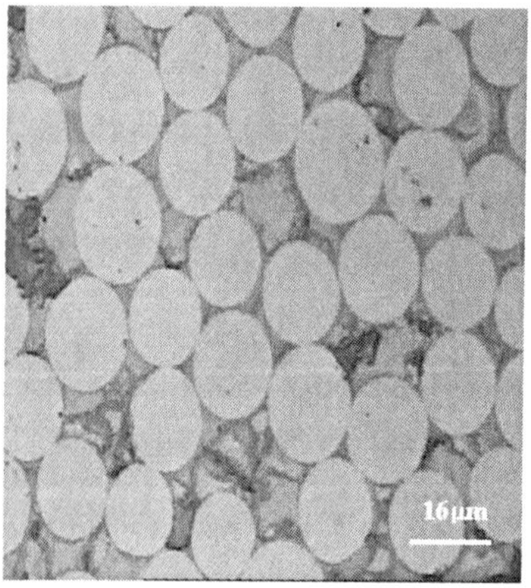

(a)

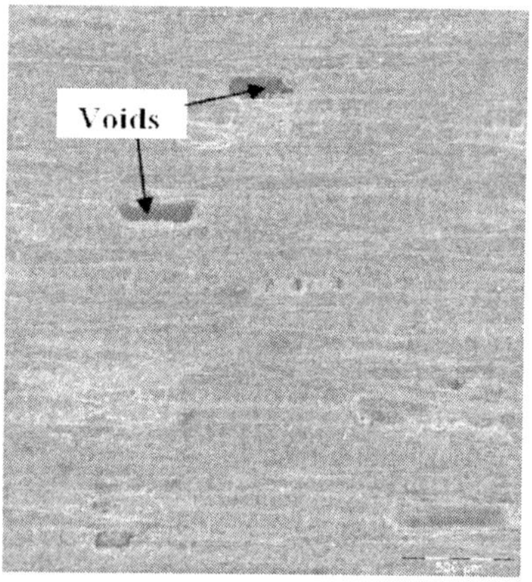

(b)

Figure 3. *SEM images of a CFCC specimen cross-section: a) fiber content, b) void content.*

Table III. *Void, fiber, and matrix percentage in manufactured CFCC.*

Material	Void %	Fiber %	Matrix %
Ni / CE	3.64	65.33	31.03

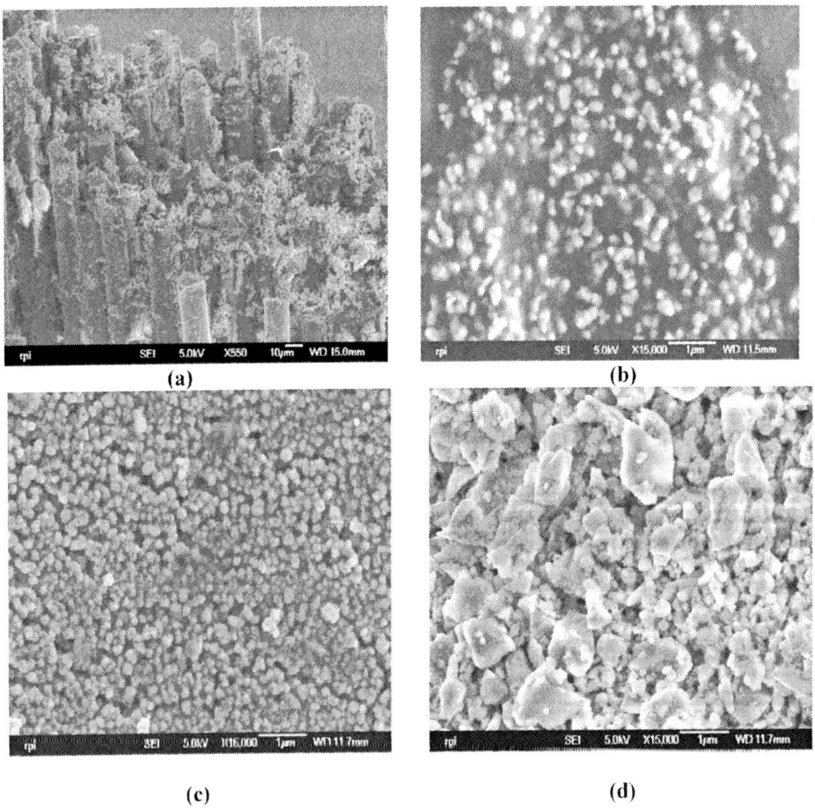

Figure 4. *SEM image of: a) Ni/CE-10 CFCC specimen, b) CERASET with 10% nanoparticles, c) CERASET with 15% nanoparticles, d) CERASET with 20% nanoparticles.*

Dispersion effects study using SEM images

Dispersion is an important factor in the performance of nanoparticle-reinforced nanocomposites. Improper dispersion leads to agglomeration of particles and these inclusions can act as defect inclusions with high stress concentrations, instead of acting as uniform reinforcements. Initial attempts to study the dispersion effects of nanoparticles in CFCCs using SEM studies did not reveal sufficient data, to arrive at a discrete conclusion (e.g., see Figure 4 (a)). Hence, to study the dispersion effects of nanoparticles in CERASET, the polymer was mixed with the required weight percentage of nanoparticles, 2-wt% of dicumyl peroxide, and 5-wt% of surfactant glycerol monooleate. The nanoparticle filled preceramic polymer was cured according to the procedure outlined to thermoset the liquid polymer using vinyl cross-linking mechanism. These monolithic polymer samples were then subjected to SEM studies and as shown in Figures 4 (b) and 4 (c), the nanoparticles were dispersed properly up to a weight percentage of 15% without any agglomeration of particles. As shown in Figure 4 (d), particles started to agglomerate at a threshold weight percentage of 20%.

Processing results for Ni/CE and Ni/CE-n CFCCs

For each sample, a gain in weight of less than 1% from previous pyrolysis/reinfiltration cycle was considered as the convergence criterion. Table 4 gives processing results for the two different types of samples (i.e, Ni/CE and Ni/CE-n). The difference in weight for the samples after the first and the last (in this case, sixth) pyrolysis/reinfiltration cycle is reported as cumulative percentage weight gain.

Table IV. *Processing results for Ni/CE and Ni/CE-n CFCCs.*

Specimen type	Weight gain (%)
Ni /CE	36.33
Ni/CE-5	33.23
Ni/CE-10	33.02
Ni/CE-15	32.27
Ni/CE-20	29.57

All CFCC samples manufactured without and with nanoparticles took the same number of pyrolysis/reinfiltration cycles to achieve the weight convergence. From Table 4, a general tendency merits mentioning that the cumulative percentage weight gain for Ni/CE-n type CFCCs decreases with increasing percentage of nanoparticles, and is lower than that of the base Ni/CE CFCC (see Figure 5). This can be attributed to better consolidation of composites with nanoparticles, without any premature closure affecting porosity and number of cycles required for weight convergence. Therefore, adding nanoparticles to the matrix did not affect the processing of the CFCCs, while in general, adding micron-size particle reduces the number of reinfiltration/pyrolysis due to a possible premature closure which results in higher porosity, lower density, and lower strength[20].

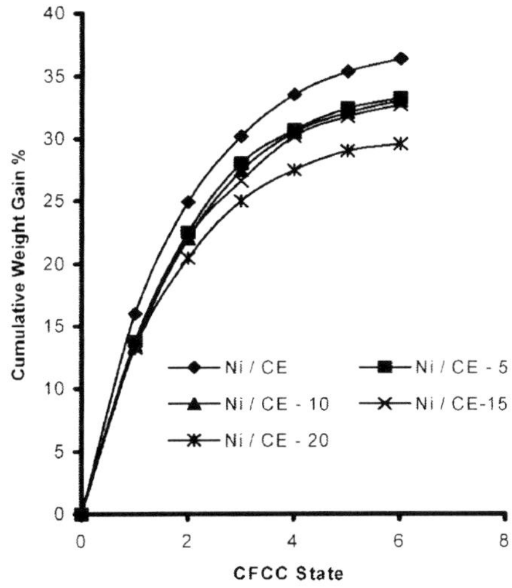

Figure 5. *Cumulative weight gain (%) for Ni/CE-n CFCCs after each pyrolysis cycle.*

FOUR-POINT FLEXURE TEST RESULTS AND DISCUSSION

This section reports and analyzes the results of the four-point bending tests. Two different types of CFCC samples, i.e., Ni/CE (base) and Ni/CE-n (nanoparticle reinforced nanocomposites), were tested with five samples for each category. Four-point bending test was chosen to quantify the performance of nanoparticles filled CFCCs. The four-point geometry is known to produce more reliable statistical data when compared with three-point bending test, particularly for brittle materials [19]. For obtaining a valid flexural strength by four-point bending test, a high support span/thickness (L/d) ratio of about 45 was used to prevent a shear failure and invalidation of the tests. All the samples tested fractured inside the uniformly stressed region of flexure specimen (i.e., between the inner loading points of the fixture), eliminating uncertainties about failures due to stress concentrations outside the uniform loading region as well as the shear failure of the specimen (see Figure 6).

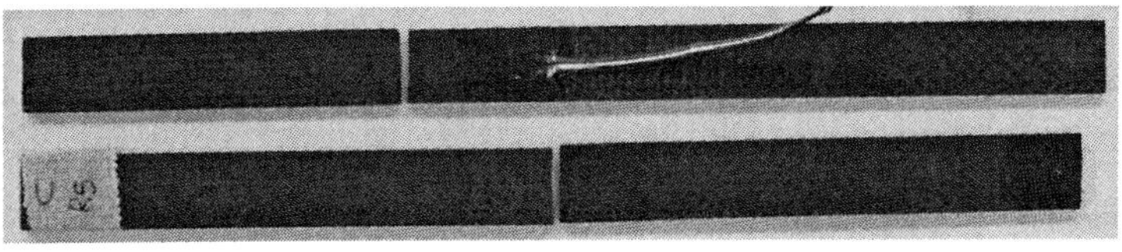

Figure 6. *Typical sample failed at its mid-span.*

Table 5 reports the average maximum load and the thickness, *d*, at the point of break from the four-point bending test for Ni/CE CFCC and for four different Ni/CE-n type CFCCs. The standard deviation of load is given in parentheses.

Table V. *Four-point flexure test results for Ni/CE and Ni/CE-n CFCCs.*

Specimen	d, mm	Load, N
Ni /CE	3.86	37.37 (2.15)
Ni/CE-5	3.4	30.64 (0.78)
Ni/CE-10	3.45	33.24 (2.35)
Ni/CE-15	3.52	42.00 (0.93)
Ni/CE-20	3.58	34.90 (0.36)

From Table 5, in the case of Ni/CE-n, the general tendency of increase in maximum load with increasing percentage of nanoparticles merits mention. The change in trend for Ni/CE-20 samples could be attributed to agglomeration of nanoparticles at that weight percentage, as revealed by SEM studies. Thus, it can be concluded that the threshold value for reinforcing yttrium oxide nanoparticles in CERASET preceramic polymer is around 15 percent of weight for nanoparticles, where proper dispersions are obtained. Hence, Ni/CE-15 CFCC takes the maximum load. Using, Equations (1) and (2), the flexural strengths and moduli for all samples were calculated, respectively, and are reported in Table 6 with their standard deviations in parentheses.

Table VI. *Flexure strength and modulus for Ni/CE and Ni/CE-n CFCCs.*

Specimen type	Strength (MPa)	Modulus (GPa)
Ni /CE	31.85 (1.06)	95.14 (3.90)
Ni/CE-5	33.30 (4.05)	88.15 (13.13)
Ni/CE-10	34.83 (2.25)	89.82 (1.26)
Ni/CE-15	42.65(0.97)	89.94(0.37)
Ni/CE-20	33.39 (0.62)	86.22 (0.95)

Analyzing the data in Table 6, a consistent reduction in modulus and the general tendency of increase in flexural strength when CFCCs are filled with increasing weight percentage of nanoparticles are observed. The peak value of the flexural strength for Ni/CE-15 could be attributed to the following trend and further decrease in strength with increase in nanoparticle weight could be attributed to agglomeration as discussed above. Figures 7 and 8 show the flexural strength enhancement and its percentage of Ni/CE-n CFCCs compared with the Ni/CE CFCC base, respectively.

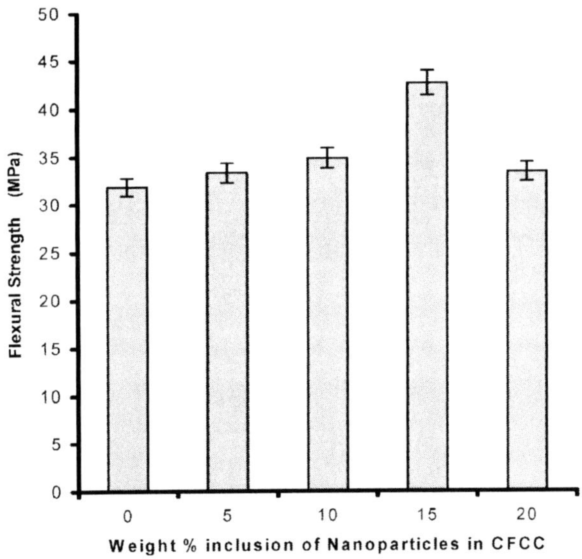

Figure 7. Flexural strengths with increasing weight percentage of nanoparticles.

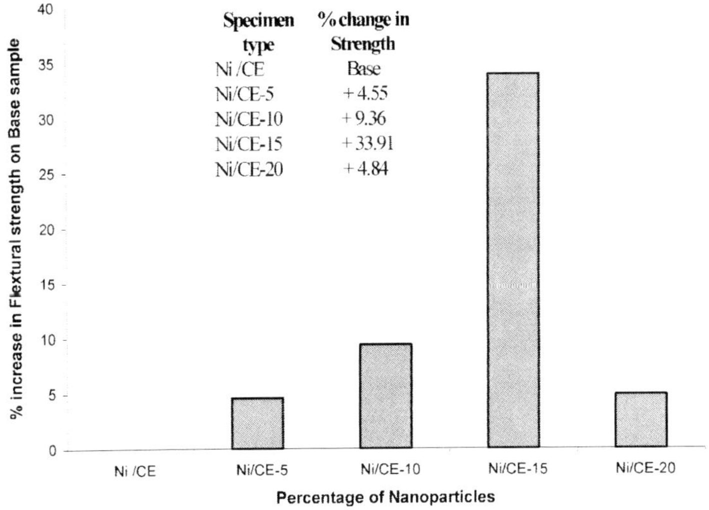

Figure 8. Percentage strength enhancement due to addition of nanoparticles to CFCCs.

Overall, Ni/CE-15 performed the best in a balance of strength and modulus. It should also be mentioned that while, in general the strength increased, the modulus decreased with the addition of nanoparticles to CFCCs (see Table 6.)

CONCLUSIONS

The effects of different weight fraction of nanoparticle inclusions on the flexural mechanical properties of continuous fiber ceramic composites (CFCCs) are experimentally investigated. We have employed a preceramic polymer pyrolysis / reinfiltration route for the manufacturing of the CFCC specimens. Five different sets of CFCC specimens are manufactured and tested. The specimens differ in the percentage weight fraction of the nanoparticle inclusions in the CFCC, the inclusions being 0% (CFCC without nanoparticle inclusion), 5%, 10%, 15%, and 20%. The results from the weight gain percentage study and SEM micrographs are used to investigate the part quality and to characterize the processing of the manufactured CFCC samples. These studies indicate good retention of shape of the specimen, compactness of the material, and reasonably uniform dispersion of the nanoparticles in the CFCCs. Four point flexure test, based on ASTM C1341 standard, is performed to analyze the effects of the nanoparticles inclusion in the CFCC nanocomposites. All the specimens fractured in the uniformly stressed region of the flexure specimen, confirming the validity of the test. It merits mentioning that the CFCC nanocomposites consistently showed improved mechanical properties compared to their CFCC composites counterpart. However, flexural modulus remained more or less the same for all different weight fraction nanocomposites, though they showed a considerable decrease compared to the base sample without nanoparticle reinforcement. Among CFCC nanocomposites, the mechanical properties (flexural strength) were observed to increase with increasing weight percentage inclusion of nanoparticles till an optimum value of inclusion (i.e., 15%). The flexural strength of the CFCC nanocomposite with 20% inclusion showed a considerable decrease compared with that with 15% inclusion. We believe that the difficulty in achieving a proper dispersion and increased chances of the nanoparticles agglomeration limits the amount of nanoparticles inclusion to a certain weight percentage to achieve increased material strengths.

ACKNOWLEDGEMENTS

The authors thank Dr. Richard Russ for helping with the CFCCs manufacturing.

REFRENCES

[1]P. M.Ajayan, L. S. Schadler, and P. V. Braun, *Nanocomposite Science and Technology*, Wiley-VCH, Weinheim (2003).

[2]V. Gudapati, V. P. Veedu, and M. N. Ghasemi-Nejhad, "Preceramic Polymer Pyrolysis Route for Processing and Mechanical Performance Evaluation of Continuous Fiber Ceramic Nanocomposites," *Composites Science and Technology*, 2005 (in press).

[3]J. Jamet, J. R. Spann, R. W. Rice, D. Lewis, and W. S. Coblenz, "Ceramic-Fiber Composite Processing via Polymer-Filler matrices," *Ceram. Eng. Sci. Proc.* **5**(7-8):677–694 (1984).

[4]J. J., Rogers, J. Semen, and Y.-F. Yu, "Silicon Carbide and Silicon Nitride Structural Ceramics Derived from a Preceramic Polymer Binder," *Ceram. Eng. Sci. Proc.*, **10**(7-8):833-836 (1989).

[5]J. Semen, and J. G. Loop, "Structural Ceramics Derived from a Preceramic Polymer," *Ceram. Eng. Sci. Proc.*, **11**(9-10):1387-1394 (1990).

[6]J. Semen, and J. G. Loop, "A Preceramic Polymer Route to Molded SiC Ceramic Parts," *Proceedings of Ceramic Engineering and Science*, **12**(9-10):1967-1980 (1991).

[7]F. I. Hurwitz, P. J. Heimann, J. Z. Gyekenyesi, J. Masnovi, and X-Y. Bu, "Polymeric Routes to Silicon Carbide and Silicon Oxycarbide CMC," *Proceedings of Ceramic Engineering and Science*, **12**(7-8):1292-1303 (1991).

[8]R. W. Rice, W. J. McDonough, G. Y. Richardson, J. M. Kunetz, and T. Schroeter, "Ceramics from Polymer Pyrolysis, Opportunities and Needs - A Materials Perspective," *American Ceramics Society Bulletin*, **62**(8):889- 892 (1983).

[9]A. J. Cornie, C. Yet-ming, D. R. Uhlmann, R. D. Mortensen, and M. J. Collins, "Processing of Metal and Ceramic Matrix Composites," *American Ceramic Society Bulletin*, **65**(2):293-304 (1986).

[10]M. N. Ghasemi-Nejhad, M. V. Chandramouli, and A. Yousefpur, "Processing and performance of Continuous Fiber Ceramic Composites by Preceramic Polymer Pyrolysis: I – Filament Winding," *Journal of Composite Materials*, **35**(24):2207-2237 (2001).

[11]M. N. Ghasemi-Nejhad, J. K. Bayliss, and A. Yousefpur, "Processing and performance of Continuous Fiber Ceramic Composites by Preceramic Polymer Pyrolysis: II – Resin Transfer Molding," *Journal of Composite Materials*, **35**(24):2239-2255 (2001).

[12]B. Wetzel, F. Haupert, and M. Q. Zhang, "Epoxy nanocomposites with high mechanical and tribological performance," *Comp. Sci. and Tech.*, **63**:2055-2067 (2003).

[13]B. Wetzel, F. Haupert, K. Friedrich, M. Q. Zhang, and M. Z. Rong, "Impact and wear resistance of polymer nanocomposites at low filler content," *Polym. Eng. and Sci.*, **42**:1919–1927 (2002).

[14]C. B. Ng, L. S. Schadler, and R. W. Siegel, "Synthesis and mechanical properties of TiO_2-epoxy nanocomposites," *Nanostructured Mater.*, **12**:507–510 (1999).

[15]Product Data Sheet KiON CERASET® Polyureasilazane, KiON Corporation, Huntingdon Valley, PA, U.S.A. (2004).

[16]Dow Corning Corporation. Information about Nicalon™ ceramic fiber (2001).

[17]Product Specification Sheet, Sigma-ALDRICH, Inc., St. Louis, MO.

[18]Material Safety Data sheet, BASF Corporation – Performance Chemicals, Mount Olive, NJ.

[19]ASTM standard: C1341-97, Standard Test Method for Flexural Properties of Continuous Fiber-Reinforced Advanced Ceramic Composites, **15.01**: 509-526 (2004).

[20]A. Yousefpour, and M. N. Ghasemi-Nejhad, "Processing and Performance of Nicalon/Blackglas and Nextel/Blackglas using cure-on-the-fly filament winding and preceramic polymer pyrolysis with inactive fillers," *Composites Science and Technology*, **61**(13):1813-1820, (2001).

INCREASE OF STRUCTUAL INTEGRITY MACHINED ALUMINA/SIC USING CRACK-HEALING

Kotoji Ando, Masato Ono, Wataru Nakao and Koji Takahashi
Yokohama National University
79-5, Tokiwadai, Hodogaya-ku
Yokohama, Kanagawa, 240-8501

Shinji Saito
NHK Spring Co., Ltd., Japan
3-10, Hukuura, kanazawa-ku,
Yokohama, Kanagawa, 236-0004

ABSTRACT

In this study, the effect of crack-heal on cracks propagated during machining was studied systematically. From the obtained results, we proposed a new method for increase in reliability of machined alumina/SiC composites by using the crack-healing treatment. As an example, we investigated the crack-healing effect on the cracks on the bottom of semicircular groove, which initiated during machining. Local fracture stress, which was reduced by machining, was found to be completely recovered by crack-healing at 1400 °C for 1 h. From the Weibull distribution, failure probability, $F(\sigma)$, of the crack-healed specimens at 1400 °C for 1h was found to be less than 4 % at the stress of 400 MPa which 90 % as-machined specimens fractured. Therefore, it was concluded that the crack-healing was effective to improve the integrity of machined ceramics.

INTRODUCTION

Ceramics have been used in various fields. In particular, alumina ceramics is expected to apply to structural component operating at high temperature in various atmospheres, because of its high temperature stability, excellent corrosion resistance and high creep resistance. However, alumina has low fracture toughness, so that countless cracks initiated at random during machining. Moreover, even if these cracks are too small not to be detected by the latest technique, large decrease in mechanical properties and reliability of ceramics component arise.[1-2]

In general, finish machining, such as polishing or rapping, and nondestructive testing, after machining was performed to remove the large cracks. Although this technique is generally very costly, it cannot eliminate minute cracks. To overcome these problems, it is expected to be the most effective method to endow ceramics with self-crack-healing ability. Self-crack-healing ability is a function which material detects surface cracks itself, and then restores it. If the whole of the countless cracks introduced during machining can be perfectly eliminated by crack-healing treatment, one can secure high reliability of ceramics component without costly finish machining.

The present authors succeeded to endow alumina with self-crack-healing ability by compounding 15vol% SiC particles. Afterward, the crack-healing behavior of the alumina has been studied systematically. From the obtained results, it was found that a semi-elliptical surface crack having surface length below 250 μm and aspect ratio of 0.9, which was introduced by indentation method, was completely healed by heat-treatment at high temperature in air. In addition, samples crack-healed have been found to have slightly higher bending strength than smooth samples, which were fabricated pursuant to Japanese Industrial standard (JIS). [3-5] However, we must know whether this function is also useful for the machined cracks or not.

In this study, the cracks, which were propagated when a semicircular groove was made at the center of specimens surface of alumina/15vol% SiC particles composites, were crack-healed under various conditions. The fracture stress of these specimens after crack-healing were investigated at RT and high temperatures. From the obtained results, the crack-healing effect on the cracks was discussed.

EXPERIMENTAL

Materials

Alumina powder (AKP-20, Sumitomo Chemical Co. Ltd, Japan) used has an average particle size of 0.5 μm. SiC powder (Ultrafine grade, Ibiden Co. Ltd., Japan) used has a mean particle size of 0.27 μm. For mixing the raw powder, the mixture of alumina powders and 15 vol% SiC powders were blended well in ethanol for 48 h using alumina balls and an alumina mill pot. Rectangular plates were hot pressed in N_2 at 1600 °C under 35 MPa for 4 h.

The sintered plates were cut into 3 mm × 4 mm × 22 mm rectangular bar specimens as shown in Fig.1 (a). The specimens were polished to a mirror finish according to JIS standard on a maximum tensile surface. The edges of specimens were beveled 45° to reduce the likelihood of

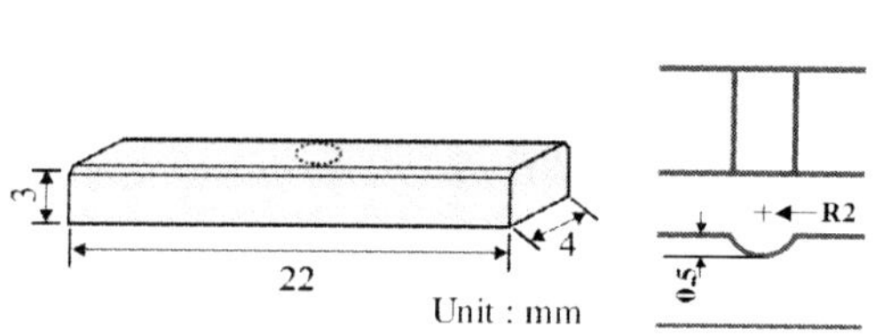

(a) (b)

Figure 1 Dimension of (a) specimen and (b) semicircular groove

Table 1 Grinding conditions

Shape of machined	Semicircular groove
Grindstone	φ4 #140 electrocoat
Number of rotations (rpm)	5000
Table feed speed (mm/min)	100
Depth of cut by one pass(μm)	10

edge-initiated failures. These specimens were called as smooth specimens.

Semicircular groove was made at the center of the smooth specimens by using a drill bonded diamond grains as shown in Fig. 1 (b). Grinding direction is perpendicular to a long side of the specimens. The other grinding conditions listed in Table 1.

Crack-healing condition and test method

These machined specimens were crack-healed in air at 1300 - 1400 °C for 1-10 h. These healed specimens were called as machined-healed specimen in this paper. Also the smooth specimens were crack-healed at 1300 °C, 1 h in air. The specimens abbreviated the heat-treated smooth specimens.

All fractured test were performed on a three-point bending system with a span of 16 mm as shown in Fig. 2. A crosshead speed is 0.5 mm/min. Test temperature is room temperature and high temperatures range from 500 °C to1300 °C.

From the maximum bending moment as specimens fractured, M_F, nominal fracture stress (σ_{NF}) of as-machined specimens was evaluated by,

$$\sigma_{NF} = M_F / Z \qquad (1).$$

where Z was section modulus of the machined specimens. The value of Z was evaluated to be 4.2.

Using stress concentration factor, α, and the following equation, one can evaluate the local fracture stress at bottom of semicircular groove,

$$\sigma_{LF} = \alpha\sigma_{NF} \qquad (2),$$

where the value of α were calculated to be 1.4 by using poisson's ratio of alumina of 2.4. [6]

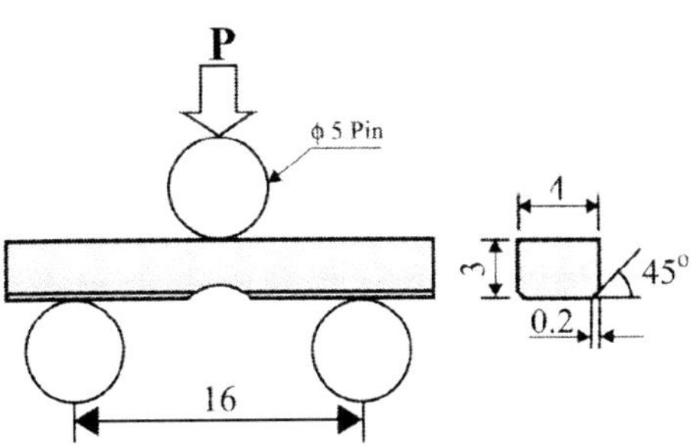

Figure 2 Three-point loading system for measuring local fracture stress at the bottom of semicircular groove and geometry of test specimen.

RESULTS AND DISCUSSIONS

Crack-healing behavior.

Prior to mention of crack-healing behavior, the shape and effect of cracks propagated during machining were discussed. Fig. 3 was shows SEM images of fracture surface in a vicinity of fracture initiation. Dotted line in this photograph indicates the tip of the crack. As shown in this figure, this crack acted as the fracture initiation. This crack was the deepest one in the observed surface cracks of this specimen. Depth of this crack was found to be approximately 30 μm and the damaged layer was also observed on the machined surface. Moreover, this as-machined specimen had the nominal fracture stress of 146 MPa. This value was approximately 400MPa lower than the value that is the fracture stress of heat-treated smooth specimens divided by α. Thus, it was found that existence of crack initiated during machining deteriorated the nominal fracture stress significantly.

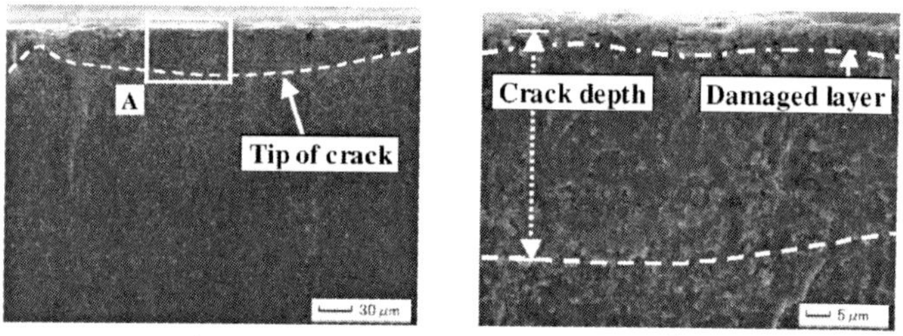

Figure 3 SEM images of surface crack on fracture surface of as-machined specimen : (a) σ_{NF}=146 MPa and (b) detail of A.

Fig. 4 shows the effect of healing temperature on the nominal fracture stress of the machine-healed specimens. Also the nominal fracture stresses of as-machined specimen were indicated as the open triangles in the left column. As the healing temperature increase, the nominal fracture stress increases largely. An average fracture stress of specimens crack-healed at 1400 °C for 1 h was found to be maximum value. In addition, Fig. 5 shows SEM image of the fracture surface in the vicinity of fracture initiation. As shown in this figure, it was found that the machine-healed specimens did not fracture from the machined cracks but from a chip on the edge. Moreover, the rest of the machined cracks cannot be observed. Therefore, these results suggest that the machined cracks were perfectly eliminated by crack-healing treatment. Additionally, the chips of the edge were confirmed to also exist in as-machined specimens. However, because the chips were exactly smaller than the machined cracks, the fracture of as-machined specimens did not initiate at this chip.

From these results, it was concluded that alumina/SiC particle composites could completely heal the whole machined cracks by crack-healing treatment at 1400 °C for 1 h.

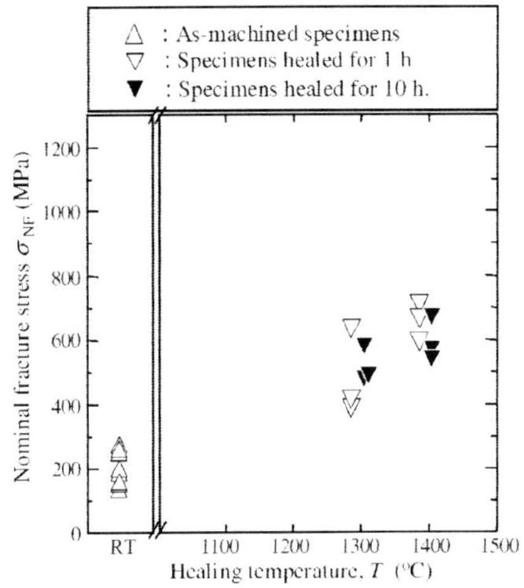

Figure 4 Effect of healing temperature on nominal fracture stress of machine-healed specimens.

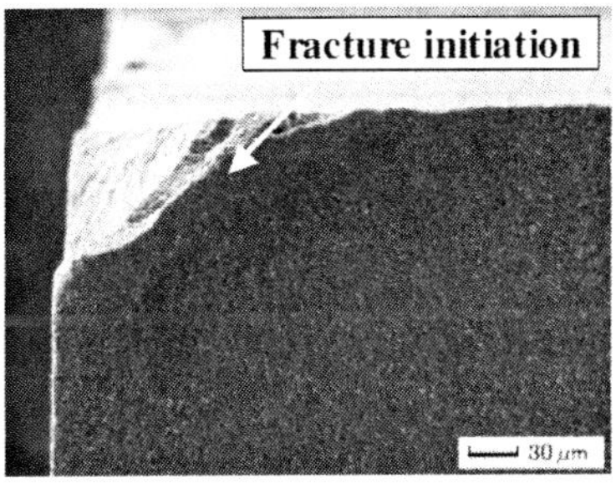

Figure 5 SEM image of the fracture surface in the vicinity of fracture initiation of machined-healed specimen

Effect of crack-healing treatment on failure provability of machined specimens

Fig. 6 shows the Weibll plots of the local fracture stresses of as-machined (◇), machine-healed (□) and heat-treated smooth specimens (●), where the fracture of the heat-treated smooth specimen initiates at the largest embedded flaw. [4] In addition, Table 2 shows Weibll modulus, which describes the width of the strength distribution, and characteristic strength, which was statistical average of local fracture stress, of three kinds of samples in Fig. 4.

The Weibll modulus and the characteristic strength of the as-machined specimens was less than that of the heat-treated smooth specimens. Thus, the maximum size and its scatter of the machined cracks were found to be larger than that of the embedded flaw.

On the other hand, the characteristic strength of the machine-healed specimens was almost equal to that of the heat-treated smooth specimens. These results suggest that machined cracks were completely healed by crack-healing treatment. As an example, consider that the as-machined and machined-healed specimens were subjected to the stress of 400 MPa, corresponding to the stress, which 90 % the as-machined specimens fractured. From the Weibull distribution, failure probability, $F(\sigma)$, of the machine-healed specimens was found to be less than 4 %. Therefore, it was concluded that the crack-healing treatment was effective to improve the integrity of machined ceramics at room temperature.

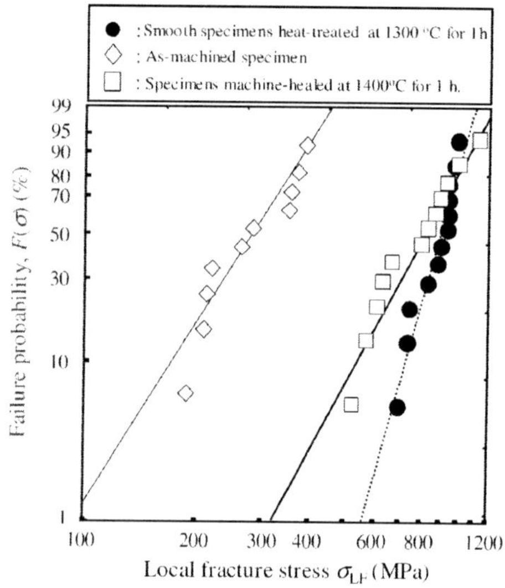

Figure 6 the Weibull plot of local fracture stress of machine-healed specimens.

Table 2 Weibll modulus and characteristic strength

	Weibll modulus	Characteristic strength
Smooth specimens heat-treated at 1300 °C for 1h	8.95	891
As-machined specimens	3.85	307
Specimens machine-healed at 1400°C for 1 h	4.53	846

Effect of test temperature on the local fracture stress of machine-healed specimens

Fig. 7 shows the temperature dependence of the local fracture stress of the machine-healed specimen (◇). For comparison, the temperature dependences of the fracture stresses of crack-healed (▲) and heat-treated smooth (●) specimens are also shown in this figure, respectively. Symbols attached with asterisks indicate the crack-healed specimens fractured outside the crack-healed part. These specimens indicate that the pre-crack was completely healed. [3-5,7]

The local fracture stresses of the machine-healed specimens were almost constant in the temperature range from room temperature to 1100 °C. The constant value of σ_{LF} was found to be approximately 700 MPa, and was higher than fracture stress of the crack-healed specimens and heat-treated smooth specimens. Above 1100 °C, σ_{LF} decreases largely as the test temperature increase. This decrease was caused by plastic deformation of alumina. Therefore, the crack-healed parts have high reliability at high-temperatures.

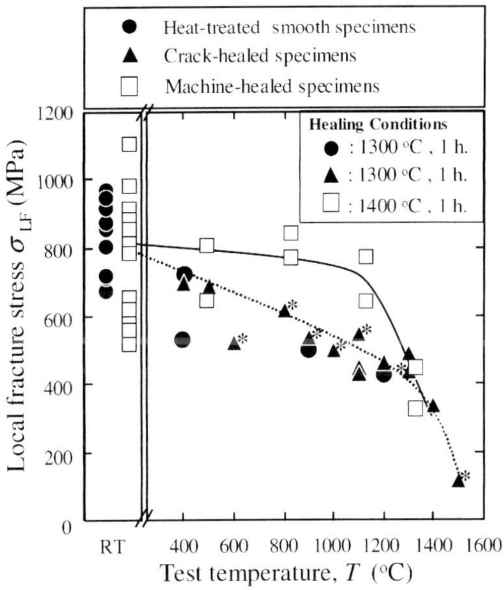

Figure 7 Temperature dependence of local fracture stress of machine-healed specimens
: * indicate the specimens fracture outside the crack healed part

CONCLUSIONS

In this study, the cracks, which were propagated when a semicircular groove was made at the center of specimens surface of alumina/15vol% SiC particles composites, were crack-healed under various conditions. The fracture stress of these specimens after crack-healing were investigated at RT and high temperatures. From the obtained results, the crack-healing effect on the cracks was discussed.

(1) Existence of crack initiated during machining deteriorated fracture stress of machined specimens significantly.
(2) Machine-healed specimens did not fracture from the machined cracks but from a chip on the edge.
(3) Alumina/SiC particle composites could completely heal the whole machined cracks by crack-healing treatment at 1400 °C for 1 h.
(4) From the Weibull distribution, the crack-healing treatment was effective to improve the integrity of machined ceramics at room temperature.
(5) The crack-healed parts have high reliability at high-temperatures.

REFERENCES

[1] A. Tsukuda, Y. Kondo, Y. Kuroshima and K. Uematsu, "Relationship between Grinding Mechanism and Residual Cracks in Single Point Grinding Test of Alumina Ceramics with Different Grain Size," *J. Ceram. Soc. Japan*, **103**, [3], 268-273, (1995).

[2] W. kanematsu, Y. yamauchi, T. Ohji, S. Ito and K. Kubo, "Formulation for the Effect of Surface Grinding on Strength Degradation of Ceramics," *J. Ceram. Soc. Japan*, **100**, [6], 775-779, (1992).

[3] K. Ando, B.S.KIM, M.C. Chu, S. Saito, and K. Takahashi, "Crack-Healing and mechanical Behavior of Al_2O_3/SiC composites at elevated temperature," *Fatigue Fract. Engng. Mater. Struct.*, **27** , 533-541, (2004).

[4] B.-S.KIM, K. Ando, M.C. Chu and S. Saito, "Crack-Healing Behavior of Monolithic Alumina and Strength of Crack-Healed Member," *J. Soc. Mat. Sci., Japan*. **52** [6], 667-673, (2003).

[5]K. Ando, B.S. Kim, S. Kodama, S.H. Ryu, K. Takahashi and S. Saito, "Fatigue Strength of An Al_2O_3/ SiC Composite and A Monolithic Al_2O_3 Subjected to Crack-Healing Treatment," *J. Soc. Mater. Sci. Jpn.*, **52** [11], 1464-1470, (2003).

[6] M.Nishida, "Stress Concentration," *Morikita Publishing Co., Ltd.*, 572-574, (1967).

[7] K. Ando, M. Yokouchi, Lee, K., K. Takahashi, W. Nakao and H. Suenaga, "Crack-Healing Behavior , High Temperature Strength and Fracture toughness of Alumina Reinforced by SiC Whiskers," *J. Soc. Mat. Sci., Japan*. **53** [6], 599-606, (2004).

INFLUENCE OF GEOMETRICAL IRREGULARITIES ON THE CREEP BEHAVIOUR OF CERAMIC FIBRES

Ian J. Davies and Rodney D. Entwistle
Department of Mechanical Engineering, Curtin University of Technology, GPO Box U1987, Perth, WA 6845, Australia

ABSTRACT

Due to the variation in fibre radius for current ceramic fibres based on the silicon carbide (SiC) system, a sufficiently long gauge length may span a region where the fibre radius varies by up to 50%. The potential effect of this variation on initial creep rate, $\dot{\varepsilon}_o$, and creep rupture time, t_r, were investigated as functions of the radius variation, r_v ($0 < r_v < 0.5$), creep stress exponent, n ($0 < n < 3$), and the creep threshold stress, σ_0 ($\sigma/\sigma_o = 1.1, 1.5, 2.0$). It was found that the initial creep rate, $\dot{\varepsilon}_o$, increased compared to that of a prismatic fibre with the same mean radius as both n and r_v increased whereas the time to rupture decreased. The creep threshold stress was found to have little effect for the values of n encountered with current SiC-based fibres. It was concluded that researchers should be aware of the variation of fibre radius within their chosen gauge lengths and account for the effects noted in this paper.

INTRODUCTION

The economic requirement for increased operating temperatures, and hence increased efficiency, of gas turbine engines in aerospace and energy generation applications has been a major reason behind the development of high performance ceramics and ceramic matrix composites (CMCs). Whilst monolithic ceramics exhibit many suitable high temperature mechanical properties, it is their relatively low fracture toughness which has proved to be a serious disadvantage for many applications. For example, the fracture toughness of monolithic silicon carbide (SiC) is typically[1,2] 3~4 MPa·m$^{1/2}$ and requires reinforcement by particles, whiskers, or fibres in order to reach a satisfactory level; the fracture toughness of SiC/SiC composites containing chopped Tyranno® Si-Al-C fibres (mean length: 706 μm)[3] has reached 6.3 MPa·m$^{1/2}$ although this is still far below the value of 40 MPa·m$^{1/2}$ obtained for a 3-D woven SiC/SiC composite containing continuous Tyranno® Si-Ti-C-O fibres.[4]

Whilst the mechanical performance of SiC-based fibres has continued to improve, one aspect of concern is the variation in radius both along individual fibres[5-7] and also between neighbouring fibres within a single fibre bundle.[8,9] Table I presents radius variation data for several SiC-based fibres with it being clear that the fractional radius variation, $r_v = \Delta r/r_0$, where Δr is the magnitude of radius variation and r_o is the mean radius, was in the approximate range $0.25<r_v<0.5$. In addition, the fibre radius was often noted to exhibit a regular variation along the fibre length with a characteristic radius variation wavelength, λ, in the range $130<\lambda<160$ mm for the fibres under investigation. It could be imagined that fractional radius variations in the range 0.25~0.5 may have the potential for significant influence on the properties of the fibres and resulting composites. One of the authors has previously investigated the influence of radius variation on the strength of fibre populations[10] and, whilst an effect was noted, the magnitude was such that it could be ignored for all intents and purposes.

Table I: Variation in radius for several ceramic fibres based on the SiC system.[5-8]

Fibre type	Investigated fibre length, 2*L* (mm)	Mean radius, r_o (μm)	Minimum / maximum radius (μm)	Fractional radius variation, r_v	Radius variation wavelength, λ (mm)	Reference
Tyranno® ZMI (Si-C-O)	500	5.5	2.78 / 7.22	0.40	130	Morimoto[5]
Tyranno® SA (SiC)	300	4.7	2.7 / 7.05	0.46	160	Youngblood *et al.*[6]
Nicalon™ (Si-C-O)	-	8.0	4 / 11	0.44	-	Zhu *et al.*[7]
Hi-Nicalon™ (Si-C-O)	250	6.9	3.9 / 10.1	0.45	160	Youngblood *et al.*[6]
Hi-Nicalon™ S (SiC)	300	6.15	4.3 / 7.3	0.24	-	Youngblood *et al.*[6]
Sylramic™ (SiC)	300	4.6	3.1 / 7.45	0.47	-	Youngblood *et al.*[6]
Tyranno® LoxM (Si-C-O)	-	4.03	2.13 / 5.72	0.45	-	Davies *et al.*[8]

However, in contrast to this, it may be envisaged that the influence of radius variation on fibre creep at elevated temperature may be non-trivial. Whilst several expressions exist for the creep rate of ceramic fibres, perhaps the simplest relationship between creep rate, $\dot{\varepsilon}$, in the "steady-state" regime and stress, σ, is given by:[11]

$$\dot{\varepsilon} = A\sigma^{n} e^{-\left(\frac{Q}{RT}\right)} \tag{1}$$

where A is an empirical constant, σ is the applied stress, n is the stress exponent, Q is the activation energy, R is the gas constant, and T is the absolute temperature. Values of n in the range $0.97 < n < 3.3$ have been noted for ceramic fibres based on the SiC system[12-16] whilst the overall creep behaviour of SiC-based fibres is known to depend on factors such as grain size, oxygen content, impurity content, and atmosphere.[20]

The creep response of ceramic fibres is often characterised through applying a constant force, F, to the fibre ends and measuring the fibre extension as a function of time.[18] Thus, for a fibre of constant radius the value of σ will be given simply by:

$$\sigma = \frac{F}{\pi r_o^2} \tag{2}$$

For the case of a fibre with a linear variation in fibre radius as illustrated in Figure 1 the stress at any point, $\sigma(x)$, will be given by:

$$\sigma(x) = \frac{F}{\pi r_o^2} \frac{1}{\left(1 + \frac{r_v}{L} x\right)^2} \tag{3}$$

whilst the initial creep rate, $\dot{\varepsilon}(x)$, can be determined from substitution of Equation (3) into Equation (1) as follows:

$$\dot{\varepsilon}_o(x) = \frac{AF^n e^{-\left(\frac{Q}{RT}\right)}}{\pi^n r_o^{2n}} \left(1 + \frac{r_v}{L} x\right)^{-2n} \tag{4}$$

The total initial creep rate for the fibre will be simply the integration of the creep rate at each position along the fibre length, *i.e.*,

$$\dot{\varepsilon}_o = \frac{AF^n e^{-\left(\frac{Q}{RT}\right)}}{\pi^n r_o^{2n}} \int_{-L}^{+L} \left(1 + \frac{r_v}{L} x\right)^{-2n} dx \tag{5}$$

From the point of view of highlighting the influence of radius variation it would be pertinent to normalise the fibre initial creep rate with regards to a fibre of constant radius, *i.e.*, $r_v = 0$, as follows:

$$\dot{\varepsilon}_o^* = \frac{1}{2L} \int_{-L}^{+L} \left(1 + \frac{r_v}{L} x\right)^{-2n} dx \tag{6}$$

where $\dot{\varepsilon}_o^*$ is the normalised initial creep rate. Equation (6) thus provides an expression for the normalised initial creep rate in the "steady-state" regime. Whilst the value of $\dot{\varepsilon}_o^*$ may be solved analytically for the case of $n = \{0, \frac{1}{2}, 1, \frac{3}{2}, \ldots, z\}$, in general it is necessary to solve Equation (6) using a numerical procedure.

An examination of literature relating to the experimental determination of creep rates in monofilament ceramic fibres indicates the fibre radius value, utilised for the calculation of creep stress, to often be calculated from the simple mean of several radii measurements. One of the present authors has recently investigated the influence of radius variation on initial creep rate for several different geometries including linear variation (Figure 1) and sinusoidal variation.[17] The main aim of the work was to determine whether the use of a simple mean radius value was justified for fibres containing a significant radius variation along their length, *i.e.*, whether the

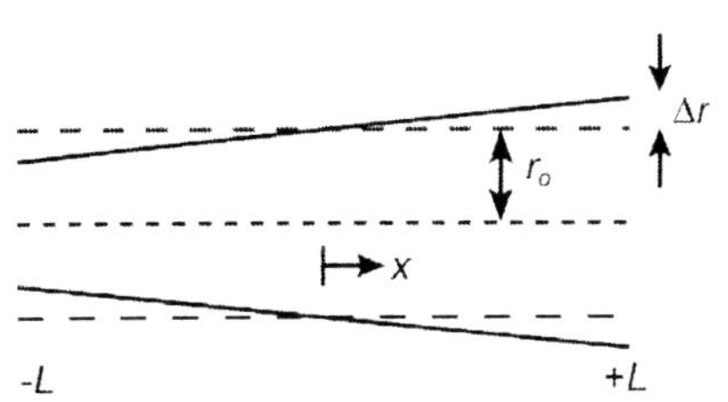

Figure 1: Schematic illustration of a fibre with a linear variation in radius along the fibre length.

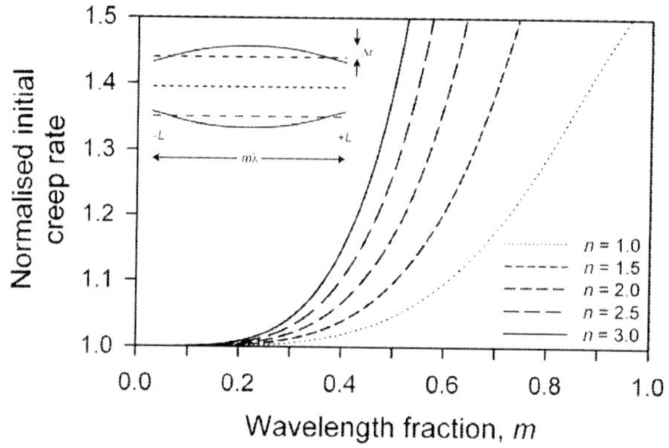

Figure 2: Influence of creep stress exponent, n, and characteristic wavelength fraction, $m = 2L/\lambda$, on the normalised initial creep rate, $\dot{\varepsilon}_o^*$, of a fibre with a sinusoidal shape geometry as shown in the figure. Adapted from data in Davies.[17]

measured creep rate is identical (or similar) to the "intrinsic" material creep rate for a fibre of constant radius. The main conclusions were that, for typical current SiC-based fibres, radius variation along the fibre length could significantly influence the measured creep rates.[17] Such a result is consistent with previous experimental work that suggested statistical scatter between the creep properties of individual Hi-Nicalon™ SiC-based fibres to be as high as 100%.[19,18] Analysis of the influence of characteristic wavelength fraction, $m = 2L/\lambda$, and n for several different sinusoidal radius variation profiles (such as the example shown in Figure 2) suggested that fibre gauge lengths for the measurement of creep rates should be as low as possible and preferably $\leq \lambda/5$, which corresponds to an upper limit of 26~32 mm for current SiC-based fibres.[17]

From consideration of the initial results[17] that radius variation has the potential to significantly influence the initial creep rate of ceramic fibres, in the present work the authors have investigated further aspects of this phenomenon.

EXPERIMENTAL PROCEDURE

Similar to the previous work[17] conducted on this topic, trends in the creep rate were deduced under various initial conditions from numerically integrating the creep rate along the length of the fibre by first dividing the fibre into, in this case, 10^4 elements of equal length and calculating the strain rate within each element. In this work the authors have only considered a linear variation in fibre radius as shown in Figure 1 whilst the ranges of n and r_v investigated were $0 < n < 3$ and $0 < r_v < 0.5$, respectively. The influence of radius variation on creep rupture time was analysed through the use of the Monkman-Grant relationship[21] as follows:

$$t_r \dot{\varepsilon}_o^m = C \tag{7}$$

where t_r is the rupture time, $\dot{\varepsilon}_o$ is the minimum creep rate (from Equation (5)), and m and C are empirical constants. A value of m = 1.2 was assumed for the present work as this value was found to be appropriate for a range of SiC fibres and matrices.[20] In order to highlight the influence of the radius variation on t_r, normalised values of the rupture time, t_r^*, were obtained from dividing t_r by the appropriate value for a fibre with constant radius.

Several researchers[13] have also noted the phenomenon of a stress threshold, below which the creep stress is zero, with the creep rate in this case being given by:

$$\dot{\varepsilon} = A\left(\sigma - \sigma_o\right)^n e^{-\left(\frac{Q}{RT}\right)} \tag{8}$$

where σ_o in the threshold stress; this equation thus reducing to Equation (1) for the simple case of $\sigma_o = 0$. In the present work the authors have investigated the influence of σ_o on the normalised initial creep rates for fibres containing a linear variation in fibre radius for the conditions σ/σ_o = 1.1, 1.5, and 2.0.

RESULTS AND DISCUSSION

Normalised Initial Creep Rate

Values of normalised initial creep rate, $\dot{\varepsilon}_o^*$, for the case of a fibre containing a linear radius variation have been presented in Fig. 3(a) for the region $0 < n < 3$ and $0 < r_v < 0.5$ whilst trends in $\dot{\varepsilon}_o^*$ versus n for constant r_v (equivalent to horizontal line profiles in Figure 3(a)) have been presented in Figure 4(a). The main point from Figure 3(a) is the general trend in increasing $\dot{\varepsilon}_o^*$ for higher n and r_v values. In order to keep $\dot{\varepsilon}_o^* < 1.05$ (*i.e.*, 5% difference relative to the constant radius case) it is necessary for $n < 0.35$ (irrespective of r_v), which is below the value of n for current SiC-based fibres,[12-16] or for $r_v < 0.08$ (irrespective of n). This latter case of $r_v < 0.08$ can be compared with the range of $0.25 < r_v < 0.5$ noted in Table I for current SiC-based fibres, suggesting as before that gauge lengths for monofilament creep testing should be as small as possible. For a difference of 10% with the constant radius case (*i.e.*, $\dot{\varepsilon}_o^* < 1.1$) then the respective values would be $n < 0.50$ and $r_v < 0.12$. For a 20% difference the values would be $n < 0.76$ and $r_v < 0.16$. It is thus clear that the potential exists for significant errors in the experimental values of creep rates for current SiC-based fibres and that trying to reduce the value of r_v, most likely by reducing the test gauge length, may be appropriate for the minimisation of experimental errors. In the longer term it would be advantageous for manufacturers to work towards minimising radius variation within individual fibres in order to improve the creep characteristics.

The influence of r_v on $\dot{\varepsilon}_o^*$ is more clearly seen in Figure 4(a) with the r_v = 0.10 curve increasing only marginally towards a value of 1.073 for n = 3; thus emphasising that small (linear) variations in the fibre radius may be tolerated, at least from the viewpoint of initial creep rates. However, radius variations above this value are shown to result in increasingly larger deviations from the constant radius case, *i.e.*, the portion of the initial creep rate attributed to the fibre geometry is considerably larger than that due to the material's intrinsic creep properties. The extreme case of n = 3 and r_v = 0.5 investigated in this work produced a value of $\dot{\varepsilon}_o^* = 6.37$, meaning that 84% of the measured initial creep rate would be due to the fibre geometry.

Whilst not shown here, it might be interesting to note that $\dot{\varepsilon}_o^*$ for the case of r_v = 0.5 was an order of magnitude higher (compared to the intrinsic material value) for n = 3.5 and two orders of magnitude higher once n reached a value of 5.5; this latter value producing large increases in $\dot{\varepsilon}_o^*$ for relatively small radius variation, *e.g.*, $\dot{\varepsilon}_o^* = 1.24$ for r_v = 0.1.

Normalised Creep Rupture Time

The influence of stress exponent and fractional radius variation on the normalised creep rupture time, t_r^*, for a fibre containing a linear radius variation has been shown in Figures 3(b) and 4(b). Note that in this work the authors have utilised the initial creep rate of the entire fibre for the value of $\dot{\varepsilon}_o$ in Equation (7). The value of $\dot{\varepsilon}_o$ would be maximum at the position of highest stress within the fibre, corresponding to the point where $r = r_o - \Delta r$, and in many situations would be considerably higher than the overall fibre creep rate. If the creep rupture time were to be determined by the maximum value of $\dot{\varepsilon}_o$ within the fibre, rather than the overall fibre

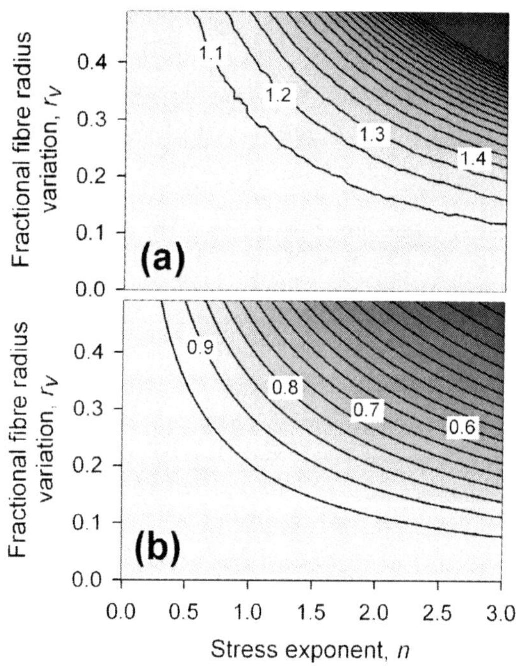

Figure 3: Influence of stress exponent, n, and fractional radius variation, r_V, on the creep properties of ceramic fibres containing a linear radius variation (m = 1.2 in Equation 7): (a) normalised initial creep rate, $\dot{\varepsilon}_o^*$, and (b) normalised creep rupture time, t_r^*.

value, then the values presented in Figures 3(b) and 4(b) would represent the upper limit of the rupture time.

As would be expected from inspection of Equation (7), essentially similar trends were observed in t_r^* as were previously noted for $\dot{\varepsilon}_o^*$ although in this case the value of t_r^* was always the same or less than that for a fibre with constant radius and decreased with both increasing n and r_V. As an example, the value $t_r^* = 0.9$ was bounded by $n = 0.48$ (irrespective of r_V) and $r_V = 0.069$ (irrespective of n). Figure 4(b) indicated that t_r^* decreased in a more constant rate with increasing r_V than did $\dot{\varepsilon}_o^*$, *i.e.*, t_r^* is less sensitive to changes in r_V than is $\dot{\varepsilon}_o^*$. The authors consider this result to be somewhat surprising and plan to further investigate this phenomenon through analysing the influence of maximum $\dot{\varepsilon}_o$ within the fibre rather than relying on the overall fibre value.

Variation in Creep Rate along the Fibre Length

It was mentioned in the previous section that the maximum creep rate within a fibre containing a linear radius distribution has the potential to be significantly higher compared to the overall fibre creep rate. Consideration of the minimum and maximum radius values within the

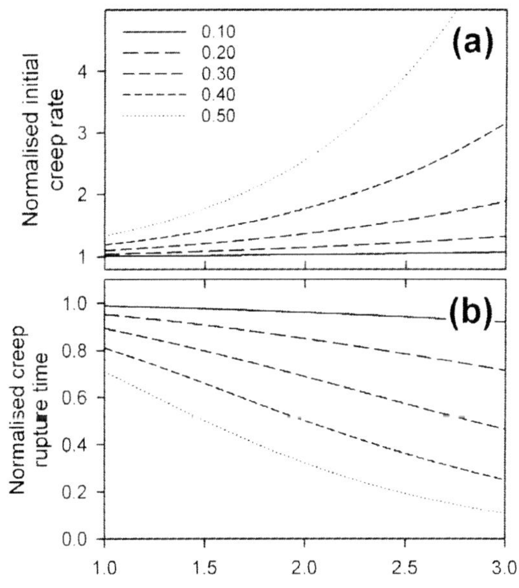

Figure 4: Influence of stress exponent, n, and fractional radius variation, r_V, on the creep properties of ceramic fibres containing a linear radius variation: (a) normalised initial creep rate, $\dot{\varepsilon}_o^*$, and (b) normalised creep rupture time, t_r^*.

fibre ($r_o - \Delta r$ and $r_o + \Delta r$, respectively) suggests that the creep rate will furthermore not be symmetrical about the fibre length midpoint. The normalised initial creep rates, $\dot{\varepsilon}_o^*(x)$, along the length of such a fibre, as a function of r_v and n, have been illustrated in Figure 5 where, by definition, all curves must pass through the fibre length midpoint (*i.e.*, $r = r_o$). As expected, increasing the radius variation, *i.e.*, r_v, tends to spread out the range of $\dot{\varepsilon}_o^*(x)$ values whilst a significant effect is also noted from changing the stress exponent from n = 1 (Figure 5(a)) to n = 3 (Figure 5(b)). In particular, the increased $\dot{\varepsilon}_o^*(x)$ values for small radius ($r < r_o$) sections of the fibre (x-axis values <0.5 in the figure) more than outweigh the decrease in $\dot{\varepsilon}_o^*(x)$ for large radius ($r > r_o$) sections of the fibre, thus resulting in an overall increase in fibre creep rate.

The ratio of the maximum to minimum $\dot{\varepsilon}_o^*(x)$ values, $\dot{\varepsilon}_o^*(x)_{max}$ and $\dot{\varepsilon}_o^*(x)_{min}$ respectively, for the n = 1 case (Figure 5(a)) was only 1.49 for r_v = 0.1 but increased to 8.88 for r_v = 0.5 whilst for n = 3 (Figure 5(b)) the respective values were 3.31 and 701; the value for r_v = 0.2 in the n = 3 case was still 11.2. These results thus indicate that the range of creep rates present within an individual fibre containing a linear increase in radius may be substantial and it

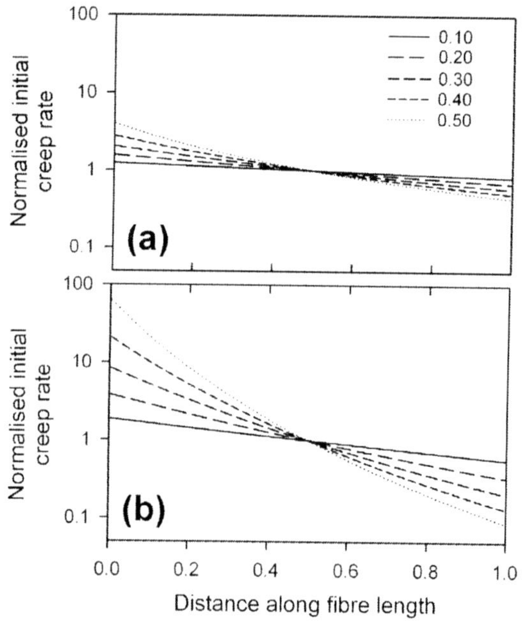

Figure 5: Variation in normalised initial creep rate, $\dot{\varepsilon}_o^*(x)$, as a function of distance along fibre length and fractional radius variation, r_v, for a fibre containing a linear variation in radius: (a) n = 1, and (b) n = 3.

might easily be imagined that the physical processes leading to creep behaviour and final rupture may vary along the length of such a fibre.

Influence of Creep Threshold Stress

As mentioned earlier, several researchers[13] have noted that creep within ceramic fibres may only occur above a threshold creep stress, σ_o. The present authors have investigated this effect on $\dot{\varepsilon}_o^*$ for creep stress ratios, σ/σ_o, of 1.1, 1.5, and 2.0, as shown in Figures 6(a) to (c), respectively. Whilst all the figures illustrate a general trend of increasing $\dot{\varepsilon}_o^*$ with increasing n and r_V, all three cases contain regions where $\dot{\varepsilon}_o^* < 1$, *i.e.*, an initial creep rate less than that expected for a fibre with a constant radius. This was particularly apparent for the low relative creep stress situations, *i.e.*, σ/σ_o = 1.1 and 1.5. Whilst this $\dot{\varepsilon}_o^* < 1$ region exists over a wide range of r_v values (although shrinking towards large r_v values as σ/σ_o increases), relatively small values of n (typically <0.7) are required for any significant effect and, as such, is not expected to occur for current SiC-based fibres where n is generally larger. The reason behind this phenomenon is simply that, for small enough n values, the loss of fibre creep in the large r region of the fibre is not made up by the increased creep in the small r region of the fibre, whereas as n increases then the small r region makes increasing larger contributions until such a point as they negate the lack of creep in the remained of the fibre.

CONCLUSIONS

Following on from previous work[17], the influence of linear radius variation on some creep properties of monofilament fibres was studied numerically. The primary variables considered were the stress exponent n ($0 < n < 3$), the radius variation r_v ($0 < r_v < 0.5$) and the creep threshold stress σ_o ($\sigma/\sigma_o = 1.1, 1.5, 2.0$). The investigation suggests the following conclusions:

(i) The normalised initial creep rate, $\dot{\varepsilon}_o^*$, is greater than that of a prismatic fibre for all combination of n and r_V. For low values of n and r_v the increase is negligible while for $n > 1.5$ and $r_v > 0.25$ the increase cannot be ignored. For values of n typical of SiC-based fibres along with radius variations reported in Table 1, increases above 100% are indicated.

(ii) Based on the Monkman-Grant relationship (Equation 7) with $m = 1.2$, the normalised creep rupture time was estimated for the same range of n and radius variation. Again, for low values of these parameters the normalised creep rupture time shows negligible decrease but drops rapidly for $n > 1.5$ and $r_v > 0.2$. The significant decrease for typical SiC-based fibre parameters suggests that the radius variation needs to be considered as the time to rupture may decrease to one third or less of that for a prismatic fibre.

(iii) The linear variation of radius with distance along a fibre results in non-linear variations in local initial creep rate, with the smaller radii region dominating the net effects.

(iv) The observed threshold creep stress appears to affect the initial creep rate of fibres with linear radius variation only for very low stress exponents, n. These lower values are not usually applicable to SiC-based fibres. Where a very low value of n does apply and σ_o is large, the initial creep rate may be decreased slightly compared to the equivalent prismatic fibre.

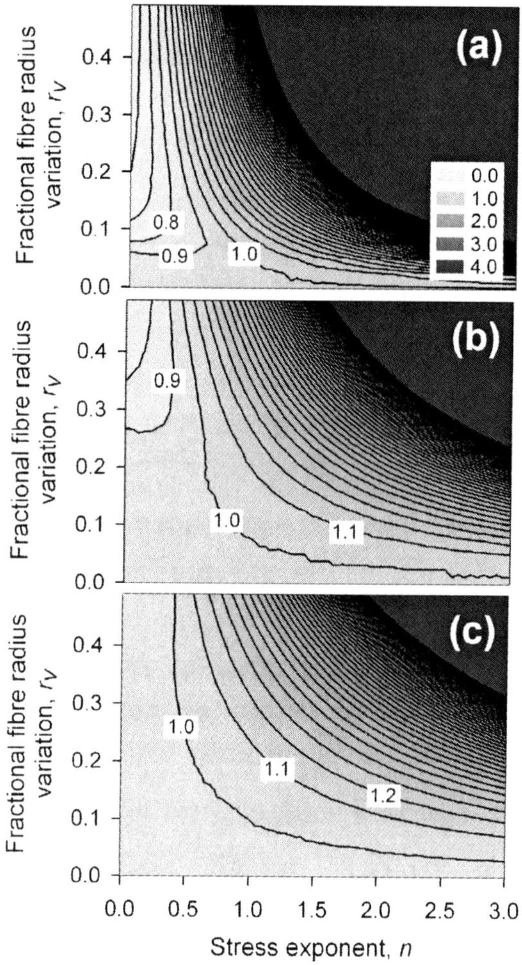

Figure 6: Influence of creep threshold stress, σ_o, on the normalised initial creep rate, $\dot{\varepsilon}^*_o$, for fibres containing a linear variation in fibre radius:
(a) σ/σ_o = 1.1, (b) σ/σ_o = 1.5, and (c) σ/σ_o = 2.0.

REFERENCES

[1]A. G. Evans and E. A. Charles, "Fracture toughness determinations by indentation", *J. Amer. Ceram. Soc.*, **59(7-8)**, 371-372 (1976).

[2]J. L. Henshall, D. J. Rowcliffe, and J. W. Edington, "Fracture toughness of single-crystal silicon carbide", *J. Amer. Ceram. Soc.*, **60(7-8)**, 373-375 (1977).

[3]K. Itatani, M. Sato, I. J. Davies, and S. Koda, "Mechanical properties of silicon carbide composite with Si-Al-C fiber and vapour grown carbon nanofiber addition", in Proceedings of the ninth Japan international SAMPE symposium and exhibition (JISSE-9), November 29th-December 2nd 2005, Tokyo, Japan, *submitted* (2005).

[4]T. Ishikawa, T. Yamamura, T. Hirokawa, Y. Hayashi, Y. Noguchi, and M. Matsushima, M., "Strength and fracture toughness properties of oxidation resistant high-temperature ceramic matrix composites", pp. 137-144 in Proceedings of the ninth international conference on composite materials (ICCM-9): Vol. 2 (edited by A. Miravette), Madrid, Spain, July 1993, Woodhead Publishing Co., Cambridge, U. K. (1993).

[5]T. Morimoto, "A representative diameter for the evaluation of brittle monofilament strength", pp. 265-272 in Proceedings of the 9th US-Japan conference on composite materials (edited by H. Fukuda, T. Ishikawa, and Y. Kogo), Mishima, Japan, July 2000, Japan society for composite materials, Tokyo, Japan (2000).

[6]G. E. Youngblood, C. R. Eiholzer, C. A. Lewinsohn, R. H. Jones, A. Hasegawa, and A. Kohyama, "Fiber diameter variation/sample preparation and analysis techniques", *Ceram. Eng. Sci. Proc.*, **20(3)**, 481-486 (1999).

[7]Y. T. Zhu, S. T. Taylor, M. G. Stout, D. P. Butt, W. R. Blumenthal, and T. C. Lowe, "On the statistical strength of Nicalon fibers and its characterization", *Ceram. Eng. Sci. Proc.*, **18(3)**, 119-126 (1997).

[8]I. J. Davies, T. Ishikawa, M. Shibuya, T. Hirokawa, and J. Gotoh, "Fibre and interfacial properties measured *in situ* for a 3-D woven SiC/SiC-based composite with glass sealant", *Composites Part A*, **30(4)**, 587-591 (1999).

[9]I. J. Davies and T. Ishikawa, "Bundle to bundle variation of mean fibre radius for Tyranno® LoxM Si-Ti-C-O fibres", *J. Maters. Sci. Letts.*, **20(6)**, 505-507 (2001).

[10]I. J. Davies, "Effect of radius variation on the mean strength of brittle fibres", *J. Maters. Sci. Letts.*, **20(12)**, 1103-1105 (2001).

[11]J. E. Dorn, "Some fundamental experiments on high temperature creep", *J. Mech. Phys. Solids*, **3**, 85-116 (1954).

[12]G. Simon and A. R. Bunsell, "Creep behaviour and structural characterization at high temperature of Nicalon SiC fibres", *J. Maters. Sci.*, **19**, 3658-3670 (1984).

[13]A. R. Bunsell and G. Simon, "Mechanical and structural characterisation of Nicalon SiC fibres up to 1300°C", *Compos. Sci. Technol.*, **27**, 157-171 (1986).

[14]R. Bodet, X. Bourrat, J. Lamon, and R. Naslain, "Tensile creep behavior of a silicon carbide-based fiber with a low oxygen content", *J. Maters. Sci.*, **30**, 661-677 (1995).

[15]R. Bodet, J. Lamon, N. Jia, and R. E. Tressler, "Microstructural stability and creep behavior of Si-C-O (Nicalon) fibers in carbon monoxide and argon environments", *J. Am. Ceram. Soc.*, **79(10)**, 2673-2686 (1996).

[16]H. M. Yun and J. A. DiCarlo, "Comparison of the tensile, creep, and rupture strength behaviors of stoichiometric SiC fibers", *Ceram. Eng. Sci. Proc.*, **20(3)**, 259-272 (1999).

[17]I. J. Davies, "Effect of variable radius on the initial creep rate of ceramic fibres", *J. Maters. Sci.*, in press (2005).

[18]J. A. DiCarlo, "Property goals and test methods for high temperature ceramic fibre reinforcement", *Ceram. Int.*, **23**, 283-289 (1997).

[19]H. M. Yun, J. C. Goldsby, and J. A. DiCarlo, "Tensile creep and stress-rupture behaviour of polymer derived SiC fibres", *Ceram. Trans.*, **46**, 17-28 (1994).

[20]J. A. DiCarlo, H. M. Yun, and J. B. Hurst, "Fracture mechanisms for SiC fibers and SiC/SiC composites under stress-rupture conditions at high temperatures", *Appl. Math. Comput.*, **152**, 473-481 (2004).

[21]F. C. Monkman and N. J. Grant, "An empirical relationship between rupture life and minimum creep rate in creep rupture tests", *Proc. ASTM*, **56**, 593-605 (1956).

[22]J. A. DiCarlo, "Creep of chemically vapour deposited SiC fibers", *J. Maters. Sci.*, **2**, 217-24 (1986).

[23]G. N. Morscher, J. A. DiCarlo, and T. Wagner, "Fiber creep evaluation by stress relaxation measurements", *Ceram. Eng. Sci. Proc.*, **12(7-8)**, 1032-8 (1991).

[24]J. A. DiCarlo and G. N. Morscher, "Creep and stress relaxation modelling of polycrystalline ceramic fibers", *ASME*, **AD-Vol-122**, 15-22 (1991).

INFLUENCE OF OXIDATION ON THE MICRO-MECHANICAL PROPERTIES OF A 3-D WOVEN SiC/SiC COMPOSITE TESTED IN AIR AT 1100 °C

Ian J. Davies and Sorn Sieng
Department of Mechanical Engineering, Curtin University of Technology,
GPO Box U1987, Perth, WA 6845, Australia

Toshio Ogasawara and Takashi Ishikawa
Advanced Composite Evaluation Technology Center (ACE Tec), Japan Aerospace Exploration Agency (JAXA), 6-13-1 Ohsawa, Mitaka, Tokyo, 181-0015, Japan

ABSTRACT

A major issue limiting the widespread utilisation of ceramic matrix composites is the poor oxidation resistance of the fibre/matrix interface, which is often comprised of pyrolytic carbon or hexagonal boron nitride. Indeed, the fibre/matrix interface shear strength, τ, is known to be highly sensitive to oxidation damage, in addition to playing an important role in the macroscopic mechanical behaviour of the composite. One potential method to increase the oxidation resistance of composite specimens has been to coat or seal the composite surface with an oxidation barrier material such as glass or pure silicon carbide (SiC). In the present work the authors have investigated the spatial distribution of micro-mechanical properties (specifically fibre strength and Weibull modulus, in addition to τ) for the case of a partially oxidised 3-D woven composite based on the SiC/SiC system creep tested in air at 1100 °C. The composite was comprised of Tyranno® LoxM Si-Ti-C-O fibres embedded within a matrix derived from the repeated polymer impregnation and pyrolysis of a precursor similar to polytitanocarbosilane. The composite had been coated (and any open porosity sealed) with a layer of glass following machining to the final test geometry. In addition to micro-mechanical properties, preliminary data was also presented for the spatial distribution of fibre pullout length within the composite fracture surface.

INTRODUCTION

Monolithic ceramics such as silicon carbide (SiC) are known to possess high strength and stiffness at elevated temperatures (*e.g.*, ≥1000 °C) which, when combined with their low density, excellent chemical and wear resistance, high hardness, and high thermal conductivity[1], makes them potential candidates for use as high temperature structural materials in aerospace and space applications. However, one of the main disadvantages of monolithic ceramics is their relatively poor fracture toughness (typically 2~3 MPa·m$^{1/2}$ for SiC[2,3]) which results in brittle fracture and wide scatter of strength data. The requirement for higher fracture toughness, K_{IC}, has led to the development of ceramic matrix composites (CMCs) in which particles, whiskers, or fibres are embedded within a ceramic matrix; values of 40 MPa·m$^{1/2}$ have been reported for a continuous Tyranno® Si-Ti-C-O fibre-reinforced 3-D woven SiC/SiC composite.[4]

A key phenomenon that directly impinges on the fracture toughness behaviour of CMCs is the interaction of microcracks with the fibre/matrix interface.[5] Ideally this interaction should result in deflection of the microcrack along the interface, debonding of the fibre/matrix interface, and subsequent pullout of fibres from the fracture surface.[6] Optimisation of mechanical properties for CMCs based on the SiC/SiC system have been achieved through the use of a weakly bonded fibre/matrix interface which is characterised by the fibre/matrix interface shear strength, τ. Thus far, the main materials utilised for the fibre/matrix interface region in SiC/SiC composites have been pyrolytic carbon (py-C)[7] and hexagonal boron nitride (h-BN)[8], both of which possess a layered structure and low sliding resistance; La-monazite ($LaPO_4$) has also been the subject of recent research.[9]

A major concern with py-C and h-BN is their relatively poor oxidation resistance above 500 °C, particularly an issue with CMCs that exhibit low-stress matrix microcracking.[10] The resulting oxidation of the fibre/matrix interface[11] and fibre surface[12-15] has the potential to significantly reduce the strength of the composite; the authors have noted strength decreases on the order of 90% following several minutes of exposure to oxidising conditions.[16] The maximum oxidation rate in SiC/SiC composites is known to occur at intermediate temperatures (*e.g.*, 500~900 °C)[17] and decreases at higher temperatures (*e.g.*, ≥1100 °C)[18] due to the sealing of surface cracks[18,19]. One method that has proved effective for increasing the oxidation resistance of SiC/SiC composites has been to seal the composite surface with a glass-based coating.[20,21] The oxidation of such a CMC can be thought to comprise of several different stages as indicated in Figure 1 with oxygen transport through the composite (stage 2), individual fibre bundles (stage 3), and at the surface of individual fibres (stage 4), proceeding rapidly once the surface layer has been breached (stage 1). Recent work by the authors on the oxidation of individual fibre bundles[22] and fibre surfaces[23] has followed on from investigations into oxidation of the entire composite for a 3-D woven SiC/SiC system.[21,24]

Whilst several researchers have investigated the oxidation properties of SiC/SiC composites containing py-C or h-BN interfaces[25-31], relatively little data is available regarding

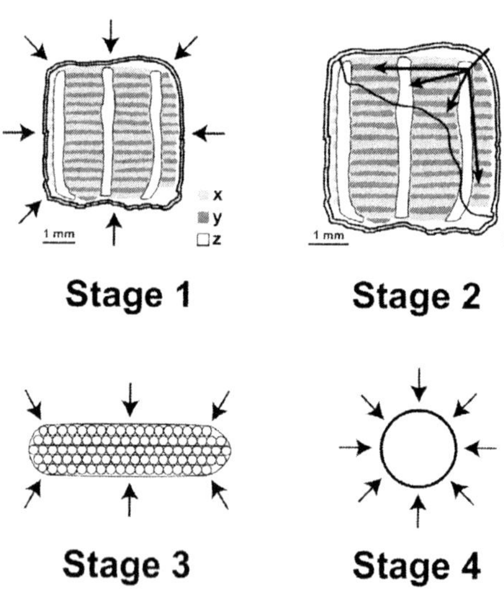

Figure 1: Schematic illustration of the different stages of oxidation within ceramic matrix composites: *Stage 1* – Oxygen ingression from the environment into the composite surface; *Stage 2* – Oxygen transport along microcracks and fibre bundles within the composite; *Stage 3* – Oxygen ingression into individual fibre bundles, and *Stage 4* – Oxidation of individual fibre surfaces.

the analysis of micro-mechanical data which has a direct influence on the macro-mechanical behaviour of the composite. For example, the tensile strength, S, of continuous fibre CMCs may be estimated using the following relationship:[32]

$$S = V_f \sigma_o \left(\frac{2}{m+2}\right)^{\frac{1}{m+1}} \left(\frac{m+1}{m+2}\right) \qquad (1)$$

where V_f is the volume fraction of fibres in the direction of loading and σ_o and m are the Weibull strength parameter and Weibull modulus, respectively, of the fibres measured *in situ* the composite.[16,20] The values of σ_o and m are related to τ through:[32,33]

$$\tau = \frac{r\lambda(m)\sigma_o}{4\langle h \rangle} \qquad (2)$$

where r is the fibre radius (taken to be 4.03 μm for Tyranno® LoxM Si-Ti-C-O fibres)[20], $\langle h \rangle$ is the mean fibre pullout length, and:[34]

$$\lambda(m) \approx 0.716 + \frac{1.36}{m^{0.6}} \qquad \text{for } m \geq 1 \qquad (3)$$

The present authors have investigated the spatial dependence of micro-mechanical properties within a partially oxidised 3-D woven SiC/SiC composite with preliminary results being presented in the current work. The aim of this investigation was, for the first time, to provide quantitative data relating to the influence of oxidation on the micromechanical properties within an entire CMC specimen.

EXPERIMENTAL PROCEDURE

The composite under investigation was based on the SiC/SiC system and comprised of surface-modified Tyranno® LoxM Si-Ti-C-O fibres (800 fibres per bundle) woven into a 3-D orthogonal configuration; the surface modification resulted in an outer 10 nm SiO_x-rich layer surrounding an inner 40 nm layer rich in carbon.[35] The total volume fraction of fibres within the composite was 0.40 with approximately 0.19 being in the direction parallel to loading. Matrix densification of the composite was achieved through the repeated polymer impregnation and pyrolysis of a precursor similar to polytitanocarbosilane. Following machining to a geometry suitable for mechanical testing,[35] the composite was repeatedly coated and impregnated with a proprietary glass sealant based on the SiO_2-Na_2O system. The specimen under investigation in the present work failed following creep testing in air at 1100 °C for 2.6 x 10^6 s (~30 days) under a creep stress of 140 MPa and equivalent to approximately 35% of the room temperature tensile strength (404 MPa).

Following failure, the specimen fracture surface was investigated using a scanning electron microscope (SEM) with an overall view of the fracture surface being presented in Figure 2. The top thirteen (13) rows of fibre bundles were selected for analysis with each row containing three fibre bundles (apart from row 13 which contained 2 fibre bundles) as shown in Figure 2(b). Individual fibre bundle was designated according to the respective row and column with, for example, bundle "3_1" being position in the first (left) column of the third row from the top. Furthermore, each fibre bundle was divided into nine (9) equally sized sections with the results of analysis being averaged for each section; there being a total of 342 sections analysed within the fracture surface. The following parameters were determined for each section: (i) mean fibre pullout length, (ii) fraction of fibres exhibiting a fracture mirror (Figure 3(a)), (iii) fibre strength, and (iv) fibre/matrix interface shear strength.

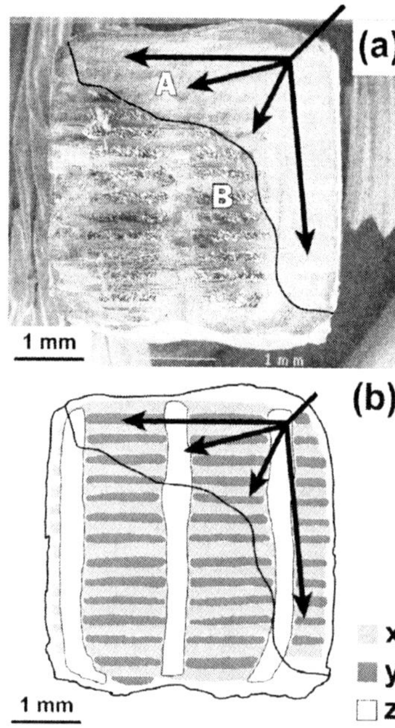

Figure 2: Fracture surface of a creep tested 3-D woven SiC/SiC specimen (140 MPa/1100 °C/2.63 x 10^6 s): (a) SEM micrograph, and (b) schematic representation of the fibre bundle positions and oxygen path (adapted from Davies *et al.*[36]).

The fibre strength for individual fibres, σ, measured *in situ* the composite was obtained from the measurement of fracture mirror radii, r_m, using the following relationship:[37-39]

$$\sigma = \frac{A_m}{\sqrt{r_m}} \tag{4}$$

where A_m is known as the "mirror constant" and found to have a value of 2.50 ± 0.09 MPa·m$^{1/2}$ for the current Si-Ti-C-O fibres measured *in situ* the composite[40,41] and similar to that proposed for nominally similar Nicalon® Si-C-O fibres (2.51 MPa·m$^{1/2}$).[42] Values of σ_0 and m were determined from nonlinear curve fitting of the standard Weibull equation to the set of ranked σ data; the resulting value of σ_0 was further normalised to a scale length, L_o, of 10^{-3} m using:

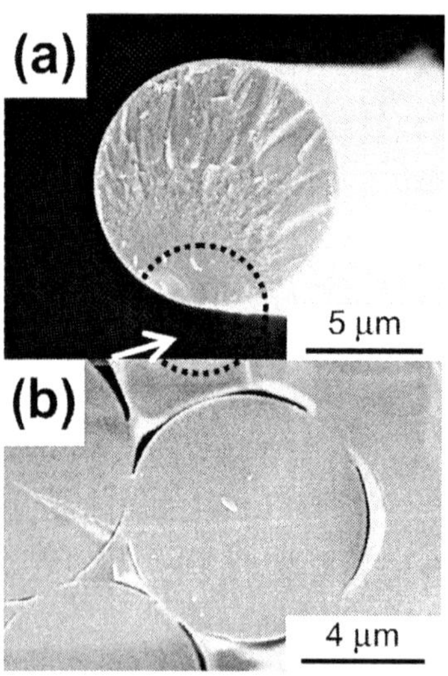

Figure 3: Scanning electron micrographs illustrating the main fibre fracture surface types noted in this work: (i) fracture mirror (the originating flaw is denoted by an arrow whilst the outer boundary of the fracture mirror is denoted by a broken line), and (ii) flat and featureless.

$$\sigma_o(L_o = 10^{-3} m) = \sigma_o\left(\frac{\delta_c}{10^{-3}}\right)^{\frac{1}{m}} \tag{5}$$

where δ_c is the scale length over which the fibre was tested and independent of the composite scale length:[32,33]

$$\delta_c = \frac{4<h>}{\lambda(m)} = \frac{rS_o}{\tau} \tag{6}$$

RESULTS AND DISCUSSION

The spatial distribution of mean fibre pullout length within the composite fracture surface has been presented in Figure 4 with $<h>$ being noted to vary between zero and 423 µm. As might be expected from preliminary inspection of Figure 2(a), most of the zero and small fibre pullout length region was concentrated in the top right corner of the specimen close to the point of oxygen ingression. The low fibre pullout lengths along the top few rows and right column would suggest the transport of oxygen to be relatively high along the x- and z-fibre bundles. The maximum value (423 µm) noted in this work was approximately half that noted for specimens tested at room temperature (810 µm),[43] suggesting that even the fibre bundles furthest away from the point of oxygen ingression to have suffered a degree of oxidation damage. In several cases, differences in $<h>$ were found to be present even within individual fibres bundles. For example, fibre bundle "5_2" exhibited a significant decrease in $<h>$ from left to right with fibre bundles in the middle column being more likely to indicate this phenomenon. Previous work by the authors[22] had highlighted a concentric distribution of fibre pullout length within a randomly picked oxidised fibre bundle (with pullout length being smallest at the fibre bundle perimeter and largest in the bundle centre) but this was not obviously apparent in the present investigation.

The fraction of fibres exhibiting a fracture mirror (Figure 3(a)) within each fibre bundle has been presented in Figure 5. Compared to the fibre pullout length data, the fraction of fibres exhibiting a fracture mirror, F_{mirror}, was relatively polarised in nature with F_{mirror} being either (close to) zero for fibre bundles in the top right of the fracture surface or else close to unity for a large number of fibre bundles towards the lower left of the fracture surface. From comparison between $<h>$ and F_{mirror} it was clear that a threshold amount of interfacial oxidation is required before F_{mirror} starts to decrease significantly. Overall there was a strong correlation between $<h>$ and F_{mirror} as previously noted for the case of a single fibre bundle.[22]

The Weibull strength measured *in situ* the composite and normalised to a scale length of 10^{-3} m has been shown in Figure 6 with the values ranging between approximately zero and 2.5 GPa; this maximum value still being significantly lower than that of 3.86 GPa for fibres in the same composite tested at room temperature.[16] The trend in normalised fibre strength was closely correlated to that of pullout length (Figure 4) with high fibre strength being associated with large mean fibre pullout length. As also noted for the mean fibre pullout length, significant variations in fibre strength were noted along the main axis (*i.e.*, from left to right) of several fibre bundles.

Finally, the spatial distribution of fibre/matrix interface shear strength has been presented in Figure 7 with values of τ being shown on a logarithm scale from a minimum value of 10 MPa to more than 5 GPa. As would be expected, the highest values of τ were noted in the top right of the specimen at the position of oxygen ingression whilst the lowest values were concentrated towards the lower left of the specimen. From inspection of Equation (2) and the mean fibre pullout (Figure 4) and fibre strength (Figure 6) data it is clear that trends in τ were mainly dominated by changes in $<h>$. Whilst the lower range of values were only slightly higher compared to that noted in specimens tested at room temperature (~5 MPa),[16] the highest values were significantly larger than values generally reported for SiC/SiC composites. The main reason for the high τ values was attributed to the observation of fibre pullout lengths on the order of several microns within heavily oxidised fibre bundles. In support of the high τ values noted in Figure 7, careful analysis of τ within a partially oxidised single fibre bundle also noted values in excess of 1 GPa.[16]

One note of caution regarding the validity of τ values for the most heavily oxidised fibres (*i.e.*, smallest $<h>$ and largest τ) is that Equation 2 assumes the presence of global load sharing conditions within the composite which may not necessarily be true for heavily oxidised fibres. As such, it may be the case that τ values for heavily oxidised fibre bundles in Figure 7 should be considered to be semi-quantitative in nature rather than quantitative.

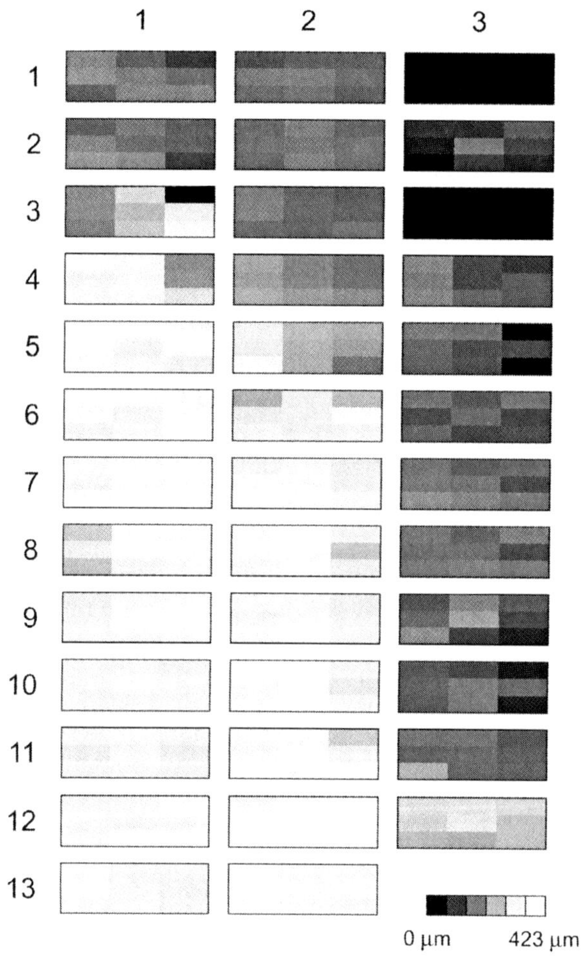

Figure 4: Spatial distribution of mean fibre pullout length, $<h>$, within a creep tested 3-D woven SiC/SiC specimen (140 MPa/1100 °C/2.63 x 10^6 s). Note that fibre pullout length data is shown on a logarithm scale.

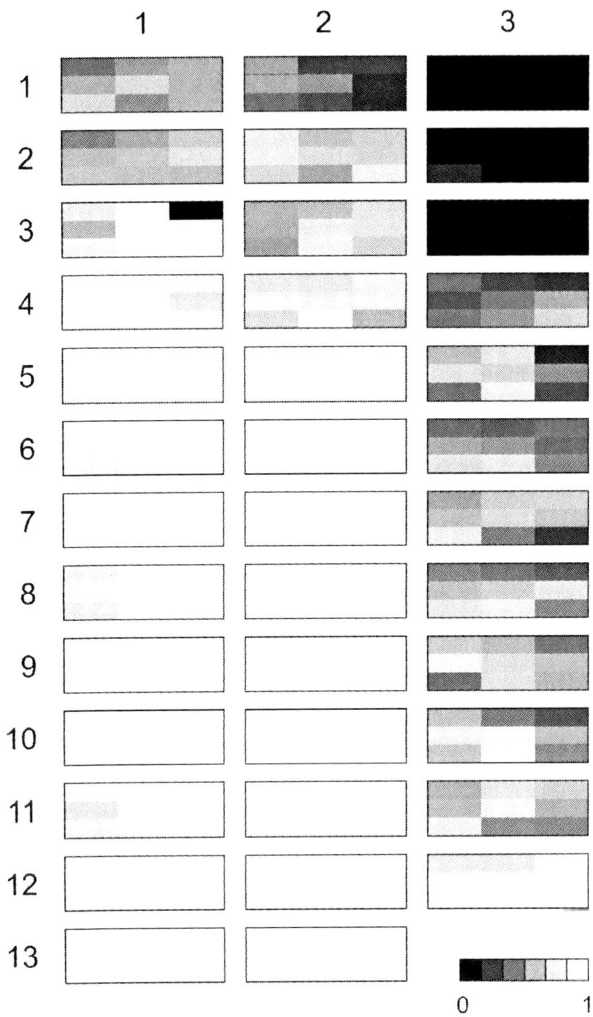

Figure 5: Spatial distribution of fraction of fibres exhibiting a fracture mirror within a creep tested 3-D woven SiC/SiC specimen (140 MPa/1100 °C/2.63 x 10^6 s).

CONCLUSIONS

In summary, preliminary results for the spatial distribution of: (i) mean fibre pullout length, (ii) fraction of fibres exhibiting a fracture mirror, (iii) Weibull strength normalised to a gauge length of 10^{-3} m, and (iv) fibre/matrix interface shear strength, have been presented for a 3-D woven SiC/SiC composite that had failed following creep testing in air at 1100 °C. To the knowledge of the authors, this is the first time that such data has been presented on such a scale for partially oxidised CMCs. It is hoped that this data may prove useful for researchers wishing to model the oxidation of complex CMCs.

ACKNOWLEDGEMENTS

The authors wish to gratefully acknowledge Dr. M. Shibuya of Ube Industries Ltd., Dr. J. Gotoh of Kawasaki Heavy Industries Ltd., and T. Hirokawa and T. Tanamura of Shikibo Ltd. for the manufacture and supply of all materials used in this study, and E. Miller of Curtin University of Technology for assistance with SEM analysis.

REFERENCES

[1]M. Srinivasan, pp. 99-159 in Treatise on Materials Science and Technology (volume 29), Academic Press, Inc., New York (1989).

[2]R. G. Munro, "Material properties of a sintered alpha-SiC", *J. Phys. Chem. Ref. Data*, **26**, 1195-1203 (1997).

[3]J. Schlichting and F. L. Riley, pp. 426-429 in Concise Encyclopedia of Advanced Ceramic Materials, ed. R. J. Brook, Pergamon Press, Oxford (1991).

[4]T. Ishikawa, T. Yamamura, T. Hirokawa, Y. Hayashi, Y, Noguchi, and M. Matsushima, M., "Strength and fracture toughness properties of oxidation resistant high-temperature ceramic matrix composites", pp. 137-144 in Proceedings of the ninth international conference on composite materials (ICCM-9): Vol. 2 (edited by A. Miravette), Madrid, Spain, July 1993, Woodhead Publishing Co., Cambridge, U. K. (1993).

[5]H. He and J. W. Hutchinson, "Crack deflection at an interface between dissimilar elastic materials", *Int. J. Solid Struct.*, **25(9),** 1053-67 (1989).

[6]A. G. Evans, M. –Y. He, and J. W. Hutchinson, "Interface debonding and fiber cracking in brittle matrix composites", *J. Am. Ceram. Soc.*, **72(12)**, 2300-2303 (1989).

[7]L. Filipuzzi, G. Camus, R. Naslain, and J. Thebault, "Oxidation mechanisms and kinetics of 1D-SiC/C/SiC composite materials: I, An experimental approach", *J. Am. Ceram. Soc.*, **77(2)**, 459-66 (1994).

[8]S. Prouhet, G. Camus, C. Labrugere, A. Guette, and E. Martin, "Mechanical characterization of Si-C(O) fibre/SiC (CVI) matrix composites with a BN-interphase", *J. Am. Ceram. Soc.*, **77(3)**, 649-56 (1994).

[9]J. B. Davis, R. S. Hay, D. B. Marshall, P. E. D. Morgan, and A. Sayir, "Influence of interfacial roughness on fibre sliding in oxide composites with La-monazite interphases", *J. Am. Ceram. Soc.*, **86(2),** 305–16 (2003).

[10]T. Ogasawara, T. Ishikawa, H. Ito, N. Watanabe, and I. J. Davies, "Multiple cracking and tensile behavior for an orthogonal 3-D woven Si-Ti-C-O fibre/Si-Ti-C-O matrix composite", *J. Am. Ceram. Soc.*, **84(7)**, 1565-74 (2001).

[11]R. J. Kerans, R. S. Hay, T. A. Parthasarathy, and M. K. Cinibulk, "Interface design for oxidation-resistant ceramic composites", *J. Am. Ceram. Soc.*, **85(11)**, 2599-632 (2002).

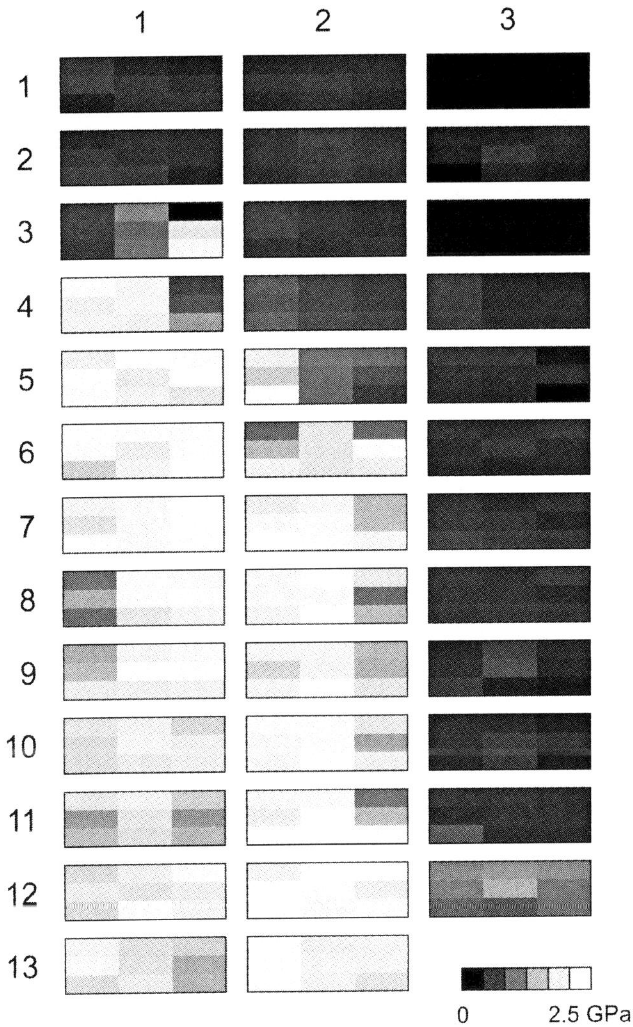

Figure 6: Spatial distribution of mean Weibull fibre strength, σ_0 (normalised to a scale length of 10^{-3} m) measured *in situ* a creep tested 3-D woven SiC/SiC specimen (140 MPa/1100 °C/2.63 x 10^6 s).

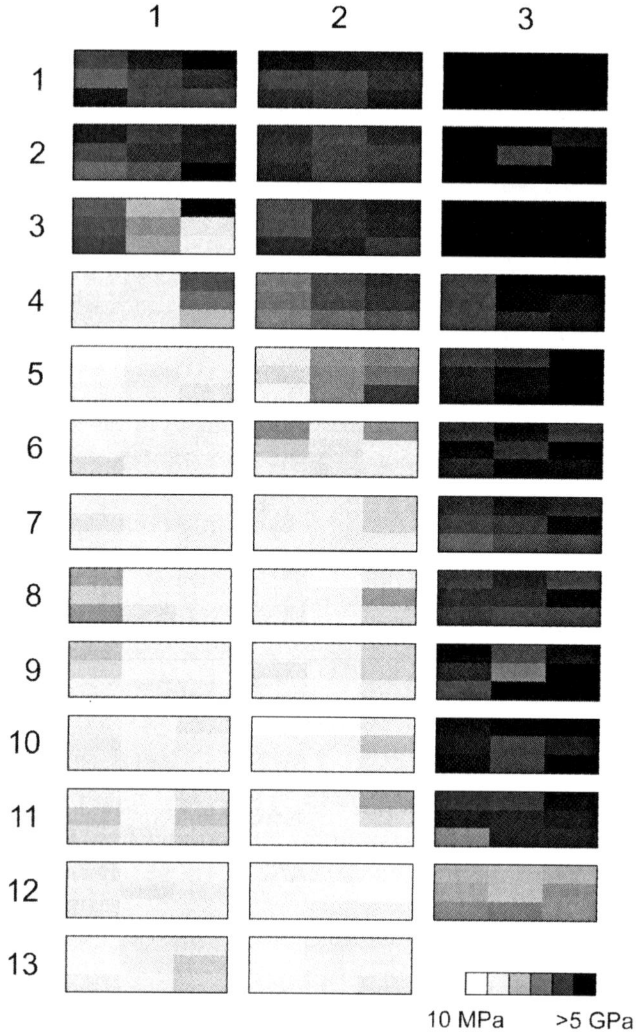

Figure 7: Spatial distribution of fibre/matrix interface shear strength, τ, within a creep tested 3-D woven SiC/SiC specimen (140 MPa/1100 °C/2.63 x 10^6 s). Note that fibre/matrix interface shear strength data is shown on a logarithm scale.

[12]K. Kakimoto, T. Shimoo, and K. Okamura, "Oxidation-induced microstructural change of Si-Ti-C-O fibres", *J. Am. Ceram. Soc.*, **81(2)**, 409-12 (1998).

[13]M. Takeda, A. Urano, J. Sakamoto, and Y. Imai, "Microstructure and oxidation behavior of silicon carbide fibres derived from polycarbosilane", *J. Am. Ceram. Soc.*, **83(5)**, 1171-76 (2000).

[14]T. Shimoo, F. Toyoda, and K. Okamura, "Thermal stability of low-oxygen silicon carbide fibre (Hi-Nicalon) subjected to selected oxidation treatment", *J. Am. Ceram. Soc.*, **83(6)**, 1450-56 (2000).

[15]T. Shimoo, T. Morita, and K. Okamura, "Oxidation of low-oxygen silicon carbide fibres (Hi-Nicalon) in carbon dioxide", *J. Am. Ceram. Soc.*, **84(12)**, 2975-80 (2001).

[16]I. J. Davies, T. Ishikawa, M. Shibuya, and T. Hirokawa, "Fibre strength parameters measured *in situ* for ceramic-matrix composites tested at elevated temperature in vacuum and in air", *Compos. Sci. Technol.*, **59**, 801-11 (1999).

[17]A. G. Evans, F. W. Zok, R. M. McMeeking, and Z. Z. Du, "Models of high-temperature, environmentally assisted embrittlement in ceramic-matrix composites", *J. Am. Ceram. Soc.*, **79(9)**, 2345-52 (1996).

[18]L. Philipuzzi and R. Naslain, "Oxidation mechanisms and kinetics of 1D-SiC/C/SiC composite materials: II, Modeling", *J. Am. Ceram. Soc.*, **77(2)**, 467-80 (1994).

[19]F. Lamouroux, R. Naslain, and J. –M. Jouin, "Kinetics and mechanisms of oxidation of 2D woven C/SiC composites: II, Theoretical approach", *J. Am. Ceram. Soc.*, **77(8)**, 2058-68 (1994).

[20]I. J. Davies, T. Ishikawa, M. Shibuya, T. Hirokawa, and J. Gotoh, "Fibre and interfacial properties measured *in situ* for a 3D woven SiC/SiC-based composite with glass sealant", *Composites: Part A*, **30**, 587-91 (1999).

[21]T. Ogasawara, T. Ishikawa, N. Suzuki, I. J. Davies, M. Suzuki, J. Gotoh, and T. Hirokawa, "Tensile creep behavior of 3-D woven Si-Ti-C-O fibre/SiC-based matrix composite with glass sealant", *J. Mater. Sci.*, **35**, 785-93 (2000).

[22]I. J. Davies, T. Ogasawara, and T. Ishikawa, "Distribution of fiber pullout length and interface shear strength within a single fiber bundle for an orthogonal 3-D woven Si-Ti-C-O fiber/Si-Ti-C-O matrix composite tested at 1100 °C in air", *J. Eur. Ceram. Soc.*, **25(5),** 599-604 (2005).

[23]J. Connell and I. J. Davies, "Modelling the influence of oxidation on the strength of ceramic fibres", in The 9th Japan international SAMPE symposium (JISSE-9), November 29th-December 2nd 2005, Tokyo, Japan, Japan chapter of SAMPE, Tokyo, Japan, *in press.*

[24]I. J. Davies, T. Ogasawara, and T. Ishikawa, "Scanning electron microscopy study of failure in glass-sealed SiC/SiC-based composite (NUSK-CMC) creep tested at 1100 and 1200 °C in air", *Adv. Composite Maters.*, **10(4)**, 357-367 (2001).

[25]C. F. Windisch, C. H. Henager, G. D. Springer, and R. H. Jones, "Oxidation of the carbon interface in Nicalon-fibre-reinforced silicon carbide composite", *J. Am. Ceram. Soc.*, **80(3)**, 569-74 (1997).

[26]E. Lara-Curzio, "Analysis of oxidation-assisted stress-rupture of continuous fibre-reinforced ceramic matrix composites at intermediate temperatures", *Composites: Part A*, **30**, 549-54 (1999).

[27]R. H. Jones, C. H. Henager, C. A. Lewinsohn, and C. F. Windisch, "Stress-corrosion cracking of silicon carbide fibre/silicon carbide composites", *J. Am. Ceram. Soc.*, **83(8)**, 1999-2005 (2000).

[28]C. Badini, P. Fino, G. Ubertalli, and F. Taricco, "Degradation at 1200 °C of a SiC coated 2D-Nicalon/C/SiC composite processed by SICFILL® method", *J. Europ. Ceram. Soc.*, **20**, 1505-14 (2000).

[29]K. L. More, K. S. Ailey, R. A. Lowden, and H. T. Lin, "Evaluating the effect of oxygen content in BN interfacial coatings on the stability of SiC/BN/SiC composites", *Composites: Part A*, **30**, 463-70 (1999).

[30]G. M. Morscher, J. Hurst, and D. Brewer, "Intermediate-temperature stress rupture of a woven Hi-Nicalon, BN-interphase, SiC-matrix composite in air", *J. Am. Ceram. Soc.*, **83(6)**, 1441-49 (2000).

[31]L. U. J. T. Ogbuji, "Pest-resistance in SiC/BN/SiC composites", *J. Europ. Ceram. Soc.*, **23**, 613-17 (2003).

[32]W. A. Curtin, "Theory of mechanical properties of ceramic-matrix composites", *J. Am. Ceram. Soc.*, **74(11)**, 2837-45 (1991).

[33]W. A. Curtin, "*In situ* fibre strengths in ceramic-matrix composites from fracture mirrors", *J. Am. Ceram. Soc.*, **77(4)**, 1075-78 (1994).

[34]W. A. Curtin, "Theory of the mechanical properties of ceramic matrix composites", pp. 1-5 in Proceedings of the Japan Society for Composite Materials - Research Presentation Meeting 1997, Tokyo Science University, May 12-13th 1997, Japan Society for Composite Materials, Tokyo (1997).

[35]T. Ishikawa, K. Bansaku, N. Watanabe, Y. Nomura, M. Shibuya, and T. Hirokawa, "Experimental stress/strain behavior of SiC-matrix composites reinforced with Si-Ti-C-O fibres and estimation of matrix elastic modulus", *Compos. Sci. Technol.*, **58(1)**, 51-63 (1998).

[36]I. J. Davies, T. Ogasawara, T. Ishikawa, and N. Suzuki, "Case study of failure in a glass-sealed SiC/SiC-based composite (NUSK-CMC) creep tested at 1100 °C in air", *Maters. Letts.*, **48(3-4),** 205-209 (2001).

[37]E. B. Shand, "Breaking stresses of glass determined from dimensions of fracture mirrors", *J. Am. Ceram. Soc.*, **42(10)**, 474-77 (1959).

[38]W. C. Levengood, "Effects of original flaw characteristics on glass strength", *J. Appl. Phys.*, **29(5)**, 820-26 (1958).

[39]D. A. Krohn and D. P. H. Hasselman, "Relation of flaw size to mirror in the fracture of glass", *J. Am. Ceram. Soc.*, **54(8)**, 411 (1971).

[40]I. J. Davies and T. Ishikawa, "Estimation of the fracture toughness of Tyranno® Si-Ti-C-O fibres from flaw size and 'fracture mirror' data measured *in situ* a 3-D woven SiC/SiC composite", *Int. J. Maters. Prod. Tech.*, **16(1-3)**, 189-96 (2001).

[41]I. J. Davies and T. Ishikawa, "Estimation of the 'mirror constant' for Tyranno® Si-Ti-C-O fibres *in situ* a 3-D woven SiC/SiC composite", *J. Am. Ceram. Soc.*, **85(3)**, 691-93 (2002).

[42]A. J. Eckel and R. C. Bradt, "Strength distribution of reinforcing fibres in a Nicalon fibre/chemically vapor infiltrated silicon carbide matrix composite", *J. Am. Ceram. Soc.*, **72(3)**, 455-58 (1989).

[43]I. J. Davies, T. Ishikawa, M. Shibuya, and T. Hirokawa, "Optical microscopy of 3-D woven SiC/SiC-based composites", *Composites Sci. Technol.*, **59(3)**, 429-437 (1999).

INFLUENCE OF EBC COATING ON TENSILE PROPERTIES OF MI SIC/SIC COMPOSITES

Ramakrishna T. Bhatt
Vehicle Technology Center
US Army Research Laboratory
Glenn Research Center at Lewis Field
21000 Brookpark Rd.
Cleveland, OH 44135

Gregory N. Morscher
Ohio Aerospace Institute
Glenn Research Center at Lewis Field
Cleveland, OH 44135

Kang N. Lee
Rolls-Royce Corporation
P.O. Box 420, Speed Code W-08
Indianapolis, IN 46206-0420

ABSTRACT

Tensile properties of notched and environmental barrier coated SiC/SiC composites were investigated. The composites were fabricated by melt infiltration (MI) and the environmental barrier coating (EBC) was deposited by air plasma spraying. Some of the EBC coated MI SiC/SiC composite specimens were exposed to static air at 1300, 1350, and 1400^{0}C for 500 h. Microstructural stability of the EBC coating and the substrate after oxidation exposure was analyzed by SEM. To evaluate notch sensitivity, some uncoated MI SiC/SiC composite specimens were also notched. Influence of EBC coating thickness, oxidation, and notch depth on room temperature tensile properties were investigated. The damage evolution during tensile deformation was monitored by acoustic emission. Results indicate that externally deposited EBC coatings, whether as-fabricated or oxidized, or machined notches do not significantly affect the tensile properties of MI SiC/SiC composites. This suggests that the composite is either notch insensitive or mildly notch sensitive. SEM analysis shows that noticeable microstructural changes occur in the EBC coating and the substrate after oxidation exposure with increasing time and temperature. At temperatures greater than 1300^{0}C, BSAS and grown silica in the EBC coating react to form a glass, and at 1400^{0}C intrinsic degradation of the composite occurs due to fiber degradation. The EBC coating degradation does not affect ultimate tensile strength (UTS) of the composites, but the fiber degradation does.

INTRODUCTION

As structural materials for high-temperature components in advanced engines for power and propulsion, fiber-reinforced ceramic matrix composites (CMC) offer a variety of performance advantages over the best metallic alloys with current structural capability to ~1100°C. These advantages are primarily based on the CMC being capable of displaying higher temperature capability, lower density (~30-50% metal density), and sufficient toughness for damage tolerance and prevention of catastrophic failure. These properties should in turn result in many important benefits for advanced engines, such as reduced cooling air requirements, simpler component design, reduced weight of support structure, improved fuel efficiency, reduced emissions, longer service life, and higher thrust. For the last two decades, SiC/SiC composites have been developed for high temperature turbine components such as blades, combustor liners, nozzle vanes, inter turbine ducts, divergent and convergent flaps, and flame holders [1, 2]. Currently, these composites are fabricated by three different processing methods; melt infiltration (MI), chemical vapor infiltration (CVI), and polymer infiltration pyrolysis (PIP) methods, but the composites fabricated by MI have been investigated the most. Although silicon-based ceramic composites exhibit excellent high temperature properties, strain capability beyond matrix cracking stress, high conductivity, and oxidation resistance in static air, they undergo surface recession when exposed to a combustion environment containing moisture at temperatures> 1100^0C [3, 4]. To avoid surface recession, a barium strontium aluminum silicate (BSAS) based environmental barrier coating (EBC) has been developed [3, 4]. Under thermal loading conditions alone, the EBC coated MI SiC/SiC components show microstructural stability after ~15,000 h exposure to 1200^0C in a combustion environment [5]. The EBC coating can have adverse or beneficial affect on mechanical properties of the composites because of the differences in elastic, fracture, and thermal expansion between the EBC coating and the substrate. Large difference in thermal expansion between the substrate and the EBC coatings causes residual stresses, which can change after long term exposure to high temperatures in oxidizing or combustion environment. The EBC coating can also act like a notch, affecting the strength of the composite. Preliminary results of edge notch study on uncoated MI SiC/SiC composite reported mild notch sensitivity on the ultimate strength [6]. However, the influence of face notches, EBC coating thickness and long-term oxidation exposure of EBC coating on the tensile properties of SiC/SiC composites have not been fully understood.

This study has two objectives. The first is to determine the influence of EBC coating thickness, and notches on tensile properties of the MI SiC/SiC composites. The second is to examine long term static oxidation exposure effects on tensile properties of EBC coated MI SiC/SiC composites.

EXPERIMENTAL PROCEDURE

The MI SiC/SiC composite panels (229 mm (L) x 152 mm (W) x 2 mm (T)) were supplied by GE Power Systems Composites, Newark, Delaware. All panels consisted of eight layers of 2-D five harness satin woven fabric of Sylramic-iBN SiC fibers, BN interface, CVI SiC coating, and SiC matrix which itself is a mixture of SiC particulate and silicon. Each panel contained ~34 vol% SiC fibers, ~6 vol% BN, ~28 vol% CVI SiC, and ~ 28 vol% (SiC+silicon) matrix. The composites were fabricated by a manufacturing method similar to that described in reference [7]. Each panel was first cut into rectangular specimens (152 mm (L) x 13 mm (W)) and then each specimen was cut into a dog-bone shape.

For EBC coating deposition, the gage section of the dog-boned specimen was first grit blasted to roughen the surface, and then deposited with EBC coating by atmospheric pressure plasma spraying. Plasma spraying (PS) is a high velocity impact deposition process in which melting, quenching, and consolidation take place in a single step. Reference [8] describes the PS process in detail. To describe the process briefly, a desired metal or ceramic powder is fed radially in the direction of a high velocity, high temperature, plasma flame. Molten droplets of the powder formed in the flame are propelled rapidly toward a substrate by the plasma flame. Upon impingement on to the substrate, the droplet instantaneously quenches and solidifies. The substrate, held in a fixture, is stationery, but the plasma gun is mounted on a robot arm which is movable. To deposit a layer, the plasma flame gun is rastered along the length and width of the substrate. The thickness of the layer can be adjusted by the distance between the plasma gun and the substrate, the rate at which powder fed into the plasma gun, and the velocity of the plasma flame. Multiple passes produce the desired thickness of the coating. The EBC coating consists of three sub-layers: silicon bond coat, a mixture of mullite and barium strontium aluminum silicate (BSAS) intermediate coat, and a topcoat of BSAS. Two different thickness of EBC coating were deposited on the composite specimens; 75 μm Si/75 μm mullite + BSAS/75 μm BSAS, and 125μm Si/200 μm mullite+BSAS/200 μm BSAS, respectively. For brevity, these two coatings are henceforth referred to as 225 μm and 525 μm EBC coatings, respectively. The 225 μm EBC coating is developed for the trailing edges of turbine nozzle vanes and blades, and the 525 μm EBC coating is developed for the leading edges of blades, nozzle vanes and combustor liners. In some specimens the EBC coating was deposited on one side of the gage section and in others on both sides of the gage section.

The EBC coating deposited by PS method contains numerous microstructural inhomogenieties such as pores and micro-cracks. During tensile loading, macro cracks developed in the EBC coating may act like notches on the MI SiC/SiC composites. Depending on notch sensitivity, the composites may show loss in tensile properties. To evaluate notch sensitivity, 50, 100, 150, or 200 μm deep notches were machined at the center of the gage section along the width of the dog-boned specimen. In some specimens notches were placed on one side and in others on both sides of the gage section.

Tensile testing was performed in an Instron 8502 universal testing machine. The tab ends were enveloped in wire mesh and gripped with pneumatic grips. Clip-on extensometers were used to measure displacement (25 mm gage length). A spring-loaded clip-on gauge was attached to the 25-mm length straight section of the dog bone specimen to monitor the displacement during the tensile test. The specimens were tested at room temperature until failure in a servo-controlled tensile testing machine equipped with self-aligning grips at a crosshead speed of 1.3 mm/min. Damage accumulation with loading was monitored by acoustic emission (AE) using a Digital Wave Fracture Wave Detector. Three wide-band (100 kHz to 2 MHz sensitivity; model B1025, Digital Wave, Englewood, CO) sensors were used to capture AE activity. Two sensors were placed on the face of the specimen outside of both extensometer knife-edges whereas the third was placed in the center of the gage section on the opposite face of the extensometer knife-edges. Only AE data that hit the middle AE sensor first, i.e., data that occurred in the gage section, were used for the matrix cracking analysis.

To determine microstructural and strength stability of MI SiC/SiC composite after oxidation exposure, dog-boned tensile specimens were coated all sides with 525 μm EBC and then exposed to air at either 1300, 1350, or 1400^{0}C for 500 h. After exposure, the residual tensile properties were determined at room temperature. Some of the failed specimens were sectioned, mounted in a mold, infiltrated with epoxy, and then polished using standard metallographic technique. The polished specimens were observed under an SEM equipped with energy dispersive spectroscopy (EDS).

RESULTS AND DISCUSSION

Figure 1 shows a cross section of a MI SiC/SiC composite coated with 525 μm EBC. The coating shows significant variation in thickness and contains three sub-layers: a layer of silicon bond coat on top of the substrate followed by an intermediate coat of a mixture of mullite and

BSAS, and then a top BSAS coat. A higher magnification photograph of figure 1 indicates pores, micro-cracks, and splats in the coating.

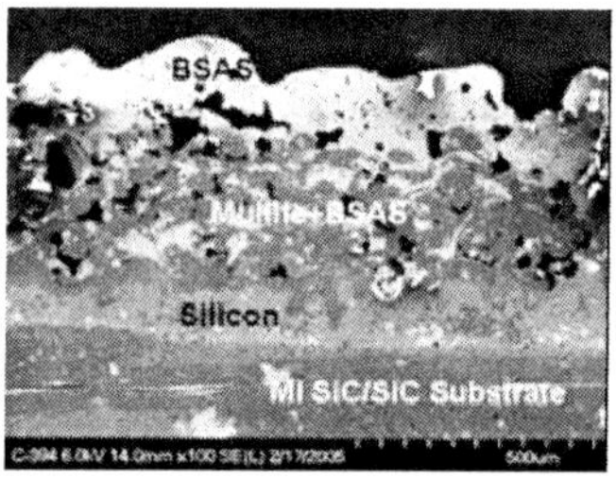

Figure1. SEM photograph of a cross-section of an EBC coated MI SiC/SiC composite showing microstructure of the coating and sub-layers.

Figure 2 shows room temperature tensile stress strain behavior of uncoated and EBC coated MI SiC/SiC composites. In all EBC coated composites, the coating was applied only on one side of the gage section. In the figure, tensile strain corresponds to average of the strain measured on the EBC and uncoated side. Tensile stress in the specimens is calculated by not including the cross-sectional area of EBC. This is valid because the EBC has low modulus and strength and does not contribute to the strength and modulus of the composites. Figure 2 indicates that except for a small difference in ultimate tensile strength, the primary modulus and the stress corresponding to deviation from linearity for the uncoated and the EBC coated composites are nearly the same. Also thickness of the EBC coating did not affect the tensile property significantly. The specimen that failed outside the gage section showed slightly lower ultimate tensile strength.

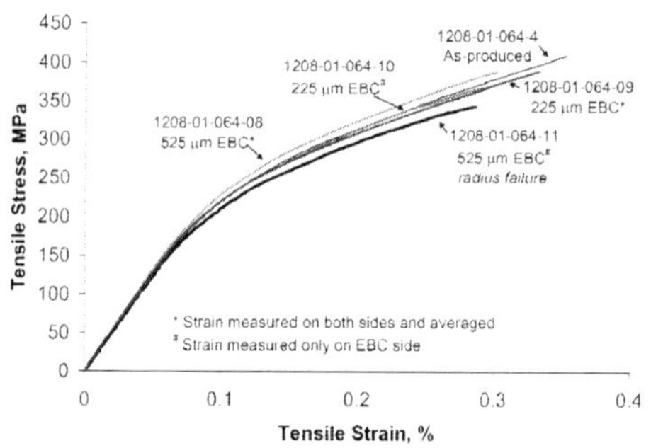

Figure 2. Room temperature tensile stress strain behavior of uncoated and EBC coated MI SiC/SiC composites showing the effect of EBC coating thickness

The AE activity occurring during tensile loading of specimens shown in figure 2 was also monitored to determine onset of matrix cracking and matrix crack density. Figure 3 shows the plot of the cumulative AE energy with increasing tensile stress. This has been shown to be an excellent measure of relative matrix crack density [9]. The differences in the cumulative AE data curves seen for the composites coated with 225 μm and 525 μm EBC are typical of specimen to specimen variation and transducer contact area/pressure. If the data is normalized by cumulative AE at 343 MPa (the failure stress of the lowest strength specimen), the relative AE activity is very similar for both cases and no real difference can be discerned.

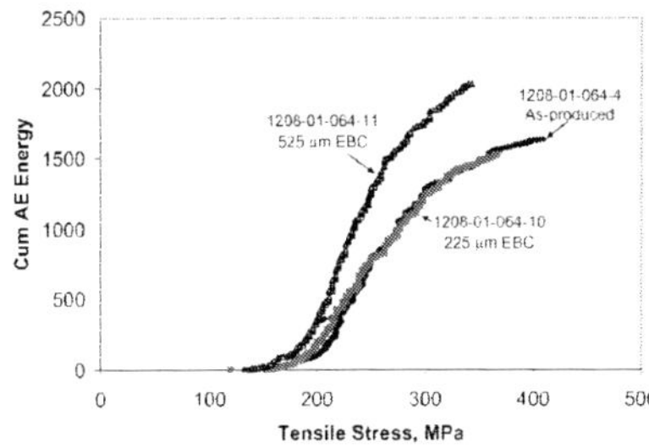

Figure3. Variation of cumulative AE energy with tensile stress for the uncoated and EBC coated MI SiC/SiC composites tensile tested at room temperature

To study the influence of notches on tensile properties, uncoated MI SiC/SiC composites were machined with through-the thickness notches to different depths. Some notches were placed only one side and others on both sides of the gage section. The room temperature tensile stress strain behavior and associated AE activity during loading of un-notched, single notched, and double notched composites are shown in figures 4 and 5. The notch definition is shown in the inset figure where t represents thickness of the specimen and a/2 represent depth of the notch.

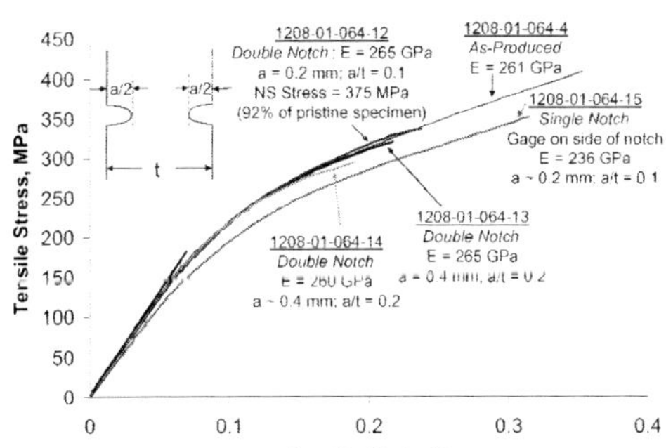

Figure 4 Room temperature tensile stress strain behavior of unnotched, single notched, and double notched MI SiC/SiC composites. The NS stress indicates net section stress.

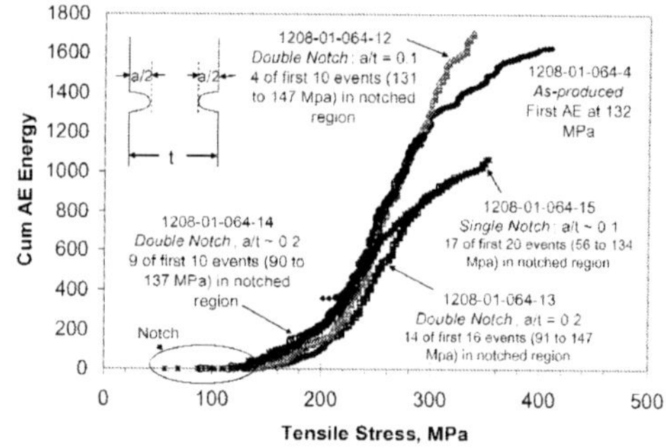

Figure 5 Cumulative AE energy with increasing stress for unnotched, single notched and double notched MI SiC/SiC composite specimens shown in figure 4.

Figure 4 indicates that as the notch depth increases ultimate tensile strength (UTS) decreases, but has no effect on the stress corresponding to deviation from linearity (DFL) and elastic modulus of the double notched specimens. Although the general shape of the stress strain behavior of the single notched specimen is similar to that of the double notched specimen, it exhibits lower modulus possibly due to within batch variation in modulus and due to the placement of the extensometer on the side of the notch. The plot of cumulative AE energy with tensile stress for the notched specimens (fig. 5) exhibits a sigmoidal shape. Detailed analysis of the AE curve shows that the damage in the form of matrix crack starts at the notch root. With increasing stress, matrix crack density increases initially and then levels off. At this stage some weak fibers also fractures. Other than the first 10 to 20 events up to stress level of 275 MPa, cumulative AE energy curves for un-notched and notched specimens are nearly the same.

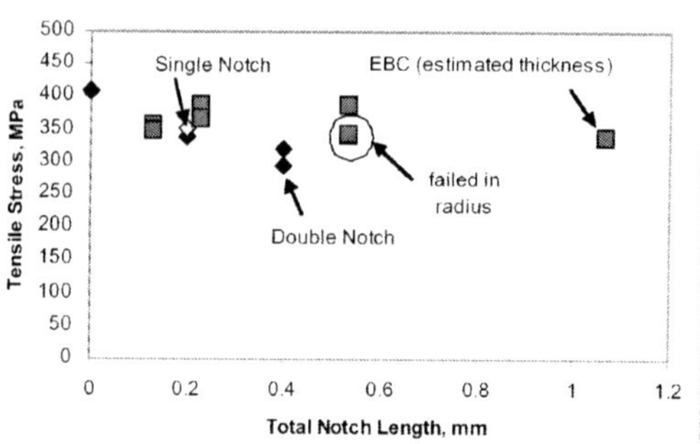

Figure 6. Variation of room temperature UTS with notch length for MI SiC/SiC composites. Filled square and diamond symbols represent EBC coated and double notch data, respectively and the filled circle represents the as-fabricated data.

Influence of notch on the room temperature UTS of MI SiC/SiC composites is shown in figure 6. The notch strength is based on gross cross sectional area of the specimen. Also included in the figure are the UTS values of as-fabricated and EBC coated specimens for comparison. Thickness of EBC coating is considered as notch length. Obviously as the notch length increases the UTS of the composite decreases linearly. This is expected because of loss of cross sectional area due to the notch. Externally applied coating up to a thickness of 0.55 mm does not degrade the strength of the composites, which indicate notch insensitivity. In figure 7, the notch strength data shown in figure 6 is re-plotted as normalized net cross sectional strength against a/t. The normalized net cross sectional strength is obtained by dividing UTS based on net cross sectional area of notched specimens with UTS of un-notched as-fabricated composite from the same batch. Data shown in the figure indicate no significant notch sensitivity for MI SiC/SiC composites

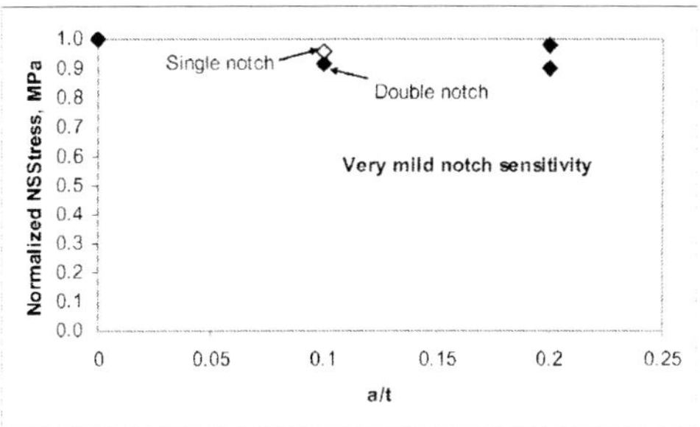

Figure 7 Variation of normalized net cross sectional strength with notch parameter, a/t, for MI SiC/SiC composites.

Microstructural stability of EBC coated MI SiC/SiC composites after 500 h oxidation exposure was investigated. For this purpose, both uncoated and EBC coated composites were exposed to air at either 1300, 1350, or 1400^{0}C for 500 h, and then their residual tensile properties were measured at room temperature. Figures 8, 9, and 10 show SEM micrographs of the cross-section of the MI SiC/SiC composites coated with 525 µm EBC that were exposed to air at 1300, 1350, and 1400^{0}C for 500 h, respectively. At 1300^{0}C, the sub layers of EBC coating are microstructurally stable except for growth of silica in the micro cracked regions of silicon bond coat. At 1350^{0}C, silica grown on the silicon bond coat reacts with the BSAS in the intermediate coat to form a glass which is consistent with the observations of Lee et al [10]. At 1400^{0}C, the glass that is formed by interaction of the silicon bond coat and the intermediate coat oozed out to the top coat of BSAS, and silicon from the (SiC+Si) matrix diffused through the CVI coating and reacted with the SiC fibers.

The residual tensile properties of EBC coated MI SiC/SiC composites exposed to air at either 1300, 1350, or 1400^{0}C for 500 h were measured at room temperature. Figures 11(a), (b), and(c) show the UTS of specimens exposed under these conditions. For comparison, included in the figures are the residual UTS values of uncoated MI SiC/SiC composites exposed to air for 100 and 500 h at similar exposure temperatures from reference [11]. The filled symbols represent

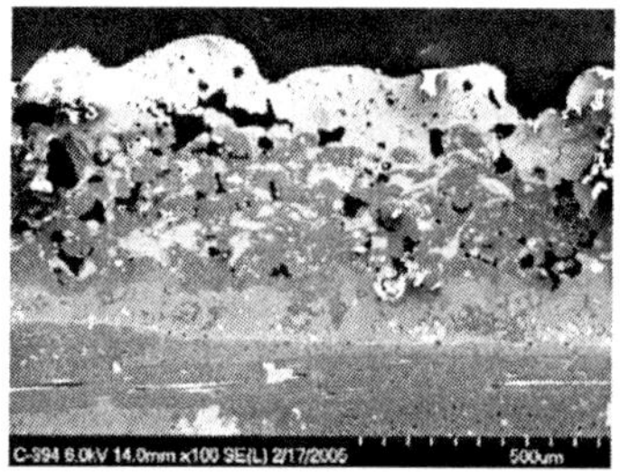

Figure 8 SEM micrograph of a cross section of an EBC coated MI SiC/SiC composite exposed to air at 1300^{0}C for 500 h

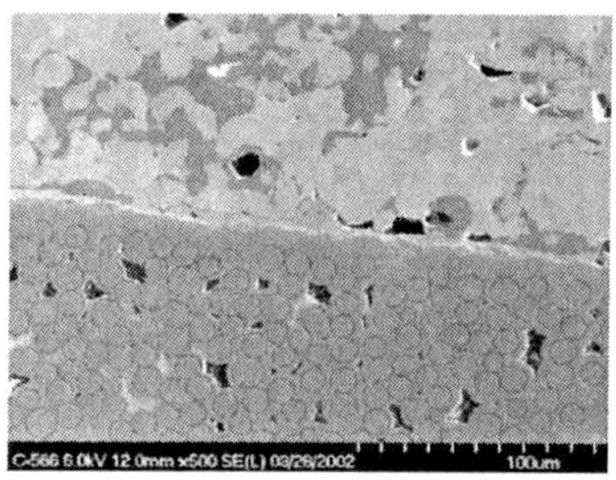

Figure 9 SEM micrograph of a cross section of an EBC coated MI SiC/SiC composite exposed to air at 1350^{0}C for 500 h showing formation of glass in EBC coating

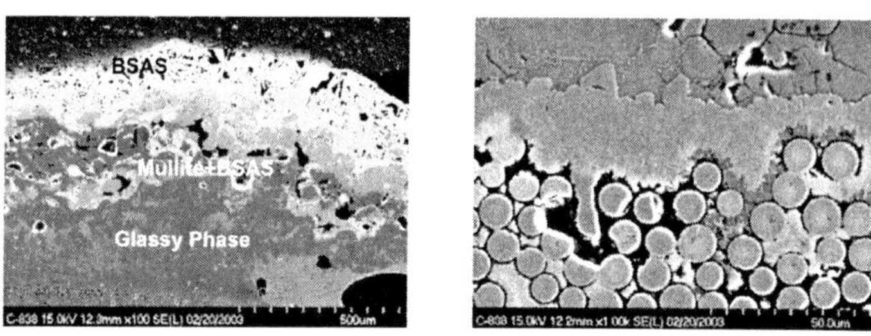

(a) (b)

Figure 10 SEM micrograph of a cross section of an EBC coated MI SiC/SiC composite exposed to air at 1400^{0}C for 500 h showing glass oozing into top coat and degradation of fibers inside the tow

uncoated composite data. It is reported in reference [11] that for fixed volume % SiC fibers, the residual UTS of MI SiC/SiC composites after long term exposure at high temperatures depends on CVI SiC content. The predominant mechanism of UTS degradation in MI SiC/SiC composites is attributed to silicon diffusion from the matrix into the SiC fibers and degrading them. As the CVI SiC content increases the residual strength also increases because of reduced diffusion of silicon into SiC fibers. Since the CVI SiC content of panels used in this study was different from those of uncoated composite data, the UTS data in the figures were plotted against CVI SiC content. Typical scatter seen in the DFL stress and UTS of as-fabricated composites are shown as bands in figure 11.

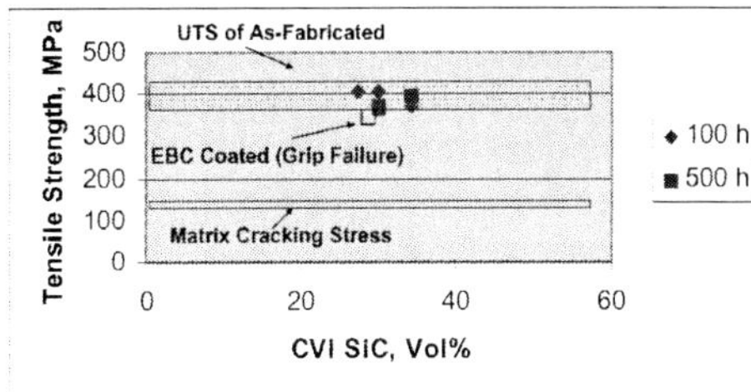

(a)

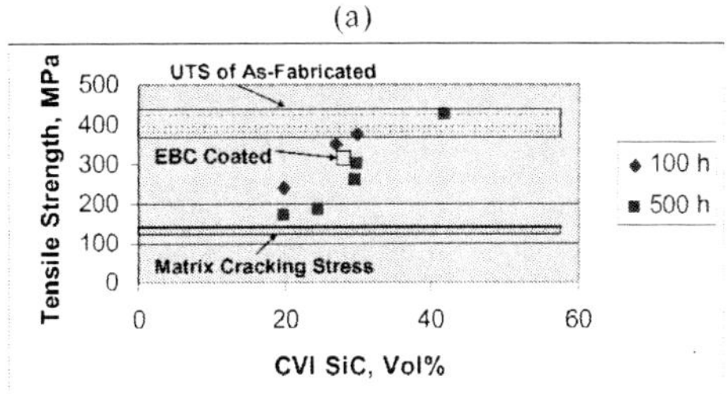

(b)

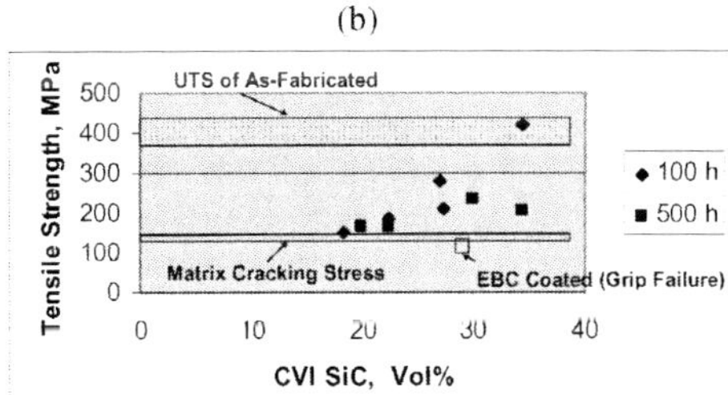

(c)

Figure 11 Room temperature ultimate tensile strength of uncoated and EBC coated MI SiC/SiC composites after exposure to air for 100 and 500 h at (a) 1300, (b) 1350, and (c) 1400^{0}C. Open square indicates EBC coated composites exposed to 500 h.

Premature grip failure is possibly due to misalignment/bending of EBC coated MI SiC/SiC composite specimens during tensile testing. Contributing factors for misalignments are specimen thickness variations and warping of the specimens after long tem exposure in static air at high temperatures.

CONCLUSIONS

Effects of notch and EBC coating on tensile properties of MI SiC/SiC composites have been investigated. Furthermore, residual tensile properties and microstructural stability of EBC coated MI SiC/SiC composite specimens exposed to air for 500 h at either 1300, 1350, or 1400^{0}C were also studied. Major findings are the following.

(1) Both EBC coating and notches on MI SiC/SiC composites show no or mild notch sensitivity effect on tensile properties.
(2) EBC coating undergoes microstructural changes beyond 1300^{0}C, but these changes do not affect tensile properties of the composites.
(3) Strength degradation seen in EBC coated MI SiC/SiC composites after long term oxidation exposure is intrinsic and caused by silicon diffusion from SiC matrix into SiC fibers.

REFERENCES

[1]D. Brewer, "HSR/EPM Combustor Materials Development Program," *Materials Science and Engineering*, **A261**, 284-291 (1999).

[2]NASA Ultra Efficient Engine Technology (UEET) Program, http://www.grc.nasa.gov/WWW/RT2000/2000/2100shaw.html

[3]K.N. Lee, "Environmental Barrier Coatings for Silicon-Based Ceramics," in *High Temperature Ceramic Matrix Composites*, W. Krenkel, R. Naslain, and H. Schneider, Eds, Wiley-VCH, Weinheim, Germany, (2001), pp. 224- 229.

[4]J.L. Smialek, R.C. Robinson, E.J. Opila, D.S. Fox, and N.S. Jacobson, "SiC and Si_3N_4 Recession Due to SiO_2 Scale Volatility under Combustor Conditions," *Adv. Composite Mater*, **8** [1], 33-45 (1999).

[5]H.E. Eaton and G.D. Linsey, "Accelarated Oxidation of SiC CMCs by Water Vapor and Protection via Environmental Barrier Approach," J. Euro. Ceram. Soc. **22**, 2741-2747 (2002).

[6]G.N. Morscher, J.Z. Gyekenyesi, and A.L. Gyekenyesi, "Mechanical Behavior of Notched SiC/SiC Composites," ASME International Gas Turbine and Aero Engine Congress, June 4-7, 2001, No. 2001-GT-461

[7]S. K. Lau, S. J. Calandra, and R. W. Ohnsorg, US Pat. #5,840,221, (1998).

[8]K.N. Lee, R.A. Miller, and N.S. Jacobson, "New Generation of Plasma-Sprayed Mullite Coatings on Silicon-Carbide," J. Am. Ceram. Soc. **78** (3) pp.705-710 (1995).

[9]G.N. Morscher, Comp. Sci. Tech., **64**, pp. 1311-1319 (2004).

[10]K.N. Lee, D.S. Fox, J.I. Eldridge, D.M. Zhu, R.C. Robinson, N.P. Bansal, and R.A. Miller, "Upper Temperature Limit of Environmental Barrier Coatings Based on Nullite and BSAS," J. Am. Ceram. Soc. **86**, (8) pp.1299-1306 (2003).

[11]R.T. Bhatt, J.A. DiCarlo, and T.R. McCue,"Thermal Stability of Melt Infiltrated SiC/SiC Composites," Ceram. Eng. Sci. Proc. **24** (3) p 295 (2003).

Design Concepts, Test Methods, and Standards

A MATERIAL DESIGN CONCEPT FOR TOUGHENED CERAMICS

Hideo Awaji, Takuya Matsunaga, and Seong-Min Choi
Nagoya Institute of Technology
Gokiso-cho, Showa-ku
Nagoya, 466-8555

ABSTRACT

In order to improve the fracture toughness of inherently brittle ceramics, we need to develop a new material design concept. One of the suitable concepts is to utilize dislocation activities in ceramics. Intra-type nanocomposites, where the dispersed nano-particles are embedded within matrix grains, have a special nano-structure. In the nano-structure, residual stresses are yielded around the second-phase nano-particles due to thermal expansion mismatch between the particles and matrix, and then dislocations are created. The dislocation activities enhance the strength and fracture toughness. In this research, the relation between the strength, the fracture toughness, and the critical frontal process zone (FPZ) size was clarified. The fracture toughness of ceramics is related closely to the critical FPZ size, because ceramics with larger FPZ size consume higher fracture energy during crack extension and indicate higher fracture toughness. Based on the toughening mechanism of nanocomposites we proposed previously, we fabricated toughened alumina-nickel nanocomposites using a soaking method which we developed recently to create the intra-type nano-structure, and found that the suitable annealing after sintering could produce toughened nanocomposites. The relation between the fracture toughness and the critical FPZ sizes for the dense alumina, as-sintered nanocomposites, and annealed nanocomposites was discussed. The results revealed that the annealed nanocomposites had the highest critical FPZ size and fracture toughness, because it was conceivable that the sessile dislocations in the critical FPZ at room temperature became nuclei of nano-cracks, created many nano-cracks, and expanded the critical FPZ size.

INTRODUCTION

Ceramics possess several superior properties as structural materials, such as high thermal resistance, chemical durability, and mechanical strength. However, ceramics have inherently low fracture toughness because of ionic and covalent bonding, and the plastic deformation due to dislocation activity is extremely limited. For instance, α-alumina single crystal has no dislocation activity around the room temperature under a pure shear stress experimented by Lagerlöf et al.[1] Therefore, dislocations in α-alumina can only be generated at elevated temperatures except for compressive stress states such as a Vickers indentation. And the frontal process zone (FPZ) ahead of a crack tip in ceramics is considered to be composed of nano-cracks rather than the kind of dislocations observed in metals[2-4]. To overcome the inherent brittleness of ceramics, a new material design concept must be developed. It is conceivable that expanding the size of the FPZ may improve the fracture toughness of ceramics, because the critical FPZ size of ceramics is quite small compared with that of metallic materials.

The intra-type nanocomposites proposed by Niihara[5] are based on a newly developed material design concept and are thought to exhibit the FPZ expansion mechanism. The intra-type nano-structure is constructed by dispersing second-phase nano-sized particles within the matrix grains. The thermal expansion mismatch between the matrix and the dispersed particles generates dislocations around the particles and only dislocations are possible to generate, because nano-sized particles yield residual stresses in nano-sized extent from the particles[4]. Also, these dislocations will become sessile dislocations at room temperature[1] and serve as origins for nano-crack nuclei in highly stressed field such as a tip of main crack. These dislocations, therefore, expand the FPZ and consequently enhance the fracture toughness of the materials. This is the nanocomposite toughening mechanism that we proposed previously[4]. To confirm this toughening mechanism in ceramics, we need to estimate the critical FPZ size.

The aim of this research is to propose an indirect technique for estimating the macroscopically defined critical FPZ size for ceramics using a local fracture criterion and linear fracture mechanics. We described a technique for estimating the critical FPZ size for ceramics in previous papers[6]. However, only a very short edge-cracked bending specimen was used in these studies, because it was not easy to analyze the stress ahead of the edge crack. In this paper, we develop a new technique for analyzing the exact stresses using the superposition concept of linear elasticity, and estimate critical FPZ sizes for several ceramics including nanocomposites.

First, we discuss the criterion of crack extension for ceramics, focusing on a local fracture criterion. Second, the exact stress at the critical FPZ size ahead of an edge-crack or an elliptical notch in a bending specimen is analyzed. Third, the relation between the strength of the notched bending specimen and the notch depth is obtained experimentally for dense alumina. Forth, the critical FPZ sizes are estimated for dense alumina and alumina-nickel nanocomposites. And finally, we clarify how to toughen brittle ceramics.

THEORY

Criteria for Crack Extension

For a crack in an infinite plate under mode I loading, the exact solution of the stress, σ_y, on the r-axis (ahead of the crack surface) is expressed as[7]

$$\sigma_y = \frac{\sigma(a_e + r)}{\sqrt{2a_e r + r^2}} = \frac{K_I}{\sqrt{2\pi r}} \frac{1 + r/a_e}{\sqrt{1 + r/2a_e}}, \qquad (1)$$

where σ represents the remote stress on the plate, a_e is the half crack length in an infinite plate, and $K_I = \sigma(\pi a_e)^{1/2}$ is the stress intensity factor (SIF). In the range of $r \ll a_e$, the following Irwin's expression is derived from Eq. (1).

$$\sigma_y \approx \frac{K_I}{\sqrt{2\pi r}}. \qquad (2)$$

Figure 1 shows the critical stress state around a long crack tip in an infinite plate, where K_I reaches the fracture toughness, K_{IC}. The solid curve in Fig. 1 indicates the exact stress distribution of σ_y on the r-axis shown by Eq. (1), the dotted curve represents Irwin's expression shown by Eq. (2), and r_0 is the critical FPZ size. In this case, the stress calculated by the exact stress distribution and the stress obtained from the Irwin's expression have almost the same value around r_0.

Griffith's energy equilibrium is, therefore, expressed using the Irwin's expression as

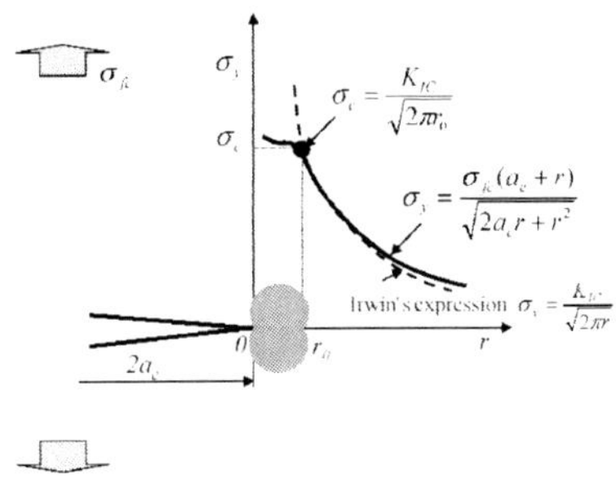

Fig. 1 Stress distribution ahead of a long crack in an infinite plate.

$$\frac{K_{IC}^2}{E'} = 2\gamma_I \ . \tag{3}$$

where the left side denotes the critical value of the maximum energy release rate and the right side is the fracture energy rate. $E' = E$ for a plane stress condition. $E' = E/(1 - \nu^2)$ for a plane strain condition. E is the Young's modulus. ν is the Poisson's ratio. and γ_I is mode I fracture energy per unit area of the crack surface.

Equation (3) is derived on the premise that the FPZ size is small enough to satisfy the condition of small scale yielding and indicates that the maximum energy release rate is determined by the SIF at the crack tip. which depends only on the outer boundary conditions such as the remote stress and the specimen geometry. as long as the condition of small scale yielding is applicable. This is called the K_{IC} criterion.

The local fracture criterion states that the crack will propagate when the stress at the characteristic distance from the crack tip reaches the critical local stress. σ_C. Therefore. if the characteristic distance is equal to the critical FPZ size. r_0. the following relation is derived.

$$\sigma_C = \frac{K_{IC}}{\sqrt{2\pi r_0}} \ . \tag{4}$$

Because K_{IC} and r_0 are material constants. σ_C becomes a material constant. too.

Next. we consider the short crack problem. where the crack length is not sufficiently long compared with the critical FPZ size. as shown in Fig. 2. In this case. the exact stress distribution and the Irwin's expression have different values at r_0. Therefore. we must use the exact stress distribution shown in Eq. (1) to estimate the critical local stress. The exact stress ahead of a crack tip in an infinite plate under a critical remote stress. σ_{fc}, can be expressed from Eq. (1) as

$$\sigma_y = \frac{\sigma_{fc}(a_e + r)}{\sqrt{2a_e r + r^2}} = \sigma_{fc} F_c(r) \ . \tag{5}$$

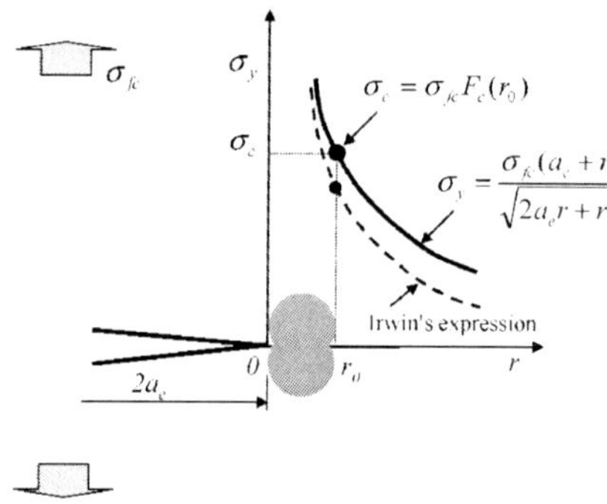

Fig. 2 Stress distribution ahead of a short crack tip in an infinite plate.

where F_c represents the non-dimensional stress distribution ahead of a crack tip in an infinite plate. The following relation is then derived at r_0.

$$\sigma_C = \sigma_{fc} F_c(r_0) \, , \tag{6}$$

where σ_C represents the critical local stress at the critical FPZ size ahead of a crack tip for any crack length in an infinite plate.

From Eqs. (5) and (6), the following relation is derived

$$\lim_{a_c \to 0} \sigma_{fc} = \lim_{a_c \to 0} \sigma_C \frac{\sqrt{2a_c r_0 + r_0^2}}{a_c + r_0} = \sigma_C \, , \tag{7}$$

which indicates that σ_C is the strength of an infinite plate with no artificial crack.

Stress Distribution Ahead of an Edge Crack Tip

We use a bending specimen instead of an infinite plate to estimate the critical FPZ size. The exact stress ahead of an edge crack is then needed to calculate the critical local stress for the bending specimen with an edge crack. Assuming the bending specimen with an edge crack, shown in Fig. 3, the imagined remote bending stress on the section of the edge crack plane is expressed as

$$\sigma(r) = \sigma_f \frac{W - a - 2r}{W - a} \, , \tag{8}$$

$$\sigma_f = \frac{3PS}{2W^2 B}$$

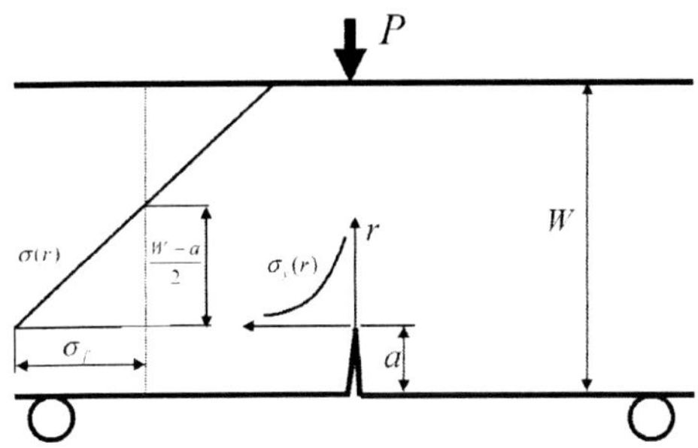

Fig. 3 Stress distribution and supposed remote stress in a bending specimen

where r represents the distance from the crack tip. σ_f is the maximum tensile stress of a rectangular shaped three-point bending specimen. W and B are the width and thickness of the specimen, respectively, a is the depth of the edge crack, P is the load, and S is the span length. The exact stress at r ahead of the edge crack tip is derived from Eqs. (5) and (8), using the superposition concept of linear elasticity as

$$\sigma_y(r) = \sigma(r)F_c(r) \ . \tag{9}$$

The critical local stress at r_0 is then derived as

$$\sigma_c = \sigma_{fc}\frac{W-a-2r_0}{W-a}F_c(r_0) \ . \tag{10}$$

Confirmation of Superposition Technique

To confirm the stress expression represented in Eq. (9) derived by the superposition concept, we analyzed the stress distribution ahead of the edge crack using a finite element

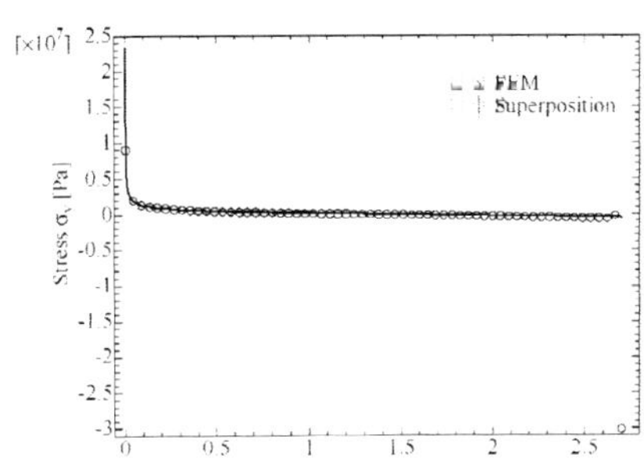

Fig. 4 Stress distributions from the crack tip to the loading point in a bending specimen.

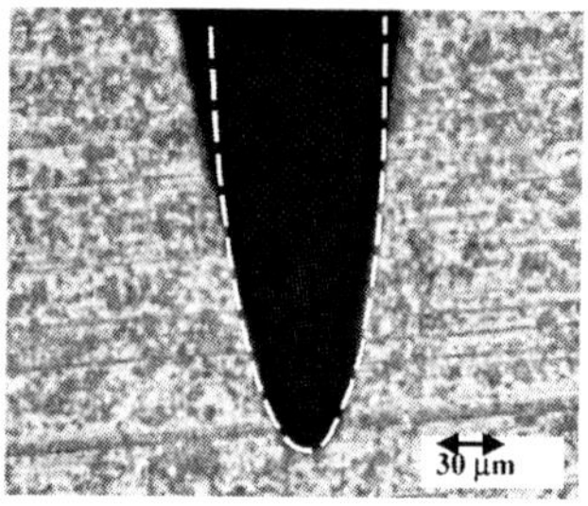

Fig. 5 Machined V-notch.

method (FEM) code. We used the ANSYS classic (ANSYS, Inc.). The size of the analyzed specimen was W = 4 mm, B = 3 mm, a = 1.3 mm, and S = 16 mm. The point load was 1 N. The smallest element around the edge crack tip was 10 μm square of 8-point iso-parametric elements.

Figure 4 shows a comparison between the superposition solution and the FEM results for the entire section of the specimen from crack tip to loading point, where some selected FEM results are shown in the figure. There is excellent agreement between the two except for the region near the loading point. This demonstrates that the proposed superposition technique does provide a good estimate of the exact stress values ahead of a crack tip in a wide range of the section of the bending specimen.

Stress Distribution Ahead of an Elliptical Notch

We used the single edge V-notched beam (SEVNB) method[7,8] to estimate the critical FPZ size of ceramics, where the V-notch radius is very sharp, around 20 μm as shown in Fig. 5. Because a machined notch can be simulated by an elliptical notch shown in Fig. 5, the stress distribution ahead of the V-notch tip can be estimated using the exact solution for an elliptical hole in an infinite plate analyzed by Williams[9]

$$\sigma_y = \frac{\sigma}{2}\left[1+\frac{2(1+f)}{\alpha^2-f}\right]+\frac{\sigma}{2}\left\{1+\frac{f^2-1}{\alpha^2-f}\left[1+\frac{f-1}{\alpha^2-f}\frac{3\alpha^2-f}{\alpha^2-f}\right]\right\}=\sigma F_e(r)$$
$$f=\frac{a_d-b}{a_d+b},\quad \alpha=\frac{r+a_d}{a_d+b}+\sqrt{\left(\frac{r+a_d}{a_d+b}\right)^2-f} \tag{11}$$

where σ_y is the stress distribution ahead of the elliptical hole tip on the r axis, σ is the remote stress, $2a_d$ and $2b$ are the lengths of the major axis and minor axis of the elliptical hole, respectively, and F_e is the non-dimensional stress distribution. Using the same technique described earlier, the critical local stress at r_0 can be derived for the bending specimen with an elliptical notch as

$$\sigma_c = \sigma_{fc}\frac{W-a_d-2r_0}{W-a_d}F_e(r_0)\ . \tag{12}$$

Calculation of Critical FPZ Size

Because the FPZ ahead of a crack tip in ceramics is considered to be composed of many microcracks rather than the kind of dislocations, the shape of the FPZ will be approximated by the area enclosed in the iso-stress contours of the maximum principal stress[10]. The stresses ahead of a crack under mode I loading is expressed

$$\sigma_{ij} = \frac{K_I}{\sqrt{2\pi r}} f_{ij}(\theta) \quad , (i, j = x, y) \ . \tag{13}$$

Then, the maximum principal stress is expressed

$$\sigma_1 = \frac{K_I}{\sqrt{2\pi r}} \cos\frac{\theta}{2}(1 + \left|\sin\frac{\theta}{2}\right|) \ . \tag{14}$$

And the radius of the iso-stress contours is derived as

$$r_1 = r_0 \cos^2\frac{\theta}{2}(1 + \left|\sin\frac{\theta}{2}\right|)^2 \ , \tag{15}$$

where r_0 represents r at $\theta = 0$ expressed as

$$r_0 = \frac{K_I^2}{2\pi\sigma_1^2} \tag{16}$$

The area enclosed in the iso-stress contours is then analyzed as

$$A = \frac{1}{2}\int_{-\pi}^{\pi} r_1^2 d\theta = r_0^2 \int_0^{\pi} \cos^4\frac{\theta}{2}(1 + \left|\sin\frac{\theta}{2}\right|)^4 d\theta \tag{17}$$

Equation (17) was calculated numerically. The results was

$$A = 3.29 r_0^2 \ . \tag{18}$$

EXPERIMENTAL PROCEDURE AND RESULTS

Materials

The materials used in this study were dense alumina[6] and alumina-nickel nanocomposites[11]. The first one was used to explain the relation between the strength of cracked specimen and crack length, and both were used to estimate the critical FPZ sizes and fracture toughness.

The polycrystalline alumina of 99.5 % purity with minor dopant of MgO was manufactured by Japan Fine Ceramics Center. The mean grain size was 2 μm with isotropic morphology and homogeneous microstructure. The fracture toughness measured using the SEPB (single-edge precracked beam) method was 3.65±0.11 $MPam^{1/2}$. Specimens 3×4×40 mm^3 in size were machined out, chamfered and finished by polishing with lapping.

Alumina-nickel nanocomposites were fabricated by our group. In order to produce intra-type nano-structure, we developed a soaking method[11]. Commercially available γ-alumina powder consists of column-like primary particles and has an agglomerated form with many nano-pores. The γ-alumina powder is soaked in nickel nitrate solution under vacuum. The γ-alumina-nickel solution composite powder is then filtered and dried, following calcinations to get nickel oxide particles in the nano-pores of the γ-alumina powder. The calcinated specimens are reduced under hydrogen atmosphere to obtain metallic nickel in the nano-pores. After sintering the powders, we obtained an intra-type nano-structure of alumina-nickel nanocomposites.

Commercially available γ-alumina powder (AKP-G015, Sumitomo Chemical, Japan) was used as a matrix and nickel nitrate with 99.9 % purity (Osaka Chemistry, Japan) as a source of nickel particles dispersed into the matrix. To improve sinterability of γ-alumina powder, we added 10 wt % of α-alumina (AKP-53, Sumitomo Chemical, Japan, average grain size 300 nm) as a seed. The mixtures were sintered by the pulse electric current sintering (PECS) method at 1250 to 1550 °C with soaking times for 5 min under 30 MPa in vacuum. The volume fraction of nickel was about 3 vol%. The specimens sintered at 1350 °C with seeds were heat-treated at 800 °C and kept for 0 to 10 min at several temperatures in order to diffuse dislocations into the matrix grains. Alumina monolith was also sintered by the PECS method at 1250 to 1450 °C.

Bending Tests for Dense Alumina

A three-point bending test was carried out on the specimens with various depths of notch machined using a sharp V-shaped diamond wheel, where the average root radius of the notch tip was about 20 μm. The SIF of the edge-cracked specimen was calculated using Wakai et al.'s equation[12]. All specimens were set as W = 3 mm (width) and B = 4 mm to compare the fracture strength of the specimens with and without the artificial notch to estimate the critical FPZ size, where the bending strength was measured using a JIS-type bending specimen.

Hereafter, we use an equivalent crack length[13] in an infinite plate to remove the influence of the shape factors, Y, on the fracture strength of the notched specimen. The equivalent crack length is necessary for determining the relationship between the crack length and strength of the notched specimen. The equivalent crack length, a_e, is defined in this case as

$$a_e = \frac{Y^2}{\pi} a , \tag{19}$$

where a represents an edge crack length and a_e is a half crack length in an infinite plate.

The experimental results are shown in Fig. 6, where the marks (□) indicate the average strength of the specimens with various depths of V-shaped notch (five specimens per each notch depth). The mark (■) indicates the average bending strength of the smooth-surface specimens (ten specimens), where the average strength was 462±18.0 MPa with the shape factor of the Weibull distribution m = 28, and the scale factor β = 471 MPa. It is observed from Fig. 9 that when the equivalent crack length is longer than about 100 μm, the strength values of the bending specimens with a notch lie on the straight line with -1/2 gradient, therefore, linear fracture mechanics is applicable in this range. By curve fitting, the fracture toughness was estimated to be 3.72 $MPam^{1/2}$, which is almost the same as the value of 3.65 $MPam^{1/2}$ obtained by the SEPB method previously mentioned.

Results for Nanocomposites

The strength was measured by a three-point bending test for specimens $2\times2\times10$ mm^3 in size and the fracture toughness was measured using a single edge V-notched beam (SEVNB) method. Table 1 shows the bending strength and fracture toughness of the alumina-nickel nanocomposites sintered with different temperatures, where the specimen sintered at 1,350 °C shows the highest fracture toughness of 5.49 $MPam^{1/2}$. To confirm the annealing effects after sintering[14], the specimens sintered at 1,350 °C were annealed under several conditions to disperse dislocations into matrix grains. The results were shown in Table 2, where the specimen annealed at 800 °C for 5 min soaking time shows quite high fracture toughness of 7.59 $MPam^{1/2}$ with high strength of 888 MPa. The fracture toughness of 7.59 $MPam^{1/2}$ is more than two times higher than that of dense alumina.

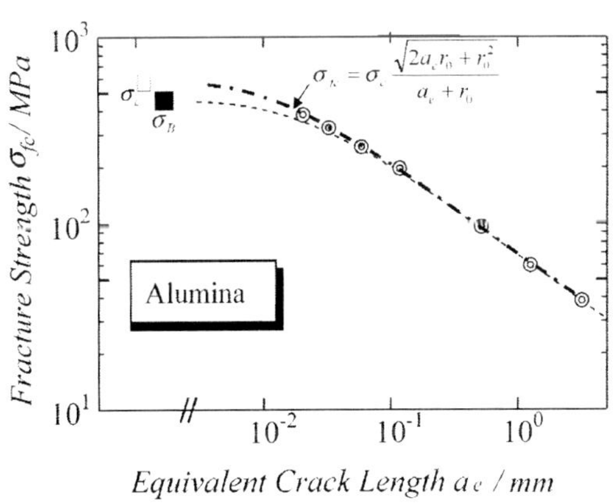

Fig. 6 Experimental result of the bending strength for the specimen with and without a V-notch.

Table 1 Experimental data of alumina-nickel nanocomposites.

Sintering temp. / C	Bending strength / MPa	Fracture toughness / MPam$^{1/2}$	Critical FPZ size, r_0 / μm
1,250	928	4.40	1.9
1,350	877	5.49	3.8
1,450	984	4.49	1.7
1,550	856	4.07	1.9

DISCUSSION

Estimation of Critical Local Stress

The value of the critical local stress, σ_c, is estimated in relation to the size effect on the strength of the bending specimen, σ_B, and the critical FPZ size. The difference between the strength σ_B and the critical local stress, σ_c, is expressed as[15]

$$\frac{\sigma_c}{\sigma_B} = \left[\frac{V_{e3}}{V_{FPZ}}\right]^{1/m} \cdot \frac{S_2}{S_1}, \tag{20}$$

where V_{e3} represents the effective volume of the three-point specimen, V_{FPZ} is the effective volume of the critical FPZ. Because the FPZ is approximated by the area enclosed in the iso-stress contours of the maximum principal stress, the stress state in the FPZ is considered to be biaxial stress state. The term, S_2/S_1, expresses the biaxial effect, where S_1 is the strength of the specimen under a uni-axial stress state, and S_2 is the strength under a biaxial stress state. The ratio S_2/S_1 is approximated to be 1 in this case.

The V_{e3} is expressed as

$$V_{e3} = \frac{V}{2(m+1)^2}, \tag{23}$$

where $V = H \times B \times S$ (H: height, B: breadth, and S: span length) is the volume of the three-point bending specimen. The V_{FPZ} is derived from Eq. (18) as

$$V_{FPZ} = 3.29 r_0^2 \times B. \tag{24}$$

The r_0 value is calculated by means of trial and error from Eqs. (20) and (4), where r_0 is firstly assumed.

Table 2

Annealing temp. / C	Soaking time / min	Bending strength / MPa	Fracture toughness / $MPam^{1/2}$	Critical FPZ size, r_0 / μm
800	0	863	6.79	6.8
	5	888	7.59	8.2
	10	561	5.16	10.0
900	0	548	4.56	7.8
	5	569	4.69	7.6
	10	555	4.37	6.8

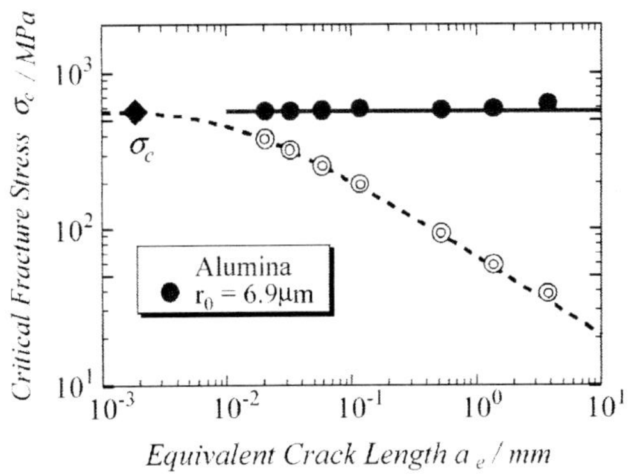

Fig. 7 Critical local stresses calculated for each equivalent crack length.

The critical FPZ size calculated from Eq. (4) for polycrystalline ceramics indicates the macroscopically defined value, because the crack does not extend uniformly along the leading edge of the crack. Some parts of the leading edge will progress earlier while others progress later depending on the microstructure (crystal orientations, grain boundaries, and defects). Several researchers have tried to observe the FPZ size directly using a transmission electron microscope (TEM).[16] However, very little dislocation activities has been observed on the growing crack surface, because of the specimen thickness effect associated with TEM observations. Therefore, dislocations move easily through the thickness of the very thin specimen and disappear. Accordingly the critical FPZ size calculated from Eq. (4) using the macroscopically defined fracture toughness and the strength of the materials is also considered to be the macroscopic value.

The critical local stress can be estimated for each notched specimen. The results for alumina are shown in Fig. 7. All the critical local stresses indicate the same value and the critical FPZ size for alumina was estimated to be 6.9 μm from Eqs. (20) and (4). The experimental results for annealed nanocomposites are summarized in Table 2.

From Eq. (4) for the long crack problem, the following relation is derived:

$$K_{IC} \propto \sigma_c \cdot r_0^{1/2} \approx \sigma_B \cdot r_0^{1/2} . \qquad (23)$$

This equation indicates that the materials with higher values of $\sigma_B \cdot r_0^{1/2}$ have higher fracture toughness. Therefore, to enhance the fracture toughness of ceramics, we need to improve both the strength and the critical FPZ size. Figure 8 shows the relation between the fracture toughness and the values of the $\sigma_B \cdot r_0^{1/2}$. Both nanocomposites have higher values of fracture toughness than that of dense alumina, and the annealed nanocomposite has the highest fracture toughness, and there is almost a linear relation between the fracture toughness and $\sigma_B \cdot r_0^{1/2}$.

CONCLUSIONS

The relation between the strength, the fracture toughness, and the critical frontal process zone size was clarified using dense alumina. Toughened alumina-nickel nanocomposites were fabricated using a soaking method and we found that a suitably annealed material showed very high fracture toughness which material had the largest critical frontal process zone size.

REFERENCES

[1]K.P.D. Lagerlöf, A.H. Heuer, J. Castaing, and J.P. Rivière, "Slip and twinning in sapphire (α-Al_2O_3)," *J. Am. Ceram. Soc.*, **77**, 385-97 (1994).

[2]R.G. Hoagrand and J.D. Embury, "A treatment of inelastic deformation around a crack tip due to microcracking," *J. Am. Ceram. Soc.*, **63**, 404-10 (1980).

[3]A.G. Evans and K.T. Faber, "Toughening of ceramics by circumferential microcracking ," *J. Am. Ceram. Soc.*, **64**, 394-98 (1981).

[4]H. Awaji, S-M. Choi, and E. Yagi, "Mechanisms of toughening and strengthening in ceramic-based nanocomposites," *Mech. of Mater.*, **34**, 411-22 (2002).

[5]K. Niihara, "New design concept of structural ceramics," *J. Ceram. Soc. Japan*, **99**, 974-82 (1991).

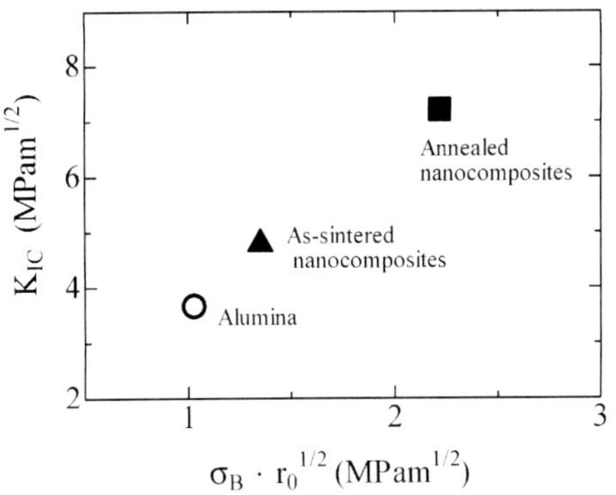

Fig. 8 Relation between the fracture toughness and $\sigma_B \cdot r_0^{1/2}$.

[6]H. Awaji, S-M. Choi, and D.D. Jayaseelan, "Indirect estimation of critical frontal process-zone size using a single-edge V-notched-beam technique," *J. Ceram. Soc. Japan*, **109**, 591-95 (2001).

[7]H. Awaji, T. Watanabe, and Y. Sakaida, "Fracture toughness measurements of ceramics by V notch technique," *Ceram. Int.* **18**, 11-17 (1992).

[8]H. Awaji and Y. Sakaida, "V-notch technique for single-edge notched beam and chevron notch methods," *J. Am. Ceram. Soc.*, **73**, 3522-23 (1990)

[9]J.G. Williams, Stress analysis of polymers, Longman, London, (1973)

[10]H. Awaji and T. Kato, "Criterion for combined mode I-II brittle fracture," *Materials Trans. JIM.*, **40**, 972-79 (1999).

[11]T. Matsunaga, U. Leela-adisorn, Y. Kobayasi, S-M. Choi, and Awaji, "Fabrication of alumina-based toughened nanocomposites," *J. Ceram. Soc. Japan*, **113**, 123-25 (2005).

[12]F. Wakai, S. Sakaguchi, and Y. Matsuno, "Calculation of stress intensity factor for SENB specimens by boundary collocation procedure," *Yo-gyo Kyokai-shi.*, **93**, 479-80 (1985).

[13]S. Usami, H. Kimoto, I. Takahashi, and S. Sida, "Strength of ceramic materials containing small flaws," *Engng.Fract. Mech.*, **23**, 745-61 (1986)

[14]S-M. Choi and H. Awaji, "Nanocomposites-a new material design concept," *Sc. And Tech. of Adv. Mater.*, **6**, 2-10 (2005).

[15]H. Awaji, Strength of Ceramic Materials, Corona Pub., Tokyo (2001).

[16]Y. Ikuhara, T. Suzuki, and Y. Kubo, "Transmission electron microscopy in situ observation of crack propagation in sintered alumina," *Phil. Magazine Letters*, **66**, 323-27 (1992).

CRACK-HEALING + PROOF TEST: A NEW METHODOLOGY TO GUARANTEE THE RELIABILITY AND THE QUALITY OF A CERAMIC COMPONENT

Masato Ono, Wataru Nakao, Koji Takahashi, Kotoji Ando
Yokohama National University
79-5 Tokiwadai, Hodogaya-ku, Yokohama, 240-8501, Japan

Masahiko Nakatani
NHK SPRING Co. Ltd.
3-10 Fukuura, Kanazawa-ku, Yokohama, 236-0004, Japan

ABSTRACT

Combination of crack-healing and proof-test is proposed as a new methodology to guarantee the reliability of a ceramics component, because crack-healing and proof test are effective techniques to overcome surface cracks and embedded flaws, respectively. In this study, the effect of crack-healing on the Weibull distribution was investigated for alumina/SiC particle composite, which has excellent crack-healing ability. The composite having a semi-elliptical surface crack was heat-treated at 1573 K for 1 h in air. Characteristic strength, which is nominal fracture stress obtained using two-parameter Weibull distribution, of the crack-healed specimens was increased 1.4 times and 2.4 times that of as-received and as-cracked specimens, respectively. Moreover, the minimum fracture stress guaranteed at high temperature for the specimens proof-tested at R.T. after crack-healing was estimated by using the measured temperature dependences of fracture stress and toughness. The fractures stress of the proof-tested specimens at elevated temperature was also measured. From this result, the minimum fracture stress at high temperature was determined. The calculated minimum fracture stress agreed well with the values measured by experiments. Therefore, it is confirmed that the combination of crack-healing and proof-test is the effective technique to guarantee the fracture stress at high temperature of the ceramics component.

INTRODUCTION

Structural ceramics such as alumina, mullite, and silicon nitride exhibit excellent heat corrosion and wear resistance. However, they are brittle and sensitive to flaws because of their low fracture toughness. The flaws in ceramics are classified into propagated surface cracks in machining and embedded flaws. In generally, the bending strengths of ceramic components were significantly reduced by surface cracks. Thus, it is necessary to conduct an accurate machining for improvement in the reliability of structural ceramics. As a result, the machining cost mounts but the efficiency is low.

Ando *et al.* have progressed in development of the structural ceramics with crack-healing ability by using oxidation of SiC[1]. When ceramics manufactured with SiC are heat treated in air at high temperature, SiC located on the crack surface reacts with O_2 in air. Then, the crack is completely sealed by the oxidation products. Moreover, the restored parts are mechanically stronger than the other parts. If crack-healing ability is used on structural components in engineering use, great benefits can be anticipated to improve reliability as well as to decrease machining and polishing costs of ceramics elements. However, the embedded flaws cannot be healed. For improvement in reliability of structural ceramics, thus, the proof test is an effective technique to overcome the embedded flaws.

Therefore, the present authors expect the combination of crack-healing and proof-test as a new methodology to guarantee the reliability of a ceramics component. For this purpose, the effect of crack-healing on the Weibull distribution was investigated using alumina/SiC particle composite, which has excellent crack-healing ability. Moreover, the minimum fracture stress, which is guaranteed at high temperature for the specimens proof-tested at R.T. after crack-healing process, was estimated by using the measured temperature dependences of fracture stress and toughness. The fractures stress of the proof-tested specimens was also measured at elevated temperatures. From the obtained results, the validity of the estimated minimum fracture stress was discussed.

PRINCIPLE

In most ceramics, the fracture strength has negative temperature dependence. Thus, the minimum fracture stress guaranteed at high temperature, σ_G, can be lower than the proof test stress at R.T.

In the previous study, Ando *et al.* have proposed the estimation method of σ_G [2]. Fracture of ceramics is of nonlinear behavior. The nonlinear fracture behavior is expressed as Eq. (1) proposed by Ando *et al.*,

$$D_C = \frac{\pi}{8}\left(\frac{K_{IC}}{\sigma_0}\right)^2 = a_e\left\{\sec\left(\frac{\pi\sigma_C}{2\sigma_0}\right) - 1\right\} \qquad (1),$$

where D_C is the limit process zone size. σ_C and σ_0 are the fracture stress of the as-cracked specimen and the plain specimen, respectively. K_{IC} and a_e is the plane strain fracture toughness and the equivalent crack length, respectively.

Figure 1 shows a schematic diagram of the relation between σ_C and a_e. The dotted line and the solid line show the relationship between σ_C and a_e at R.T. and high temperature, respectively. If the proof test is carried out at R.T., the maximum in the residual equivalent crack size, a_e^R, can be expressed as following Eq. (2),

$$a_e^R = \frac{\pi}{8}\left(\frac{K_{IC}^R}{\sigma_0^R}\right)^2 \left\{\sec\left(\frac{\pi\sigma_p^R}{2\sigma_0^R}\right) - 1\right\}^{-1} \qquad (2),$$

where σ_p^R is the proof test stress at R.T.. and the superscript R indicate the value at R.T..

Since embedded flaws can not be healed, the value of a_e^R is not changed in spite of heating to high temperature. σ_G can be expressed as following by expanding Eq. (2) about R.T. and high temperature,

$$\sigma_G = \frac{2\sigma_0^T}{\pi} \arccos\left\{ \left(\frac{K_{IC}^T}{K_{IC}^R} \right)^2 \left(\frac{\sigma_0^R}{\sigma_0^T} \right)^2 \left\{ \sec\left(\frac{\pi\sigma_p^R}{2\sigma_0^R} \right) - 1 \right\} + 1 \right\}^{-1} \quad (3),$$

where the superscript T indicates the value at elevated temperature. Obtaining the temperature dependence of K_{IC} and σ_0, one can estimate σ_G.

EXPERIMENTS

Material preparation

Alumina powder (AKP 30, Sumitomo Chemical Co. Ltd., Japan) used in this study has an average particle size of 0.3 μm. SiC powder (Ultrafine, Ibiden Co. Ltd., Japan) used has an average particle size of 0.27 μm. The mixture of alumina powder and 20 vol% SiC powder was blended well in alcohol for 48 h. Then, the slurry was dried. Rectangular plates of 90 mm × 90 mm × 7 mm were sintered in N_2 gas for 2 h via hot press under 35 MPa at 1923 K. The sintered plates were cut into the 3 mm × 4 mm × 36 mm rectangular specimens bar that were polished to mirror finished on one face, as shown in Fig. 2(a). This study includes investigating effect of proof test on embedded flaws as one of the purposes. Thus, many embedded flaws were consciously introduced to the test specimens by abating the sintering temperature from the optimized temperature.

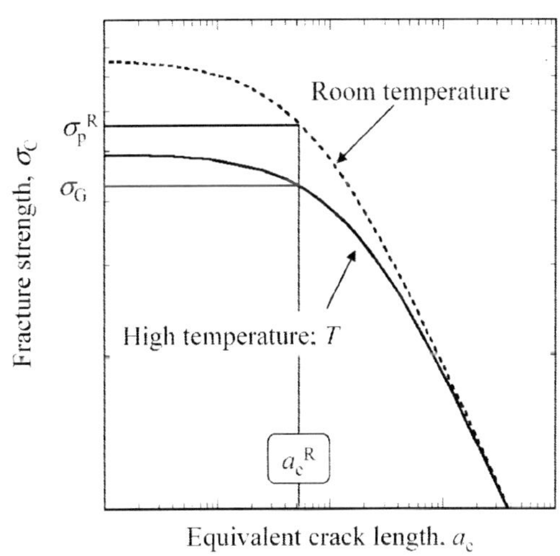

FIGURE 1 The effect of equivalent crack length on fracture strength at room temperature and high temperature.

Experimental procedure
Crack-healing treatment

A semi-elliptical surface crack of 100 μm in surface length was made at the center of the tensile surface of specimens with a Vickers indenter using a load of 19.6 N. The ratio of depth (a) to half the surface length (c) of the crack, aspect ratio, was $a/c = 0.9$. The specimens were crack-healed in air at 1573 K for 1 h.

Proof tests

The crack-healed specimens were subjected to the proof-test at R.T. The proof test stress was adapted to be 435 MPa and 530 MPa, corresponding to the values under which that 30 % and 60 % crack-healed specimens were fractured, respectively. The cross-head speed in loading the stress of the proof test was 0.5 mm/min and the proof test stress was not held. The specimens survived through the proof test are called shortly the proof-tested specimens in this paper.

Fracture testing

All fracture tests were conducted using a three point loading system with a span of 30 mm as shown in Fig. 2 (a). Test temperatures are R.T. and high temperatures from 873 K to 1373 K. Fracture toughness was measured at R.T and high temperatures from 873 K to 1473 K by SEVNB method. The V notch shape; depth = 1.2 mm, notch angle = 20°, notch root radius = 0.4 mm is shown in Fig. 2(b).

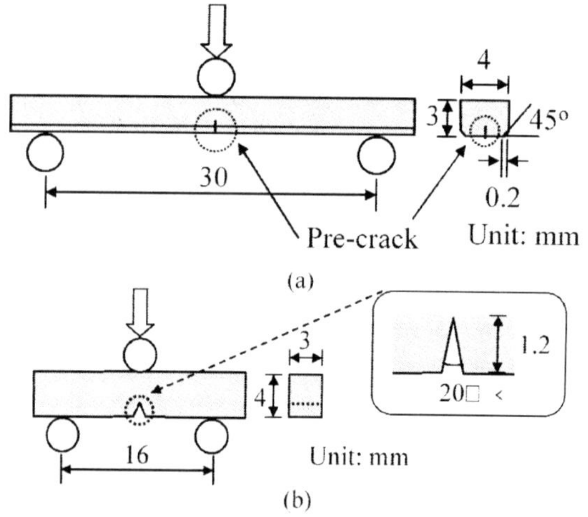

FIGURE 2 (a) Three-point loading system and geometry of test specimen. (b)Three-point loading system and geometry of the single edge V-notched beam specimen.

RESULTS AND DISCUSSIONS

Mechanical property

Figure 3 shows the Weibull plot of the fracture stress, s, corresponding to the bending strength, at R.T., where X axis is the $\ln(s)$ and Y axis is the $\ln(\ln(1/(1-F(s)))$; $F(s)$ is the fracture probability obtained by the median rank method. The open square indicates that of the crack-healed specimens. Also the open triangle and circle indicate that of the as-cracked specimen and the specimens that are made pursuant to JIS standard, abbreviated smooth specimens, respectively.

The dispersion of the fracture stress was expressed as two-parameter Weibull distribution, as shown in Eq. (4),

$$F(s) = 1 - \exp\left\{-\left(\frac{s}{\beta}\right)^m\right\} \qquad (4),$$

where m and β are Weibull modulus and characteristic strength, respectively. The data sets were subjected to a least square fitting using Eq. (4). From the obtained results, the values of β and m of the crack-healed specimens, the as-cracked specimens, and the smooth specimens were illustrated in Fig. 3.

The Weibull modulus m of the crack-healed specimen was smaller than that of the as-cracked specimen. Figure 4 (a) and (b) show the typical fracture initiations of the as-cracked and the crack-healed specimens, respectively. All as-cracked specimens fractured across the Vickers indentation, while fracture of all crack-healed specimens occurred outside of the crack-healed zone. Therefore, since the embedded flaws as the fracture initiation of the crack-healed specimens had many variable sizes, the fracture stresses exhibit a large scatter. For improvement in the reliability and the quality of the structural ceramics, thus, it is necessary to remove the specimens including large embedded flaws by proof test, even if surface cracks were completely eliminated by crack-healing.

The characteristic strength of the crack-healed specimens is higher value than that of as-cracked specimens as well as that of the smooth specimens, because cracks propagated during machining, which existed in the smooth specimens, were also crack-healed completely. As an example of this behavior, consider that the smooth specimens and the as-cracked specimens were subjected to the tensile stress of 435 MPa, which correspond to $F(s)$ = 30 % in the Weibull distribution of the crack-healed specimens. 80 % smooth specimens must fracture, and all as-cracked specimens must fracture. Therefore, the crack-healing have contributed to the improvement in the working stress of the structural ceramics.

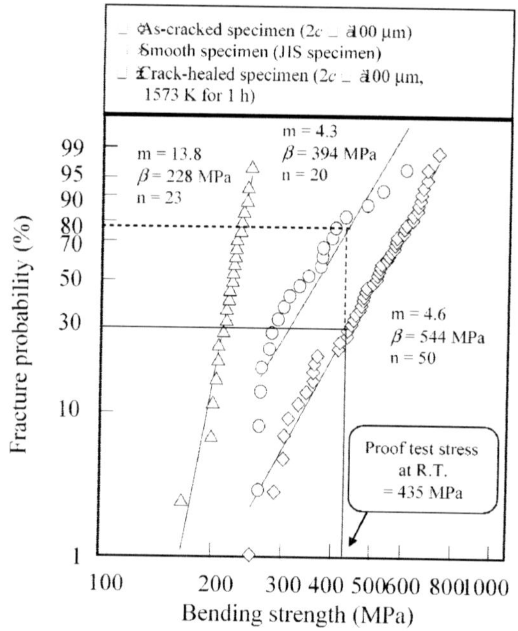

FIGURE 3 Effect of crack-healing on fracture probability.

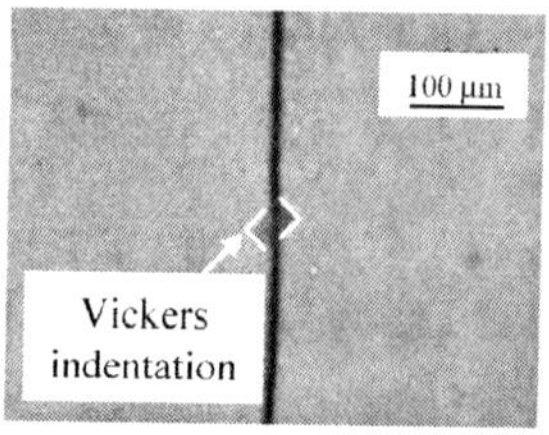

(a)

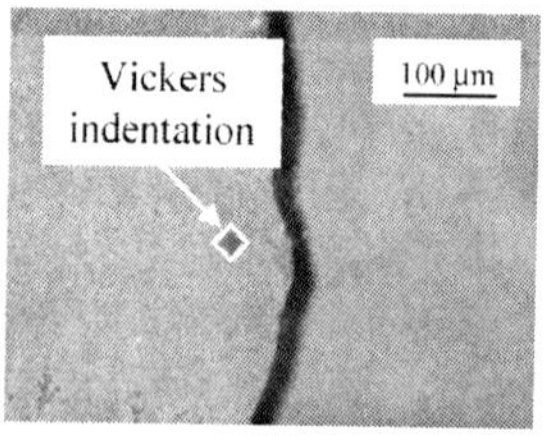

(b)

FIGURE 4 Photograph of fracture pattern: (a) specimen fractured from the Vickers indentation. (b) Fracture occurred outside of the crack-healed zone.

Temperature dependence of minimum fracture stress guaranteed

Figure 5 shows the temperature dependences of the intrinsic bending strength, σ_0, and the fracture toughness, K_{IC}, required for the estimation of the minimum fracture stress guaranteed, $\sigma_G{}^T$. The values of σ_0 were defined as the average of 5 % highest strengths at these temperatures. The temperature dependence of σ_0 was found to be almost linear and negative up to 1373 K. K_{IC} at R.T. is about 3.2 MPa·m$^{1/2}$ and is almost constant from 873 K to 1473 K, as 2.6 MPa·m$^{1/2}$.

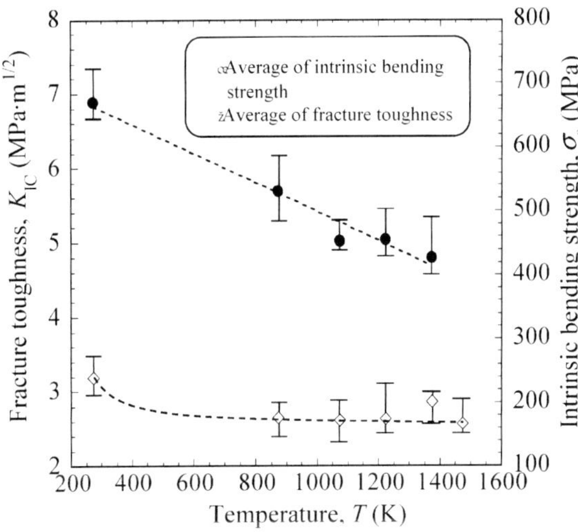

FIGURE 5 Temperature dependence of intrinsic bending strength and fracture toughness.

Figure 6 (a) shows the temperature dependence of σ_G evaluated in the case $\sigma_p{}^R$ = 435 MPa as solid line, in which the fracture stresses of the crack-healed specimens and the proof-tested specimens, $\sigma_p{}^R$ = 435 MPa, were also shown as the open and closed diamonds, respectively. At 873 K, the minimum value of the measured fracture stress of the crack-healed specimen was smaller than that of the proof-tested specimen. Clearly, the specimens including large embedded flaws could be removed by proof test at R.T. Moreover, the fracture stress of the proof-tested specimen was confirmed to have negative temperature dependence. All specimens except one tested at 1373 K have the higher fracture stress than $\sigma_G{}^T$ at the representative temperature.

Figure 6 (b) shows the same temperature dependences in the case $\sigma_p{}^R$ = 530 MPa. As shown in this figure, this estimation was accepted to be correct in spite of using the different proof test stress. Figure 7 shows the same temperature dependences in the case crack-healing condition is temperature of 1373 K and time of 50 h. Also this estimation was accepted to be correct in spite of using the different crack-healing condition.

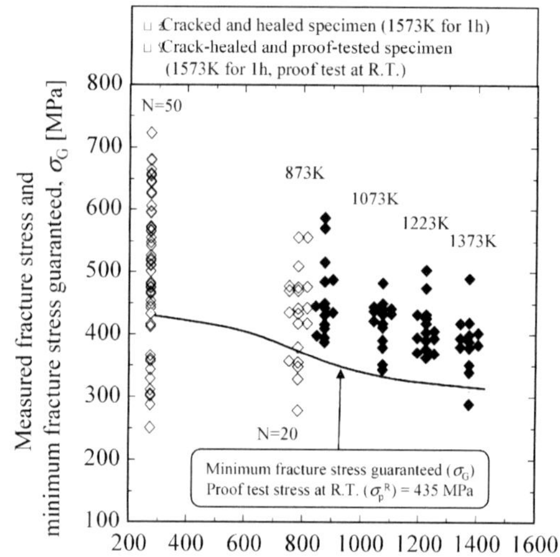

(a); σ_p^R = 435 MPa

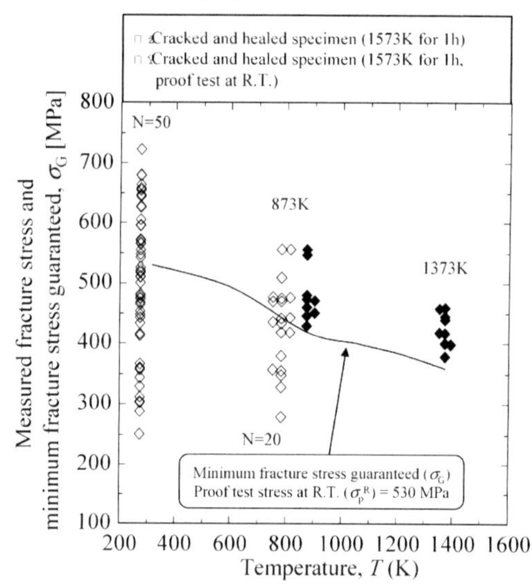

(b); σ_p^R = 530 MPa

FIGURE 6 The comparison between minimum fracture stress guaranteed and measured fracture stress as a function of test temperature; (a) Proof test stress at R.T. = 435 MPa, (b) Proof test stress at R.T. = 530 MPa.

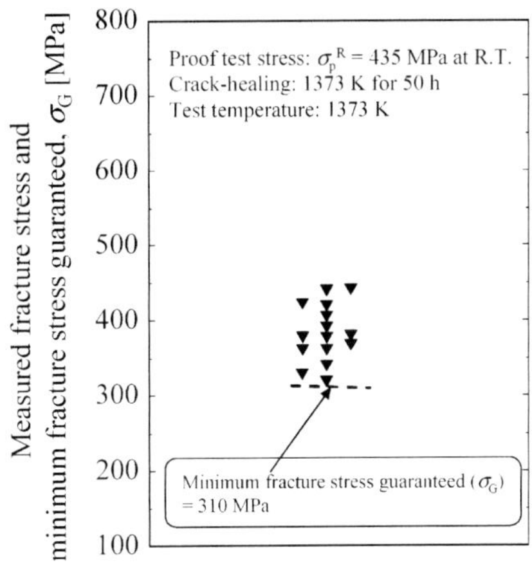

FIGURE 7 The comparison between minimum fracture stress guaranteed and measured fracture stress at 1373 K.

On the other point of view, it is interesting to investigate whether this estimation is reversible or not. Thus, the fracture stresses were investigated at R.T. for the specimens proof-tested at 1073 K, where the proof test stress was the minimum fracture stress guaranteed at 1073 K. The obtained results were shown in fig. 8. The minimum value of the measured fracture stress of the proof-tested specimens agreed well with the objective stress of 435 MPa.

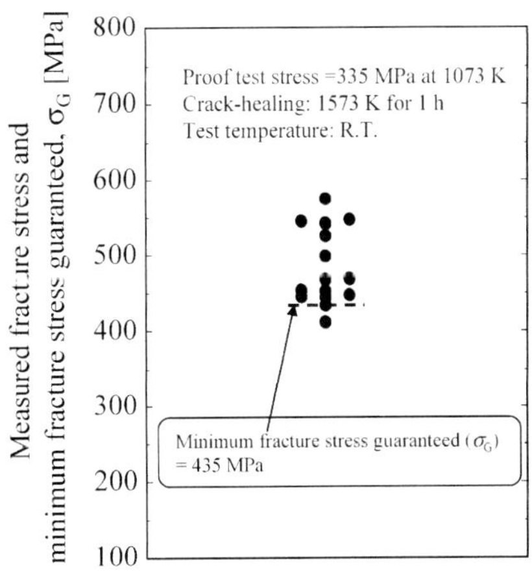

FIGURE 8 The comparison between minimum fracture stress guaranteed at R.T. and measured fracture stress at R.T. (Proof test stress; σ_G = 335 MPa at 1073 K).

CONCLUSIONS

Combination of crack-healing and proof-test is proposed as a new methodology to guarantee the reliability of a ceramics component. Moreover, this method was expanded to high temperature.

The characteristic strength of the crack-healed specimens is higher than that obtained for the as-cracked specimens as well as that of the smooth specimens, because cracks introduced by indentation method and propagated during machining were crack-healed completely.

The estimated minimum fracture stress agreed with the minimum value measured for the proof-tested specimens at several temperatures. Moreover, even if the proof test stress and the crack-healing condition differ, this estimation was proved to be correct.

It is concluded that the combination of crack-healing and proof-test is the effective technique to guarantee the fracture stress at high temperature of the ceramics component.

REFERENCES

[1]M. C. Chu, S. Sato, Y. Kobayashi and K. Ando, "Damage Healing and Strengthening Behavior in Intelligent Mullite/SiC Ceramics," *Fatigue Fract. Engng. Mat. Struct.*, **18** , 1019, (1995).

[2]K. Ando, Y. Shirai, M. Nakatani, Y. Kobayashi and S. Sato, "(Crack-healing + Proof test): a new methodology to guarantee the structural integrity of a ceramics component," *J. Euro. Ceram. Soc.*, **22**, 121, (2002).

DEVELOPMENT OF ELEVATED FRACTURE TOUGHNESS TESTING METHOD STANDARD FOR ADVANCED CERAMICS

Michael G. Jenkins
University of Detroit Mercy
Department of Mechanical Engineering
4001 W McNichols Road
Detroit, MI 48221 USA

Jonathan A. Salem
NASA-Glenn Research Center at Lewis Field
Life Prediction Branch
21000 Brookpark Road
Cleveland, OH 44135 USA

ABSTRACT

Since 1999, several national (e.g., ASTM, JIS) and international (e.g., ISO, CEN) full-consensus fracture toughness test method standards for advanced ceramics have been approved. Currently, these standards employ three different types of notch/crack geometries (SEPB, SCF and CNB) and have been verified at ambient temperature using certified SRMs. However, while these standards provide baseline information on the fracture resistance of advanced ceramics, many future applications (e.g., gas turbines, solid oxide fuel cells, etc.) for ceramics require information on fracture resistance at elevated temperatures. Various notch/crack geometries and their applicability for elevated fracture toughness are reviewed and evaluated. Limited experimental data are presented to illustrate the applicability of these methods.

INTRODUCTION

Catastrophic failure is often the characteristic behavior of ceramics in the presence of cracks or crack-like flaws under ambient conditions. Ceramics and glasses (as well as some hardened metal alloys such as bearing or brake steels) are examples of engineering materials that behave as brittle solids under ambient conditions. Of course, under non ambient conditions materials that normally fail in a ductile manner may fail in a brittle manner (e.g., carbon steel at temperatures less than their nil-ductility) and those materials that normally fail in a brittle manner may exhibit nonlinear behaviour and fail in a ductile manner (e.g., ceramics containing glassy secondary phases at temperatures greater than the glass softening temperature.). Because catastrophic failure occurs without warning and can occur in any engineering material under the "suitably-wrong" conditions, it is important to characterize the fracture behavior of brittle materials such as ceramics and glasses in order to produce engineering designs that can accommodate this phenomenon. Therefore, the potential dire consequences of catastrophic fracture under design conditions requires an understanding of the fracture of ceramic and glasses not only under ambient conditions but also at higher temperatures where many of their design applications lie.

Although Jenkins and Salem[1] provided a relatively recent detailed overview of the background, implementation and application of fracture resistance testing of brittle solids, nearly all of the standardized methods [2-7] for determining fracture resistance of ceramics in place at present are related to room-temperature conditions. At the time of the article, they noted that part of the debate surrounding the development of various test methods for determining the fracture resistance of brittle ceramics was (and still is) whether the test method should measure the intrinsic fracture resistance, R_o, or the fully-developed fracture resistance R_∞ To address this, the operational aspects of the test method must take into account intrinsic versus extrinsic flaws (or cracks), the degree of crack extension prior to measurement of the fracture resistance, the resulting linearity (or nonlinearity) of the loading curve, features on the fracture surfaces, and other clues indicating brittle or non brittle fracture. Typically, brittle materials exhibit low values of fracture toughness ($K_{Ic} = R_o$)* such as <1 MPa √m for many glasses and <10 MPa √m for many monolithic ceramics. These low fracture toughnesses combined with moderate strengths mean that from a Griffith approach, intrinsic flaw sizes may only be on the order of 1 to 50 μm which often lead to low, broad strength distributions. From a fracture mechanics approach, low fracture toughness combined with very short detectable macro crack sizes severely limit the design stress allowables in components comprised of brittle materials.

Therefore, the real utility of measuring the fracture resistance of brittle materials may not be in its direct application to crack or flaw-based design (e.g., when does $K_I = K_{Ic}$?) through either Griffith or fracture mechanics approaches. Instead, measures of fracture resistance, (e.g., K_{Ic}) may find their greatest utility when combined with other material properties in relative comparison indices such as those that describe wear or brittleness. For example, one possible proposed abrasive wear model [8,9] for ceramics is:

$$\dot{V} \propto P_{1.5} K_c^{-0.5} H^{-0.625} \left(\frac{E}{H}\right)^{0.8} \tag{1}$$

where V is the volume loss rate of material, P is the applied force, K_c is the fracture resistance, H is the hardness and E is the elastic modulus. Although the exponents in Equation 1 are specific to the application and material, the model shows that to minimize abrasive wear at high forces, P, the material must possess high K_c, high H, and low E.

As another example, a possible proposed erosive wear model for ceramics has the form:[10,11]

$$\varepsilon_v \propto v^2 \frac{\pi d^2}{12} \left(E\rho / K_c^2\right) \tag{2}$$

where ε_v is volume of material removed, v is impact velocity of the particle with diameter d, E is elastic modulus, ρ is density of the particle, and K_c is fracture resistance. For high velocities of impacting particles, the erosive wear of the material is minimized by high Kc, low E and low ρ.

One last example is the susceptibility of a material to cracking rather than deformation under the action of a concentrated stress can be defined in terms of its brittleness:[12, 13] B =

$$B \propto \left(\frac{HE}{K_c^2}\right) \tag{3}$$

where B is the brittleness. In this case, in order to decrease the tendency to brittle behavior (i.e.,

* Note although the term K_{IC} as applied in this discussion s often used to denote fracture toughness without regard to how it is measured, ASTM E399 "Standard Test Method for Plane Strain Fracture Toughness of Metallic Materials" specifically reserves the use of the term to refer to the plane strain fracture toughness of the material. Other symbols (e.g, K_{Ipb} for SEPB, K_{Ivb}, for CNB and K_{Isc} for SCF, etc) are used to more correctly denote operationally specific measures of fracture resistance (see ASTM C1421, ref 2).

cracking rather than plastic deformation), B needs to be decreased by increasing H, E, and K_{Ic}.

These examples of relative comparison indices show that quantifying the fracture resistance of ceramics has direct and important applications other than those involving conventional fracture mechanics. In addition, as indicated in Equations 1 to 3, depending on the application, the exponent on the fracture resistance term can have a significant influence on the resulting value of the index being calculated. Thus, accurate and precise measures of the fracture resistance of brittle materials is of paramount importance for making sensible predictions of material behavior using relative comparison indices. Given that the current standardized methods for determining fracture toughness of ceramics are room temperature specific whereas many of the applications of advanced ceramics are at elevated temperatures, a need exists to develop a test method for the determining the elevated fracture toughness of advanced ceramics.

In this paper, the salient aspects of room temperature test methods are first presented. Then the necessary requirements for ensuring acceptable elevated temperature results are discussed and a proposed elevated temperature fracture toughness test method for ceramics is detailed. Finally, examples of the application of the test method are provided and a summary/conclusion is given.

ROOM TEMPERATURE FRACTURE TEST METHODS

Standardized methods

As of mid-2005 there are five national and international consensus standards for fracture toughness testing ceramics, all at room temperature: ASTM C1421[2], JIS1607[3], ISO15732[4], ISO18756[5], ISO24370[6]. A sixth international document on fracture toughness testing of ceramics is currently in draft form, CEN TC184 Draft.[7] Some salient aspects of each of these standards are discussed in the following.

ASTM C1421 "Standard Test Methods for Determination of Fracture Toughness of Advanced Ceramics at Ambient Temperature,"[2] includes three different test methods in one standard: single edge pre-cracked beam (SEPB), surface crack in flexure (SCF) and chevron notch beam (CNB) which are illustrated in Figure 1. Technical and historical summaries of ASTM C1421 have been published previously.[13-15]

JIS 1607 "Testing Methods for the Fracture Toughness of High Performance Ceramics,"[3] includes two test methods in one standard: SEPB and indentation fracture (IF) which are illustrated in Figure 2. Discussion of JIS 1607 and its similarities and differences to other standards is contained in Ref 1. It should be noted that while the SEPB method contained in JIS107 is similar to other standards (e.g., ASTM 1421 and ISO15732) the IF method is not accepted in any other national or international standard. There is considerable controversy surrounding the result of the IF method (including JIS1607 that states that the result of the IF method cannot be strictly considered a fracture toughness).

ISO15732 "Fine ceramics (advanced ceramics, advanced technical ceramics)-Test Method for Fracture Toughness of Monolithic Ceramics at Room Temperature by Single Edge Pre-Cracked Beam (SEPB) Method"[4] includes only one method, SEPB, that is a harmonization of SEPB methods contained in ASTM C1421 and JIS 1607.

ISO 18756 "Fine ceramics (advanced ceramics, advanced technical ceramics) - Determination of fracture toughness of monolithic ceramics at room temperature by the surface crack in flexure (SCF) method,"[5] includes only one method, SCF and primarily reflects details of the SCF method in ASTM C1421.

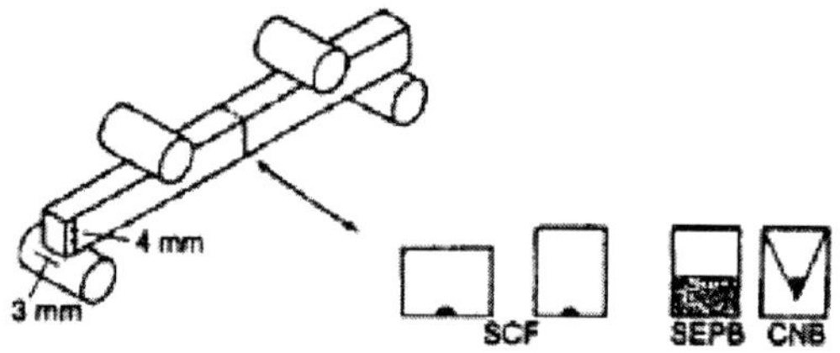

Figure 1 Three test method configurations (SCF, SEPB, CNB) contained in ASTM C1421 (Ref 2)

ISO 24370, "Fine ceramics (advanced ceramics, advanced technical ceramics) - Test method for fracture toughness of monolithic ceramics at room temperature by chevron notched beam (CNB) method,"[6] also contains only one method, CNB. However, only one standardized test specimen geometry is allowed in ISO 24370 rather than the four test specimen geometries (as well as corresponding stress intensity factor (SIF) solutions) allowed in ASTM C1421. In addition, ISO 24370 also gives provisions for generalized test specimen dimensions and a generalized SIF solution.

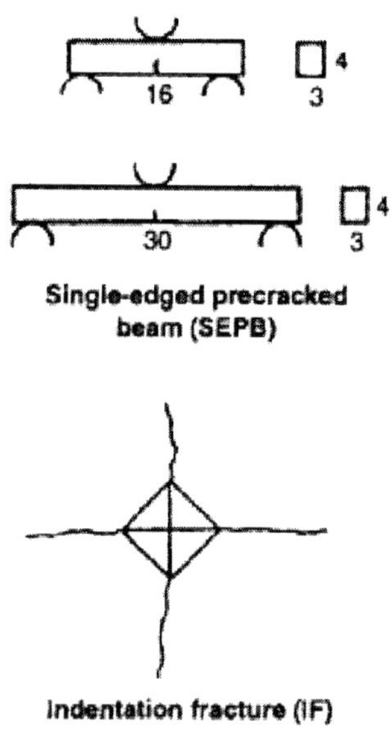

Figure 2 Two test method configurations (SCF and IF) contained in JIS1607 (Ref 3)

CEN TC184 Draft "Advanced Technical Ceramics- Monolithic Ceramics-Test Methods for Determination of Apparent Fracture Toughness"[7] is a seven-part document in which the first part provides a summary of six test methods as well as guidance to the user on the choice and application of each test method. The six test methods are illustrated in Figure 3 and include SEPB, CNB, SCF, IF, indentation strength (IS), and single edge V-notched beam (SEVNB). Per the Vienna agreement between CEN and ISO, CEN may adopt ISO 15732 and ISO 18756 directly as CEN standards. However, CEN will probably not adopt ISO 24370 directly as a CEN standard because it lacks sufficient generality to allow any test specimen dimensions regardless of previous user and analytic experience.

It is noteworthy that considerable effort has been expended in validating the three methods (SEPB, SCF, and CNB) contained in ASTM C1421, JIS R1607, ISO15732, ISO 18756 and ISO 24370. One such effort[13] involved multiple billets of a standard reference material (SRM2100) tested per the ASTM C1421. For single billet C of the SRM at room temperature, the SCF method gave K_{Isc}=4.58 ± 0.16 (26 valid tests), the CNB method gave K_{Ivc}=4.60 ± 0.13 (26 valid tests) and the SEPB method gave K_{Isc} =4.58 ± 0.10 (19 valid tests). This consistency of fracture toughness results for a single material validated the "sharp crack" methods of ASTM C1421, indicating that genuine material variability (e.g., billet-to-billet, or batch-to-batch) can be discerned and quantified.

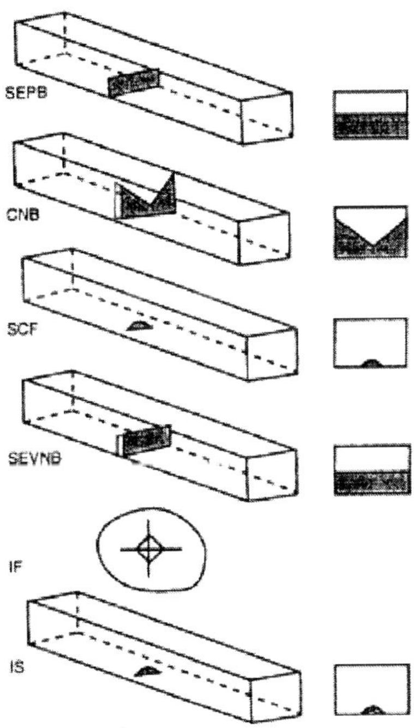

Figure 3 Six test method configurations (SCF, SEPB, CNB, SEVNB IF, IS) contained in CEN TC184 Draft (Ref 7)

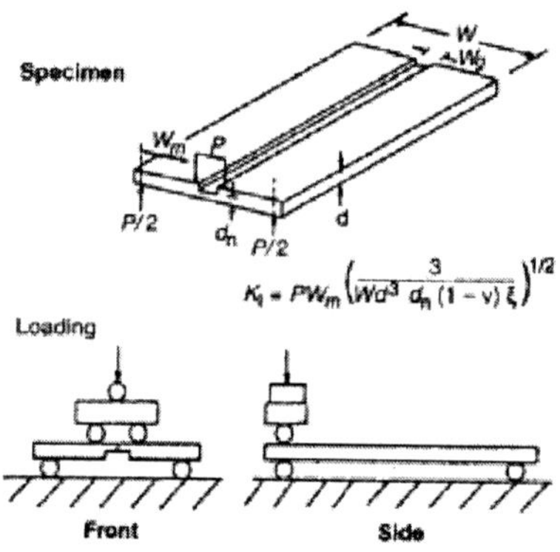

Figure 4 Illustration of DT test method including one possible SIF solution (Ref 1)

Other test methods

Many other successful methods for determining fracture toughness of ceramics and glasses and not yet standardized have been introduced and refined over the past several years. Some of the more popular methods are discussed in the following sections. Note that these discussions are limited to "sharp crack" methods because of the limitation of methods using "blunt notches."

The double torsion (DT) method was initially strongly promoted[1] as a standard test method for fracture toughness of ceramics. This method (see Figure 4) allows the use of variety of basic test specimen shapes and test configurations although the most common is a long, thin plate with a side groove cut along its length to guide the crack. There is some evidence that with good alignment, better results are obtained without the groove. Ball bearing supports help to minimize friction and improve alignment. Various analyses have shown that the applied SIF is independent of crack length over the dimensionless crack length range of 0.55 to 0.65 a/W. Although the three-dimensionality of the crack front in the DT test specimen has lead to some controversy about whether existing solutions for the SIF really represent a mode I condition, evidence exists that K_{Idt} for various brittle materials compares well to other sharp-crack mode I methods. Obvious advantages of the DT method include the simple loading configuration, its constant SIF geometry, simple test specimen geometry, stable crack propagation configuration. Some disadvantages include the relatively large volume of test material (compared to a simple flexure bar) required for a single test, the more complicated arrangement of ball bearings, etc. for supports, the curved crack front and concerns about the three-dimensionality of the crack tip. It is worthy of note that interest in the DT method is on the rise in determining the fracture resistance of thin membranes such as electrolytes for solid oxide fuel cells (SOFCs).

The double cantilever beam (DCB) test method has been part of fracture mechanics in various forms since the inception of the discipline, including its variations in ASTM E399 "Standard Test Method for Plane Strain Fracture Toughness of Metallic Materials."[16] A unique aspect of the DCB test method applied to ceramics and glasses is the method of load application which can have a variety of forms as shown in Figure 5. The applied force method can be particularly troublesome in brittle materials which are susceptible to tensile forces. However, the

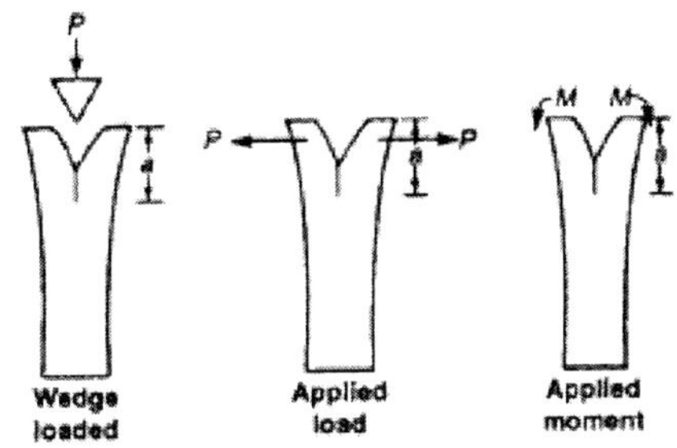

Figure 5 Three different methods of loading for the DCB method (Ref 1)

compressively-loaded wedge-loading and the compressively-loaded applied-moment methods are suited to testing brittle materials. Often side grooves are used to guide the crack longitudinally. Some advantages of the DCB method are that for some cases the SIF is independent of crack length (e.g., tapered DCB or applied moment DCB), test specimen preparation can be simple, material usage is efficient and the loading configuration can be simple. The primary disadvantage is that a sharp crack needs to be introduced to avoid problems of initiation and propagation of a crack from the blunt notch.

Chevron-notched short rod and short bar (CNSR and CNSB) test methods (see Figure 6) have many of the advantages of the CNB test method: simple notch preparation, in-situ (automatic) crack initiation at the tip of the chevron, and inherent stable crack propagation in the chevron. However, an additional advantage of the CNSR and CNSB geometries is the efficient use of material since the volume of material needed to support a flexural test specimen is avoided. A disadvantage of the CNSR and CNSB test methods is the tensile loading required for the fracture test as shown in Figure 6.

ELEVATED TEMPERATURE FRACTURE TEST METHODS

Background

In the development of an acceptable standardized fracture test method, the following conditions have been deemed necessary[1]: 1) all test specimens must have atomistically sharp pre-cracks at the onset of fracture; 2) the pre-cracks must be well characterized, measurable at the

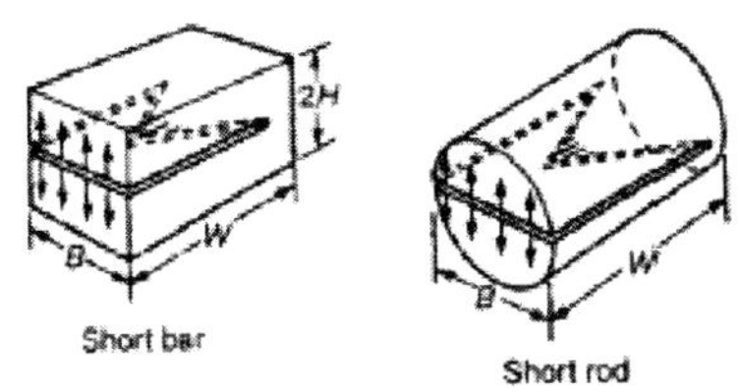

Figure 6 Examples of CNSB and CNSR test specimens (Ref 1)

conclusion of the fracture measurement; 3) the crack geometry must have a valid SIF; 4) plane strain conditions must exist at the crack tip for the measurement of plane strain fracture toughness. The three successful methods (SEPB, SCF, CNB) that have been standardized (ASTM C1421, JIS R1607, ISO15732, ISO 18756 and ISO 24370) meet these conditions for measuring fracture toughness at room temperature.

Since these same necessary conditions should apply to elevated temperature fracture tests, it would seem reasonable to extend the three successful methods to elevated temperatures. Decreasing mechanical properties and performance with increasing temperature are consistent observations for most materials. Complications of tests at elevated temperature include temperature control, environmental effects (both material and test apparati, and time-dependent behavior (e.g. creep or oxidation).

Moreover, in fracture testing, a particularly troubling complication is the effect of material/environmental interactions on the crack tip conditions. A good example of this is crack tip blunting which may more easily occur after the trouble of introducing an atomistically sharp pre-crack in fracture test specimens. Special caution is advised in extending methods which start with sharp crack introduced at ambient temperatures. These methods include SEPB, SCF, IS, and pre-cracked SENB, DCB, etc.. Methods which form cracks in-situ will probably not experience the "flaw healing" phenomenon. These methods include all the chevron-notched test specimens (CNB, CNSB, CNSR), SEVNB, and DT.

Proposed Test Method

The utility and extensive experience with the CNB method makes it an attractive choice for elevated temperature fracture toughness testing. Extensive experience with different test specimen geometries, dimensions and test configurations lead to the introduction of four standardized CNB configurations within ASTM C1421 (see Figure 7). Validated SIF solutions for each configuration are also contained in ASTM C1421. Therefore, the proposed elevated temperature fracture toughness test method uses the four CNB configurations directly.

A major consideration in elevated temperature testing is that one often ends up testing the the fixturing rather than the test material. However, this is not a major concern when using the CNB configurations of ASTM C1421 because the four and three-point flexure test setups mandated in the standard are commercially available in both room temperature (high strength, hardened steel) and elevated temperature (typically sintered silicon carbide) versions. The standard requires rollers at the loading points to minimize friction and binding effects as well as articulation (either semi or full) to promote uniform force application.

Elevated temperatures are produced by a furnace capable of establishing and maintaining a constant temperature during each testing period. The overall variation in temperature during the test must be within ±2°C determined using a temperature readout device with resolution of 1°C or less. The furnace system must minimize thermal gradients in the test specimen, so that no more than 5°C differential exists from end-to-end in the test specimen.

Typically, the test specimen temperature is monitored by a thermocouple with its tip located no more than 1 mm from the midpoint of the test specimen. A separate thermocouple can also be used to control the furnace chamber if necessary, but the test specimen must be the reported temperature of the test. The calibration of the thermocouple and temperature of the must be maintained per existing standards.

The furnace may have an air, inert, or vacuum environment, as required. If an inert or vacuum chamber is used, and it is necessary to direct force through a bellows, fittings, or seal, then the user must verify that losses or errors do not exceed 1 % of the expected failure forces.

Proceduraly, there are two ways in which the test specimen is brought to temperature: cold-furnace loading and hot-furnace loading. In both situations, placement and alignment of the test specimen on the test fixturing takes place in ambient conditions although each method has

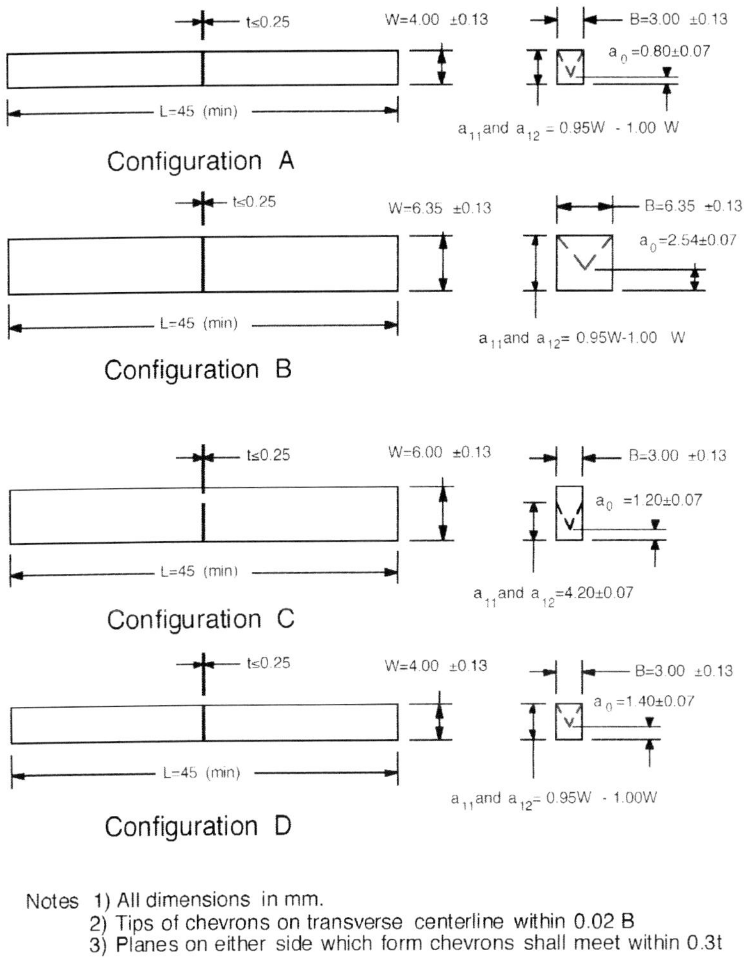

Figure 7 CNB test specimen geometries from ASTM C1421 proposed for elevated fracture toughness testing (Ref 2)

advantages and disadvantages related to furnace heating and cooling. Cold-furnace loading has the advantage of having the test specimen and fixturing fully enclosed by a static furnace but has the disadvantage of having a lengthy ramp-up to temperature following by a prolonged hold time during which the furnace and fixtures reach thermal equilibrium. Hot-furnace loading has the advantage of eliminating the lengthy ramp-up to temperature but can introduce thermal shock to the fixturing and test specimen as the hot-furnace encompasses the test setup.

For cold-furnace loading, test specimens may be loaded onto the test fixtures in a cooled furnace. Some means of ensuring alignment of the system during subsequent heating to test temperature must be provided. The furnace must be raised to the test temperature at a constant heating rate that must be stated in the report. Temperature overshoot (over the test temperature) must be strictly controlled and must be no more than 5°C. The temperature must be held constant (reported soak time) for the time necessary for the test specimen/furnace reach equilibrium.

For hot-furnace loading, test specimens may be loaded directly into a hot furnace. This must be conducted in a fashion so as to minimize or eliminate thermal shock damage to the test specimen and test fixturing. Temperature overshoot (over the test temperature) shall be strictly controlled and shall be no more than 5°C. The temperature must be held constant (soak time) for the necessary and reporting soak time for the test specimen and furnace to come to equilibrium.

Once the test temperature has been reaching the fracture tests are conducted similar to CNB tests in ASTM C1421. Test mode is in displacement (stroke) control at rates between 0.0005 to 0.005 mm/s to achieve a rate of SIF between 0.1 and 2.75 $MPa\sqrt{m}/s$.

A valid test must meet the following general requirements in addition to the specific requirements of the particular test:

a) Test machine shall have provisions for autographic recording of force versus deflection or time, and the test machine shall have an accuracy in accordance with Practice E4.

b) Test fixtures shall i) have inner and outer rollers free to roll as required, ii) have roller pins with a hardness of 40 Rockwell C or greater, iii) have rollers that have lengths at least three times the test specimen dimension, B, diameters of 4.5 ± 0.5 mm, with each roller parallel to each other within 0.015 mm over either the length of the roller or a length of 3B or greater, iv) be capable of maintaining the test specimen alignment to the tolerances specified.

c) Dimension-measuring devices shall be accurate and precise to the level required in the appropriate annex with all applicable dimensions measured and reported.

d) Test specimen shall be aligned such that the plane of the crack shall be centered under the center roller within 0.5 mm of S_o for three-point loading and shall be located within 1.0 mm of the midpoint between the two inner rollers, S_i for four-point loading.

e) Test rate shall be such that one of the test rates shall result in a rate of increase in stress intensity factor between 0.1 and 2.75 $MPa\sqrt{m}/s$.

Post-test measurements are made so as to examine the chevron notch at sufficient magnification (~30 X). The tip of the chevron must be on center within 0.02 x test specimen thickness, and the centerline of the notch grooves on either side of the tip shall meet within 0.3 of the notch thickness. The fracture surface must be examined to determine how well the crack i) followed the chevron notch plane and ii) separated the test specimen into two pieces. If the "crack follow" through the chevron section is poor, the crack may have deviated substantially farther into one half than the other and the test may be invalid.

Post test interpretation must be made of the test record to ensure that it exhibits a smooth (nonlinear) transition through the maximum force prior to final fracture. If the test specimen exhibits a sudden drop in force from the initial linear portion for the test record not followed by a subsequent force increase, the test is unstable and invalid. The relevant maximum test force,

P_{max}, is determined from the test record. In some cases the test specimen may overload slightly at crack initiation. In the calculations, the maximum stable force, P_{max} is used. If there is evidence of environmentally-assisted slow crack growth then it is advisable to run additional tests in an inert environment. Alternatively, additional tests may be done in laboratory ambient conditions at faster or slower test rates than those specified in order to determine the sensitivity to test rates. Testing rates that differ by two to three orders of magnitude or greater than those specified are recommended, although at rates > 0.5 mm/s, stability may be difficult to detect.

Several recommendations are made regarding crack stability, machining and sensitivity of test system to stable fracture behaviour. In some instances a stable crack will not initiate from the tip of the chevron, resulting in test specimen overload (i.e., a force greater than that to produce stable fracture) or underload (i.e., a force less than that to produce stable fracture) and catastrophic fracture from the chevron tip. If this occurs, a simple compression-compression fatiguing procedure to damage the chevron tip, thereby promoting stable initiation and growth of a crack, can be used. The test specimen is placed in the test fixture upside down and the crack tip loaded in compression, several times, to approximately three times the estimated fracture force expected for the normal position. Machining of the chevron notch can influence the scatter in the results. Thinner, or more precise notch thicknesses seem to decrease scatter and initiate stable crack growth more readily. Actuator displacement (stroke) may not be as sensitive to changes of fracture behavior in the test specimen as measurements taken on the test specimen itself, such as back-face strain, load-point displacement, or displacement at the crack plane. In very stiff materials, use of back-face strain is recommended for detection of stable fracture.

Fracture toughness, K_{Ivb} is calculated as:

$$K_{Ivb} = Y^*_{min} \left[\frac{P_{max} \left[S_o - S_i \right] 10^{-6}}{BW^{3/2}} \right] \qquad (4)$$

where

K_{Ivb}	is the fracture toughness ($MPa\sqrt{m}$)
Y^*_{min}=	is the minimum stress intensity factor coefficient as determined from separate equations for each test specimen geometries
P_{max}	is the relevant maximum force (N)
S_o	is the outer span (m)
S_i	is the inner span (m)
B	is the side to side dimension of the test specimen (m)
W	is the top to bottom dimension of the test specimen (m)

ELEVATED TEMPERATURE FRACTURE TEST RESULTS

"Flaw healing" can be a particular problem at elevated temperatures for precracked fracture test methods. However, even the CNB method may exhibit difficulties in initiating and propagating cracks under certain conditions. An in-situ toughened silicon nitride was fracture tested using CNB, SEPB and fine-notched (50 μm thickness) SENB. The force-displacement curves for the CNB tests are shown in Figure 8. The stable curves at 25 and 1200°C provide valid test results per ASTM C1421. However, the highly nonlinear behaviour at 1371°C, (stable and valid per ASTM C1421), was invalid test because a crack never developed at the tip of the chevron. The highly nonlinear behavior (Figure 8) was attributed to "plastic hinging."[1]

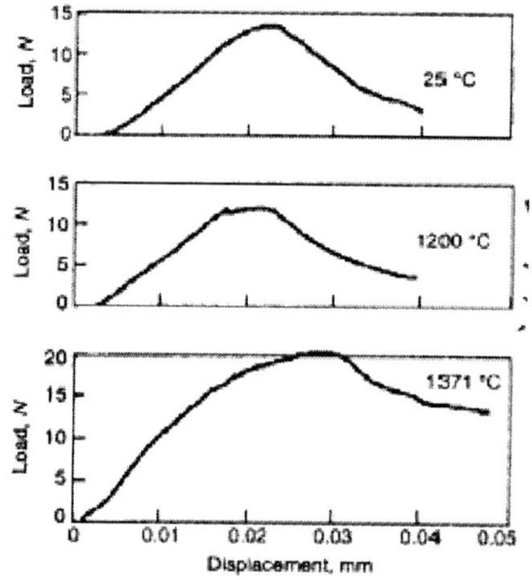

Figure 8 Force-displacement for stable fracture in CNB tests of a in-situ toughened silicon nitride. The highly nonlinear test at 1371°C is invalid because of "plastic-hinging" (Ref 1)

Comparison of fracture toughness results[1] from CNB, SEPB and sharp notch SENB are presented in Fig. 9. Note that under the proper conditions at 1371°C (fast heat up and short hold times) the SEPB method, despite the sharp precrack gives results similar to the sharp notch SENB method. The CNB method at 1371°C under these test conditions tends to over estimate fracture toughness as expected from the nonlinear loading curve shown in Fig. 8

However, a demonstration[1] of the flexibility of the application of the CNB method is shown in Figure 10 which shows fracture toughness (with error bars) for seven advanced ceramics (from aluminum oxide to silicon carbide to silicon nitride) and eight temperatures (from 20 to 1400°C) in ambient air. With the exception the non linear behavior of the in situ toughened silicon nitride, all results are predicable and consistent.

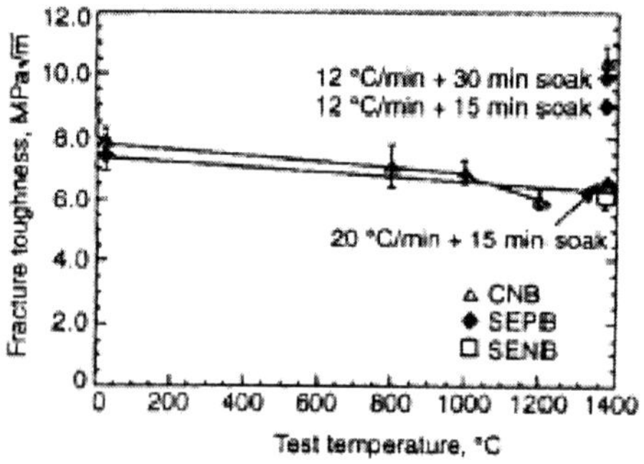

Figure 9 Comparison of fracture toughness for an in-situ toughened silicon nitride as a function of temperature for SEPB, SENB and CNB test methods (Ref 1)

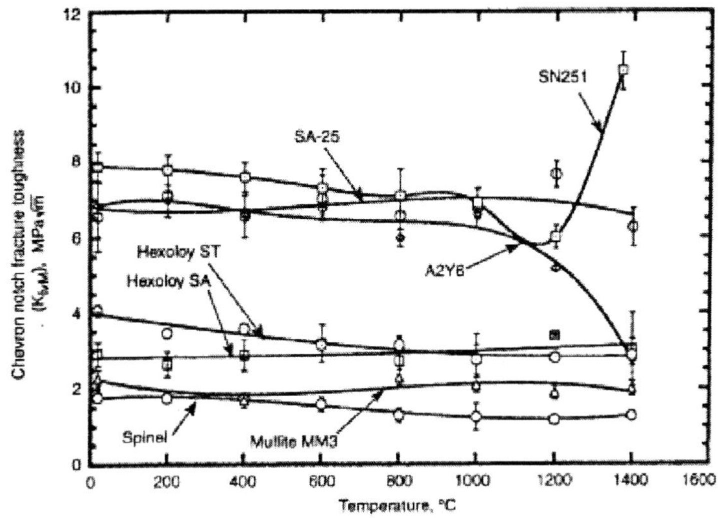

Figure 10 Fracture toughness results as a function of temperature for seven advanced ceramics determined using a CNB method (Ref 1)

SUMMARY/CONCLUSIONS

The fracture of ceramics and glasses, while conceptually (and mathematically) rather simple and straight forward, is far from simple in practice. "Real" materials often contain residual stresses, microstructures, and geometric features which can complicate test methods intended to extract what appears to be a fundamental material property: fracture resistance. Further complicating test methods are differences in fracture behaviour because of the type of flaw: intrinsic (a.k.a., micro) flaws often provide information on intrinsic (i.e., fundamental without influence of microstructure or history of crack propagation (i.e., R-curve) fracture resistance whereas extrinsic (a.k.a. induced) can be used to measure either the intrinsic or the steady state fracture resistance but can lead to large scatter in the results if the influence of the microstructure of history of crack propagation are not accounted for.

It is clear that the driving force behind test methods to determine fracture resistance is not necessarily for its use in determining the material's resistance to extrinsic flaws. Instead, the need is to accurately and precisely determine the fracture resistance of ceramics and glasses for use in relative comparison indices for evaluation of material performance in fracture resistance related (but not fracture resistance driven!) applications.

The several extant room temperature standard test methods for fracture testing ceramics and glasses (e.g., ASTM C1421, JIS R1607, and ISO 15732, ISO 18756, ISO 24370) have concentrated on maximizing repeatability and reproducibility of the fracture resistance measurements at the onset of brittle fracture. Minimizing the effects of R-curve behavior, environmentally-assisted crack growth and variability in testing are primary concerns.

At elevated temperature, methods which involve the creation of sharp pre-cracks at room temperature may become unusable at certain elevated temperatures as crack healing and crack tip blunting negate the effects of pre-cracks. The CNB method uses in-situ crack formation seems to provide better ways to measure either fracture toughness or R-curve behavior at elevated temperatures. The CNB method together with good control of temperature and test machine has shown applicability to a wide variety of advanced ceramics at temperatures up to 1400°C in air.

REFERENCES

[1] M.G. Jenkins and J. A. Salem, "Fracture Resistance Testing of Brittle Solids, ", pp. 665-678 in ASM Handbook Vol8 Mechanical Testing and Evaluation, ASM International, Metals Park, Ohio (2001)

[2] ASTM C1421 "Standard Test Methods for Determination of Fracture Toughness of Advanced Ceramics at Ambient Temperature," Annual Book of ASTM Standards, Vol 15.01, ASTM International, West Conshohocken, Penn (2005)

[3] JIS R1607 "Testing Methods for the Fracture Toughness of High Performance Ceramics," Japanese Industrial Standards Committee, Tokyo, Japan (1990)

[4] ISO15732 "Fine ceramics (advanced ceramics, advanced technical ceramics)-Test Method for Fracture Toughness of Monolithic Ceramics at Room Temperature by Single Edge Pre-Cracked Beam (SEPB) Method", ISO, Geneva, Switzerland (2003)

[5] ISO 18756 "Fine ceramics (advanced ceramics, advanced technical ceramics) - Determination of fracture toughness of monolithic ceramics at room temperature by the surface crack in flexure (SCF) method," ISO, Geneva, Switzerland (2003)

[6] ISO 24370, "Fine ceramics (advanced ceramics, advanced technical ceramics) - Test method for fracture toughness of monolithic ceramics at room temperature by chevron notched beam (CNB) method," ISO, Geneva, Switzerland (2005)

[7] CEN TC184 Draft "Advanced Technical Ceramics- Monolithic Ceramics-Test Methods for Determination of Apparent Fracture Toughness", CEN, Brussels, Belgium, (1999)

[8] J. Larsen-Basse, "Abrasive Wear of Ceramics," pp. 99-118 in Friction and Wear of Ceramics, S. Jahanmir, ed., Marcel Dekker, Inc., New York (1994).

[9] A. Evans and D. Marshall, "Wear Mechanisms in Ceramics," pp. 439-452 in Fundamentals of Friction and Wear of Materials, D. Rigney, ed., ASM, Materials Park, Ohio, (1981)

[10] Ritter, "Erosion Damage in Structural Ceramics," *Mater. Sci. and Eng*, **71** 195-201, (1985)

[11] J. B. Qui nn an d G. D. Quinn , "Hardness and Brittleness of Ceramics," *Ceram. Eng. and Sci . Pro c.*, **17** [3] 59 -68, (1996)

[12] J. B. Quinn and G. D. Quinn, "Indentation Brittleness of Ceramics: A Fresh Approach," *J. Mat. Sci.*, **32,** 4331-4346 (1997)

[13] G.D. Quinn, M. G. Jenkins, J. A. Salem and I. Bar-On, "Standardization of Fracture Toughness Testing of Ceramics in the United States," *Journal of Korean Ceramic Society*, **4** [4] 311-322 (1998)

[14] I. Bar-On, G. D. Quinn, J.A. Salem and M. G. Jenkins , "Development of ASTM C1421-99 Standard Test Methods for Determination of Fracture Toughness of Advanced Ceramics," pp. 315-335 in Fatigue and Fracture Mechanics: 32nd Volume, ASTM STP1406, ASTM International, West Conshohocken, Penn (2001)

[15] M. G. Jenkins, J. A. Salem, G. D. Quinn, I. Bar-On, "Development, Verification and Implementation of a National Full-Consensus Fracture Toughness Test Method Standard for Advanced Ceramics", pp. 49-75 in Fracture Resistance Testing of Monolithic and Composite and Brittle Materials, ASTM STP 1409, ASTM International, West Conshohocken, Penn (2002)

[16] ASTM E399 "Standard Test Method for Plane Strain Fracture Toughness of Metallic Materials," Annual Book of ASTM Standards, Vol 03.01, ASTM International, West Conshohocken, Penn, (2005)

COMPARISON OF STATIC FATIGUE SCG MODELS FOR DENSE CERAMICS

Michael G. Jenkins
University of Detroit Mercy
Department of Mechanical Engineering
4001 W McNichols Road
Detroit, MI 48221 USA

Jonathan A. Salem
NASA-Glenn Research Center at Lewis Field
21000 Brookpark Road
Cleveland, OH 44135 USA

ABSTRACT

Long-term durability of dense ceramics in clean energy applications depends on the material's susceptibility to slow crack growth (SCG) in various environments and temperatures. Typically, the SCG behaviour is described in terms of a *v-K* diagram which establishes the relationship between the applied stress intensity factor, *K*, and the crack growth velocity, *v*, in a given environment. Various models have been employed to describe this *v-K* behaviour with varying degrees of success. For example, the power law relation is commonly applied as an average of various stages of SCG behaviour. However, other models (e.g., fracture mechanics, exponential, etc) have been employed with mixed results. In this paper, the applicability of *v-K* SCG models (in addition to power law) is reviewed. Issues related to mathematical form of the model, a priori assumptions and predictive utility are examined. A recently approved full-consensus national standard (ASTM C1576) for determining slow crack growth parameters is discussed in light of these models.

INTRODUCTION

Dense ceramics along with glasses often exhibit time-dependent failure that arises from the growth of inherent and induced flaws from subcritical initial sizes to critical sizes sufficient to cause failure. Because cyclic or dynamically-applied forces are not required to produce crack extension, the phenomenon is often referred to as "static fatigue." However, other names for the phenomenon include subcritical crack growth or slow crack growth (SCG), stress corrosion, delayed failure, and environmentally induced (assisted) fatigue.

The SCG phenomenon depends on the chemical composition and temperature of the environment surrounding the ceramic or glass material as well as the applied stress intensity factor due to the flaw size, orientation, type and loading. For relatively small stress intensity factors (i.e., much less than the fracture toughness of the material), a fatigue limit (or stress intensity factor threshold) below which crack growth arrests has been measured for static loading[1-4] Since this threshold occurs at very low crack growth rates or velocities, it is generally ignored for engineering applications.

SCG is often more extensive in glass and glassy polycrystalline ceramics than in single crystals (e.g., sapphire). Therefore the structural applications of glasses and ceramics require consideration of static fatigue and SCG behaviour. In this regard various predictive models have been developed [5-7] to describe SCG behaviour and various test methods [8-12] have been developed and applied to generate data and determine SCG parameters. These models have taken many forms and their utility and applicability can vary depending on the material, operating conditions, application, and design time.

With the recent introduction of ASTM C1576[12] on static fatigue testing, the debate over the applicability of various SCG models was revisited. In this paper, the basis of SCG and SCG parameters is first discussed. Then some of the more well-known models are presented and discussed. The applicability of SCG models to actual design and some specifics of ASTM C1576 are noted. Finally a summary and some conclusions are drawn.

SLOW CRACK GROWTH BEHAVIOUR

Environmentally induced (static fatigue) in ceramics and glasses is a chemically activated atomic reaction that is enhanced by stress and temperature. Crack growth occurs by breakage of the bonds at the crack tip. The introduction of water to the environment can have a severe effect on the fatigue parameters. For example, moisture in air is often sufficient to cause crack growth in ceramics and glasses subjected to stress.

For aqueous solutions, the rate of crack growth increases with increasing concentration of OH^- ions, that break the metal-oxygen-metal bonds in silicate glass and oxide ceramics. For non-oxide ceramics not containing a glassy phase (e.g. sintered α-SiC) the presence of room temperature water vapor may have little effect on crack growth rates. Acid-base pairs have the strongest influence on crack growth.[13,14]

Michalske et al. [14] developed an atomistic model addressing the role of mechanical strain in accelerating chemical reactions between crack tip bonds in silica and environmental molecules. The strain field at the crack tip deforms the Si-O bond angle thereby resulting in bond defects that increase the bonds chemical activity. Chemisorption of the attacking species results in bond breakage. The absolute rate of crack extension depends on the combined rates of active bond formation and chemical attack at strain-induced defects.

Additional studies[13] of sapphire in water, ammonia, hydrazine and acteonitrile indicated the same mechanism (dissociative chemisorption) but with different details that make the fatigue parameters more sensitive to the specific environment.

In contrast to the SCG mechanism in sapphire, that in magnesium fluoride (ionically bonded) appeared to be ionic solvation of strained ions at the crack tip. The actively corrosive environments (in this case, water, methanol and acetonitrile) reduced the electrostatic attraction between oppositely charged ion pairs at the crack tip.

At elevated temperatures in corrosive environments such as water or pressurized stream, silicon nitrides undergo severe pitting or grain loss, depending on the oxide additives and grain morphology.[15,16] For silicon nitride with amorphous silica at the grain boundaries, integranular corrosion results in the Si_3N_4 grains falling out. However, for silicon nitrides with even small amounts of oxide additives, the leachability of the grain boundary is substantially changed. Depending on the additives, pits form under either a protective or non-protective corrosion layer.[16] The implication for environmentally induced crack growth is that harsh environments involving elevated temperature steam, such as turbines, will result in substantially accelerated crack growth.

SCG MODELS

Macroscopic observation of the behavior of cracks extending due to environmentally induced fatigue indicate three regions of growth, as illustrated in Figure 1. Region I dominates the time-to-failure during long lifetimes, and design and test methodologies focus on it. For relatively low velocities, a threshold, K_{Ith}, has been measured. In Region II the crack growth rate depends on the diffusion rate of the active species to the crack tip, and it strongly dependents on concentration changes in the environment and mechanical effects that limit exposure of the crack tip. In Region III the crack growth rate becomes sufficiently fast that it is independent of the environment chemistry. Conceptually the maximum fracture resistance of the material has been reached.

Lifetime predictions are quite sensitive[5] to the form of the crack growth law used to relate crack velocity, $v = da/dt$, and applied stress intensity factor K_I. A variety of functions, have been used as discussed in the following sections.

Power-law Form

The widely-used empirically-based power-law relation for crack velocity above the fatigue limit can be expressed as either Equation 1 or 2:

Ref (17-19) $$v(K_I) = AK_I^n \qquad (1)$$

Ref (20-22) $$v(K_I) = A\left(\frac{K_I}{K_{Ic}}\right)^n \qquad (2)$$

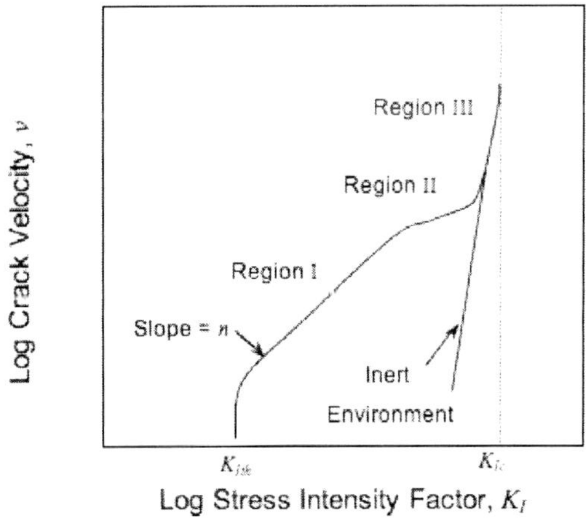

Figure 1 Generalized plot of crack velocity ($v=da/dt$) as a function of stress intensity factor (K_I) on a logarithmic scale for ceramics and glasses exhibiting SCG (Ref 7)

where $v(K_I)$ =crack velocity, da/dt
a =crack length
t =time
K_I= mode I stress intensity factor, $K_I = Y\sigma\sqrt{a}$ (note Y=geometry correction factor and σ=applied normal stress)
K_{Ic}= fracture toughness of the material
A= material/environment dependent SCG parameter
n = material/environment dependent SCG parameter

The attractiveness of Equations 1 and 2 is that substitution of either Equation 1 or 2 into the equation resulting from the differentiation of the stress intensity factor equation with respect to time gives an equation that when integrated ends up providing a equation for life time, t_f at a given applied stress, σ_a. The SCG parameters, A and n, can be determined by linear regression of either static force or dynamic force test results as discussed in a following section.

It is interesting to note that SCG behaviour is often time dependent such that Equations 1 and 2 can be modified accordingly:

$$v(K_I) = v_o K_I^n \quad \exp\left[-\left(\frac{E^*}{RT}\right)\right] \tag{3}$$

$$v(K_I) = v_o \left(\frac{K_I}{K_{Ic}}\right)^n \quad \exp\left[-\left(\frac{E^*}{RT}\right)\right] \tag{4}$$

where v_o is an SCG parameter, E^* is activation energy, T is the absolute temperature and R is the universal gas constant.

Exponential Form

Various exponential forms of the crack velocity relation have been proposed. These relations have been based on factors that may involve chemically-assisted corrosion/reaction,[23] diffusion-controlled stress rupture,[24] thermally-activation,[25] constant crack-tip chemical reaction,[1] kinetic crack growth,[26] and other factors[27]. Some proposed forms of the exponential crack-velocity relations include:

Ref (1, 23) $$v(K_I) = A\exp\left(n\frac{K_I}{K_{Ic}}\right) \tag{5}$$

Ref (26) $$v(K_I) = A\,exp\left\{n\left(\frac{K_I}{K_{Ic}}\right)^2\right\} \tag{6}$$

Ref (25) $$v(K_I) = A\left(\frac{K_I}{K_{Ic}}\right)\exp\left\{n\left(\frac{K_I}{K_{Ic}}\right)\right\} \tag{7}$$

Ref (27) $$v(K_I) = A\left(\frac{K_I}{K_{Ic}}\right)\exp\left\{n\left(\frac{K_I}{K_{Ic}}\right)\right\}^2 \tag{8}$$

Ref (26) $$v(K_I) = A\left(\frac{K_I}{K_{Ic}}\right)exp\left\{n\left(\frac{K_I}{K_{Ic}}\right)^2\right\} \tag{9}$$

where $v(K_I)$ =crack velocity, da/dt
a =crack length
t =time
K_I = mode I stress intensity factor, $K_I = Y\sigma\sqrt{a}$ (note Y=geometry correction factor and σ=applied normal stress)
K_{Ic}= fracture toughness of the material
A = material/environment dependent SCG parameter (not the same as A used in Eqs 1, 2)
n = material/environment dependent SCG parameter (not the same as n used in Eqs 1, 2)

The exponential crack velocity relations for Equations 5 to 9 can not be as easily manipulated[5] to result in the same type of simple analytical expressions for fatigue life as can be done with the power-law crack velocity relations of Equations 1 and 2. The primary complication is that the exponential form cannot be integrated analytically but rather must be integrated numerically and therefore a simple linear regression of actual test data cannot be performed to obtain the SCG parameter, A and n.

A comparison[5] of the functions for the prediction of long-term constant force data from short-term dynamic force data indicated the exponential forms may actually better predict the fatigue lives than the power-law relation of some materials. Although the exponential form has a theoretical basis,[26] the crack velocity in Region I is usually described by a power-law relation because it fits short-term test data well, and makes it apparent which parameters most affect the life of a component[22], thereby aiding material selection.

SCG MODEL APPLICABILITY

Indirect Test Methods

Strength-based methods or "indirect" methods employ smooth test specimens, such as flexural beams or tensile specimens, and estimate fatigue properties from strengths measured over different time intervals. Strength-based methods directly sample the flaw distribution within or on the surface of the test specimens. The cracks thus develop from at least some of the same sources as in a component of the same material. However, the results are subject to the scatter inherent in the strength distribution of the material, and thus a relatively large number of tests are required. Further, strength based methods do not measure the actual crack size, but infer it from the measured strength.

Two basic strength test methods and accompanying relations have been developed for determining the fatigue parameters and behavior of ceramics and glasses: static force application and dynamic force application.

Static Force Application In this type of test, tensile test specimens or flexural test specimens (e.g. beams or plates) are subjected to a constant force in the environment of interest. The time-to-failure and the initial applied stress are recorded and used to determine the fatigue parameters according to any of the Equations 1 to 9. The generalized slow crack growth behavior for static loading is shown in Figure 2. There are at least two national or international standards for determining SCG parameters for static force application tests (JIS R1632[9] and ASTM C1576[12]). One of these, ASTM C1576, is discussed specifically in a separate section.

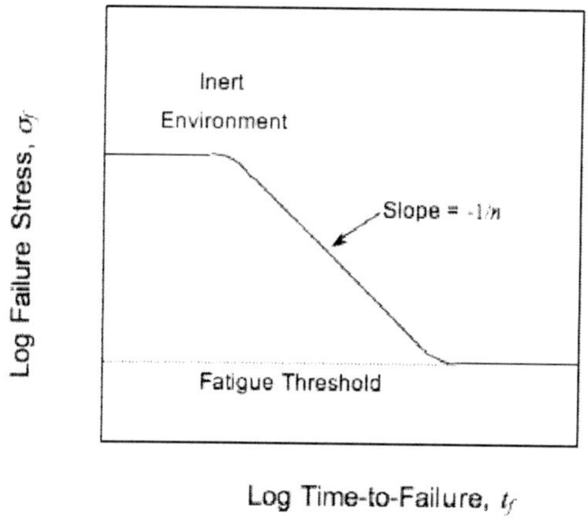

Figure 2 Generalized failure stress as a function of time-to-failure on a logarithmic scale for ceramics and glasses failing due to SCG (Ref 7)

The stress at fracture or fatigue strength, σ_f, is related to the time-to-failure, t_f:

Ref (28,29)
$$t_f = \frac{B}{\sigma_f^2}\left[\left(\frac{\sigma_i}{\sigma_f}\right)^{n-2} - 1\right] \tag{10}$$

where B = parameter associated with A, n, K_{Ic}, crack geometry and loading configuration.
σ_i = the "inert" strength (i.e., strength in non-corrosive conditions with no crack extension)

In many ceramics and glasses, $n > 10$, such that Equation 10 can be simplified to:

$$t_f = B\sigma_i^{n-2}\sigma_f^{-n}. \tag{11}$$

The fatigue parameter n can be determined from the slope of a log t_f versus log σ_f plot by writing Equation 11:

$$\log t_f = -n \log \sigma_f + \log D \tag{12}$$

where

$$\log D = \log\left[B\sigma_i^{n-2}\right] \tag{13}$$

Once the slope and intercept of Equation 12 are estimated, the parameters n and D, and the standard deviation SD_n and SD_D can be estimated[7] by a truncated Taylor series expansion of the random variable of interest:

$$n = -\alpha \tag{14}$$
$$SD_n \approx SD_\alpha \tag{15}$$
$$D = 10^\beta \tag{16}$$
$$SD_D \approx (ln\,10)(SD_\beta)(10^\beta) \tag{17}$$

where α =slope estimated by linear regression of the data represented by Equation 12
β =intercept estimated by linear regression of the data represented by Equation 12.

In some cases (e.g., special purpose design codes), the fatigue parameters B and A are necessary. The parameters B and A along with their standard deviations can be estimated[7]:

$$B = 10^{\beta}\sigma_i^{\alpha+2} \tag{18}$$

$$SD_B \approx 10^{\beta}\sigma_i^{\alpha+2}\sqrt{2\,ln(10)SD_{\beta}^2 - \frac{\alpha+2}{\sigma_i^2}SD_{\sigma_i}^2 + ln(\sigma_i)^2 SD_{\alpha}^2} \tag{19}$$

$$A = \frac{2K_{Ic}^2}{BY^2(n-2)} = \frac{-2K_{Ic}^2}{10^{\beta}\sigma_i^{\alpha+2}(\alpha+2)Y^2} \tag{20}$$

$$SD_A \approx A\sqrt{(ln\,10)^2 SD_{\beta}^2 + \frac{(\alpha+2)^2}{\sigma_i^2}SD_{\sigma_i}^2 + \left(ln\,\sigma_i + \frac{1}{\alpha+2}\right)^2 SD_{\alpha}^2} = \tag{21}$$

Note that plots of static fatigue data are usually made with the applied stress appearing as the dependent variable, however, the dependent variable actually is the time-to-failure and it must be minimized in the regression analysis.

Dynamic Force Application In this type of test, tensile test specimens or flexural test specimens (e.g. beams or plates) are subjected to a constant stress (force) rate in the environment of interest. The failure stress and the applied stress (force) rate are recorded and used to determine the fatigue parameters according to any of the Equations 1 to 9. Figure 3 illustrates the observed strength as a function of stress-rate for dynamic force application. There are at least three national or international standards for determining SCG parameters for static force application tests (ENV843-3,[8] ASTM C1368[10] and ASTM C1465[11]). A detailed error analysis of the dynamic force application test has been reported.[30]

Typically, a batch of forty or so test specimens are subjected to four or so stress rates that cover at least three orders of magnitude, with a minimum of number of tests at any of the rates. The stress at fracture or fatigue strength, σ_f, is related to the stress-rate, $\dot{\sigma}$, such that:

Ref (29, 30)
$$\sigma_f = \left[B(n+1)\sigma_i^{n-2}\dot{\sigma}\right]^{1/n+1} \tag{30}$$

where B = parameter associated with A, n, K_{Ic}, crack geometry and loading configuration,
σ_i = the "inert" strength (i.e., strength in non-corrosive conditions with no crack extension)
The fatigue parameter n can be determined from the slope of a log σ_f versus log $\dot{\sigma}$:

$$\log\sigma_f = \frac{1}{n+1}\log\dot{\sigma} + \log D \tag{31}$$

where

$$\log D = \frac{1}{n+1}\log\left[B(n+1)\sigma_i^{n-2}\right] \tag{32)}$$

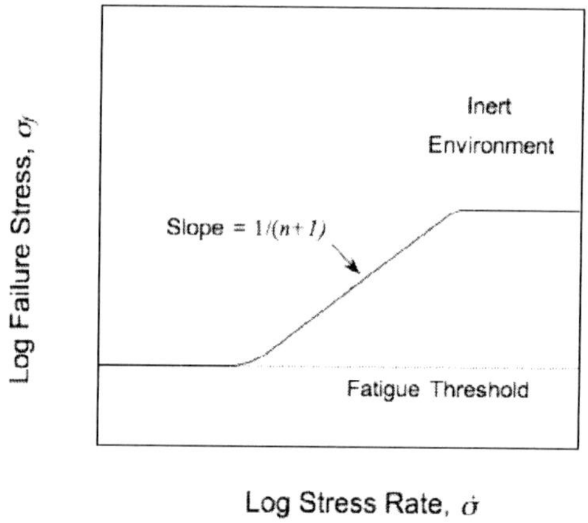

Figure 3 Generalized fracture strength as a function of stress-rate on a logarithmic scale for ceramics and glasses (Ref 7)

The dependent variable is the fracture stress and regression is preformed to minimize fracture stress variation. Once the slope and intercept of Equation 31 are estimated[7], the parameters n and D, and the standard deviation SD_n and SD_D are estimated from

$$n = \frac{1}{\alpha} - 1 \tag{33}$$

$$SD_n \approx \frac{SD_\alpha}{\alpha^2} \tag{34}$$

$$D = 10^{\beta} \tag{35}$$

$$SD_D \approx (ln\,10)(SD_\beta)(10^{\beta}) \tag{36}$$

where α =slope estimated by linear regression of the data represented by Equation 31
β=intercept estimated by linear regression of the data represented by Equation 31.

In some cases (e.g., special purpose design codes), the fatigue parameters B and A are necessary. The parameters B and A (from Equation 2) along with their standard deviations can be estimated[7]:

$$B = \frac{\alpha\, 10^{\beta/\alpha}}{\sigma_i^{1/\alpha - 3}} \tag{37}$$

$$SD_B \approx \frac{10^{\beta/\alpha}}{(\sigma_i)^{\frac{1}{\alpha}-3}} \sqrt{(ln\,10)^2 SD_\beta^2 + (1-3\alpha)^2 \frac{SD_{\sigma_i}^2}{\sigma_i^2} + (\alpha - \beta\, ln\,10 + ln\,\sigma_i)^2 \frac{SD_\alpha^2}{\alpha^2}} \tag{38}$$

$$A = \frac{2K_{Ic}^{2}(\sigma_i)^{\frac{1}{\alpha}-3}}{10^{\beta/\alpha}(1-3\alpha)Y^2} \quad (39)$$

$$SD_A \approx A\sqrt{4\frac{SD_{K_{Ic}}^2}{K_{Ic}^2} + \frac{(\ln 10)^2}{\alpha^2}SD_\beta^2 + \left(\frac{1}{\alpha}-3\right)^2\frac{SD_{\sigma_i}^2}{\sigma_i^2} + \left(\alpha - \beta\ln 10 + \ln\sigma_i - \left(\frac{1}{\alpha}-3\right)^{-1}\right)^2\frac{SD_\alpha^2}{\alpha^2}} \quad (40)$$

Direct Test Methods

Fracture mechanic methodologies or "direct" methodologies generally employ test specimens with relatively large, induced cracks. Crack growth data are typically determined directly by observation of the crack or by devices that monitor test specimen compliance, such as clip gages and strain gages. One exception to this is the double torsion (DT) test specimen, which has a relatively constant stress intensity factor, K_I, over a wide range of crack lengths. Thus the K_I can be calculated without observation of the crack length. When used in this fashion, the method is effectively an indirect, long crack method.

Two general types of fracture mechanics test specimens are employed: line-loaded or flexural specimens. Line-loaded specimens (Figure 4) such as the DT (double torsion), double (cantilever beam) DCB, or (compacted tension) CT allow cracks to be extended over large distances. Flexural specimens (see Figure 5) such as notched and precracked beams can be scaled so that crack length is comparatively small or large.

The main advantage of fracture mechanics test specimens is that large amounts of data can be derived from a single test specimen, and the result are not subject to the scatter associated with the natural flaw distribution that is sampled by strength techniques. However, because the cracks are large in comparison to those developed naturally in smooth test specimens, the fatigue behavior may be different. Although the same range of stress intensity factors is measured, the rates at which resistance develops, and possibly the rate at which fatigue damages the fracture resistance mechanism, are very different. Thus techniques that use different crack length scales may result in different fatigue parameters. Additional details regarding direct methods are contained in Ref 7.

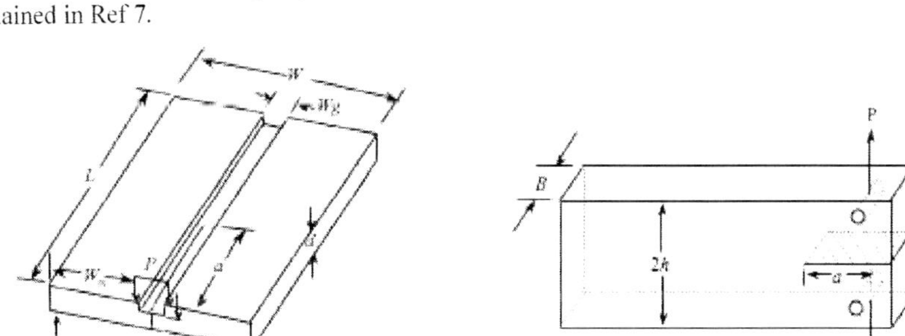

Figure 4 Two types of line-loaded fracture mechanics test specimens a) double torsion and b) compact tension (Ref 7)

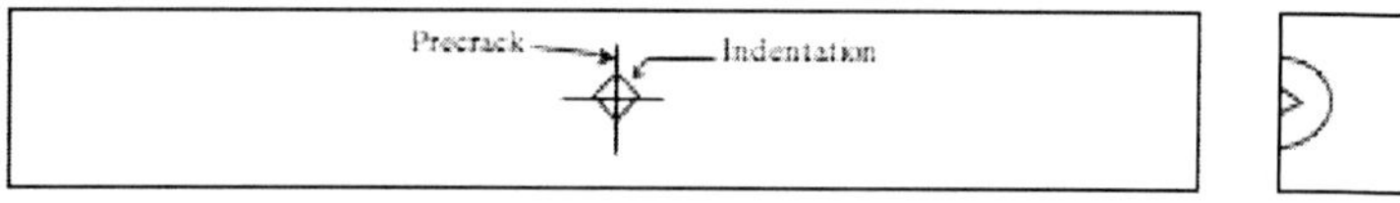

Figure 5 Flexural test specimen with a precrack formed by Vickers indentation (Ref 7)

ASTM C1576

ASTM C1576[12] "Standard Test Method for Determination of Slow Crack Growth Parameters of Advanced Ceramics by Constant Stress Flexural Testing (Stress Rupture) at Ambient Temperature," which directly addresses testing under static force conditions to determine SCG parameters was approved for publication in early 2005 after three years of drafting, debating, balloting, revising, and editing. Although two other standards[10, 11] on dynamic force testing had been approved for publication by ASTM several years earlier, ASTM C1576 is a significant contribution to the standards arena. ASTM C1576 joins JIS R1632[9] to provide two national/international standards on static force testing for SCG parameters.

ASTM C1576 "covers the determination of SCG parameters of advanced ceramics by using constant stress flexural testing in which time to failure of flexure test specimens is determined in *four-point* flexure as a function of constant applied stress in a given environment at ambient temperature. In addition, test specimen fabrication methods, test stress levels, data collection and analysis, and reporting procedures are addressed. The decrease in time to failure with increasing applied stress in a specified environment is the basis of this test method that enables the evaluation of slow crack growth parameters of a material. The preferred analysis in the standard is based on a power law relationship between crack velocity and applied stress intensity factor. Alternative analysis approaches are also discussed for situations in a detailed appendix where the power law relationship is not applicable"[12].

The approach used in ASTM C1576 "assumes that the material displays no rising R-curve behavior, that is, no increasing fracture resistance (or crack-extension resistance) with increasing crack length. The existence of such behavior cannot be determined from this test method. The analysis further assumes that the same flaw type controls all times-to-failure."[12]

The standard brings several interferences to the attention of users. These interferences include the need to control the test environment (due to chemical nature of SCG) over the long periods of time of the test, the possibility of a "strength plateau" at short test times, warning that some materials with high resistance to SCG may require more tests at high stresses because a significant portion may fail during loading, a concern about specifics of SCG and its relation to surface preparation of test specimens and the need to document fabrication history, and a statement that the analysis contained in the standard assumes a single flaw population unless fractography shows otherwise.

The actual test method of ASTM C1576 refers specifically to another ASTM standard on flexural testing advanced ceramics at room temperature (ASTM C1161[31]). Three flexural test specimen sizes are allowed including the most popular type B (3x4x50 mm). Semi and fully articulating four-point flexural test fixtures are mandated.

Ten test specimens must be tested at each stress level. Although the choice of range and number of applied stress levels (or applied force levels) not only depends on test material, but also affects the statistical reproducibility of SCG parameters, the standard mandates that at least four applied stresses covering at least four orders of magnitude in time be chosen. The standard gives guidance on how to choose stress levels based on fast fracture strength distributions of the test material. Basic information obtained from the test method include applied stress (fracture stress) as a function of time-to-failure.

Detailed formulations are provided for calculating the statistics (mean and standard deviation) of the SCG parameters, D_s and n, assuming power-law SCG behaviour. Additional information is provided on how to determine statistics related to time-to-failure.

Finally, several appendices include additional information on i) how to interpret SCG behaviour that may not follow a power-law description, ii) how to estimate scatter in time-to-failure from static force SCG testing with respect to scatter in time-to-failure from dynamic force SCG testing, and iii) simplified method for predicting time-to-failure in static force tests (static fatigue) from dynamic force (dynamic fatigue) test results.

SUMMARY/CONCLUSIONS

Interpretation of SCG behaviour in ceramics and glasses is based on multiple phenomenological models involving relationships between crack extension velocity and the applied stress intensity factor. The more popular of these models are related to simple "power law" formulations because these lead to closed-form analytical solutions for determining fatigue parameters and the resulting relations for time-to-failure. None-the-less, other models, typically those involving "exponential" formulations have been shown to better represent actual SCG data even though these formulation rely on more complex numerical methods to determine the SCG parameters.

Fatigue testing of ceramics and glasses is performed by either indirect or direct methods. Indirect or strength methods employ smooth tensile or flexure specimens and infer the fatigue parameters from strength measurements without crack length measurements. Direct methods employ either long cracks or short cracks, and the crack length is measured by observation of the crack or by inference from devices such as strain gages and electrical resistance grids. Long crack test specimens include fracture mechanics specimens such as the DT, DCB, CT and SEPB. Short crack methods employ surface cracks formed by indentation, or surface cracks that develop naturally on the surface of a smooth test specimen.

Structural ceramic and glass components that are designed to have long lives will fail from small cracks developed over long times. The cracks will develop from either inherent processing flaws or from damage generated in component machining and handling (e.g. machining cracks). Cyclic loading, though not required to induce crack growth in glasses and many ceramics, tends to accelerate fatigue crack growth. Thus, the measurement of fatigue parameters should be done with tests employing realistic crack sizes, environments and the applicable load histories. As a result, the development of standardized test methods (e.g., ENV843-3, JIS R1632, ASTM C1368, ASTM C1465 and ASTM C1576) has revolved around the use of small, inherent flaws

REFERENCES

[1]S.M. Weiderhorn and L.H. Boltz, "Stress Corrosionn and Static Fatigue of Glass," *J. Am. Ceram. Soc,* **53**., 547-548 (1970)

[2] B.J.S. Wilkins and E. Dutton, "Static Fatigue Limit with Particular Reference to Glass," *J. Am. Ceram. Soc,* **59,** 108-112 (1976)

[3] T.A. Mchalske, "The Stress Corrosion LimitL Its Measurement and Implications," *Fracture Mechanics of Ceramics, Vol 5*, Plenum Publishing 1983.
[4]T. Fett, K. Germerdonk, A. Grossmuller, K. Keller and D. Munz, "Subcritical Crack Growth and Threshold in Borosilicate Glass, *J. Mater. Sci*, **26**, 253-257 (1991).
[5]K. Jakus, J.E. Ritter, Jr, J.M. Sullivan, "Dependency of Fatigue Predictions on the Form of the Crack Velocity Equation," *J. Am. Ceram. Soc*, **64**, 372-374 (1981)
[6]S.R. Choi, N.N. Nemeth and J.P. Gyekenyesi, "Slow Crack Growth of Brittle Materials with Exponential Crack-Velocity Formulation-Part 3: Constant Stress and Cyclic Stress Experiments," NASA/TM-2002-211153/PART3, NASA-Glenn Research Center, Cleveland, Ohio, 2002
[7]J.A. Salem and M.G. Jenkins, "Fatigue Testing of Brittle Solids" ASM Handbook, Vol 8 Mechanical Testing and Evaluation, ASM, Materials Park, Ohio, 2000
[8]ENV 843-3 "Advanced Technical Ceramics-Monolithic Ceramics-Mechanical Properties at Room Temperature, Part 3: Determination of Subcritical Crack Growth Parameters from Constant Stress Rate Flexural Strength Tests," CEN, Brussels, Belgium, 1997.
[9]JIS R 1632 "Test Methods for Static Bending Fatigue of Fine Ceramics," JSA, Tokyo, Japan, 1998
[10]C1368 "Standard Test Method for Determination of Slow Crack Growth Parameters of Advanced Ceramics by Constant Stress Rate Flexural Testing at Ambient Temperature, ASTM, West Conshohocken, Pennsylvania, 2005.
[11]C1465 "Standard Test Method for Determination of Slow Crack Growth Parameters of Advanced Ceramics by Constant Stress Rate Flexural Testing at Elevated Temperature, ASTM, West Conshohocken, Pennsylvania, 2005
[12]C1576 "Standard Test Method for Determination of Slow Crack Growth Parameters of Advanced Ceramics by Constant Stress Flexural Testing (Stress Rupture) at Ambient Temperature," ASTM, West Conshohocken, Pennsylvania, 2005
[13] T.A. Michalske, B.C. Bunker and S.W. Freiman, "Stress Corrosion of Ionic and Mixed Ionic/Covalent Solids," *J. Am. Ceram. Soc.*, **69**, 10, 721-724 (1986)
[14] T.A. Michakske and B.C. Bunker, "Slow Fracture Model Based on Strained Silicate Structures," *J. Appl. Phys.*, **56**, 10, 2686-2694 (1984)
[15] T. Yishio and K. Oda, "Aqueous Corrosion and Pit Formation of Si_3N_4 Under Hydrothermal Conditions," *Cerm. Trans.*, **10** 367-386 (1990)
[16] M.K. Ferber, H.T. Lin and J. Keiser, " Oxidation Behavior of Non-Oxide Ceramics in a High-Pressure, High-Temperature Steam Environment," Mechanical, Thermal and Environmental Testing and Performance of Ceramic Composites and Components, ASTM STP 1392, ASTM International, West Conshohocken, PA, 2000.
[17] S.M. Weiderhorn, "Prevention of Failure in Glass by Proof-Testing," *J. Am. Ceram. Soc.*, **56**, 4, 227-228 (1973)
[18]A.G. Evans and S.M. Weiderhorn, "Proof Testing of Ceramic Materials – An Analytical Basis for Failure Prediction," *Int. J. Fract.*, **10**, 3, 379-392 (1974)
[19]A.G. Evans and E.R. Fuller, "Crack Propagation In Ceramic Materials Under Cyclic Loading Conditions," *Met. Trans.*, **5**, 1, 27-33 (1974)
[20]B.J.S. Wilkins and L.A. Simpson, Errors in Estimating the Minimum Time-to-Failure in Glass," *J. Am. Ceram. Soc. – Discussion and Notes*, pp. 505 (1974)
[21]W.P Minnear and R.C. Bradt, "(K-V) Diagrams for Ceramic Materials," *J. Am. Ceram. Soc. – Discussion and Notes*, **58**, 7-8, 345-46 (1975)

[22] G. Sines, "Rationalized Crack Growth and Time-to-Fracture of Brittle Materials," *J. Am. Ceram. Soc. – Discussion and Notes*, **59**, 7-8, 370-371 (1976)

[23] W.B Hillig,.; and R.J Charles, "Surfaces, Stress-Dependent Surface Reactions, and Strength. High Strength Materials" Ch. 17, John Wiley & Sons, Inc., New York, NY, pp. 682–705.

[24] R.J Charles, "Diffusion Controlled Stress Rupture of Polycrystalline Materials," *Metall. Trans. A*, **7A**, 8, 1081–1089 (1976)

[25] J.C. Pollet and S.J. Burn, "Thermally Activated Crack Propagation—Theory," *Int. J. Fracture*, **13**, 5, 667–679 (1977)

[26] B.R. Lawn "An Atomistic Model of Kinetic Crack Growth in Brittle Solids," *J. Mater. Sci.*, **10**, 469–480 (1975)

[27] E.M. Lenoe, and D.M. Neal, "Assessment of Strength-Probability-Time Relationships in Ceramics" ARPA Order 2181, AMMRC–TR–75–13 (1975).

[28] J.E. Ritter, Jr., "Assessment of Reliability of Ceramic Materials," pp. 227-251 in Fracture Mechanics of Ceramics, Vol 6, Plenum Press, NY 1983.

[29] J.E. Ritter, Jr., "Engineering Design and Fatigue Failure of Brittle Materials," pp. 667-685 in Fracture Mechanics of Ceramics, Vol 4, Plenum Press, NY 1978.

[30] J. E. Ritter, N. Bandyopadhyay, and K. Jakus, "Statistical Reproducibility of the Dynamic and Static Fatigue Experiments," *Am. Ceram. Soc. Bull.* **60**, 798 (1981)

[31] C1161 "Standard Test Method for Flexural Strength of Advanced Ceramics at Ambient Temperature," ASTM, West Conshohocken, Pennsylvania, 2005

NEXT GENERATION NUCLEAR POWER AND THE NEED FOR TEST STANDARDS, DESIGN CODES, AND DATA BASES FOR CERAMIC MATRIX COMPOSITES

Michael G. Jenkins
University of Detroit Mercy
Department of Mechanical Engineering
4001 W McNichols Road
Detroit, MI 48221 USA

ABSTRACT

Innovative and novel applications of CMCs are critical to the success of Next Generation Nuclear Power (NGNP) plants. Regularity requirements demand the implementation of test standards, design codes and data bases for CMCs as part of the licensing and approval of this new power plants. Significant progress on standards, codes and databases for heat engine and aero applications has been made since 1991, however new requirements for CMC applications in NGNP require additional efforts. The current (2005) state of standards, design codes and databases for CMCs is reviewed and discussed. Issues related to development, verification and use of new standards, design codes, and database NGNP applications are presented.

INTRODUCTION

Ceramic matrix composites (CMCs) are candidate advanced materials for a variety of industrial applications including tubing, nozzles, vanes, supports, guides in heat recovery equipment, heat engines and nuclear power assemblies. [1-3] In gas turbine applications, CMC components can result in substantial reductions in emissions and savings in the cost of power generation[2] by allowing increased combustion temperatures (see Figure 1). In nuclear power assemblies, CMC components may be the only material of choice because of their low thermal expansion, resistance to degradation from irradiation and excellent retention of mechanical properties at elevated temperatures under high pressures.[3] (see Figure 2)

CMC materials possess many of the critical characteristics of advanced ceramic monoliths (e.g., low density, high stiffness, elevated temperature capability, etc.) while exhibiting increased "toughness" over their monolithic counterparts. [1-2] Continuous fiber-reinforced CMCs exhibit greatly increased "toughness" (i.e., nonlinear energy dissipation during deformation) and therefore provide the inherent damage tolerance and increased reliability that are critical in many engineering applications where the brittleness of conventional advanced ceramics leads to the unacceptable utility of these materials.[1]

The uniqueness of CMCs often makes conventional testing and design methodologies inappropriate, just as the characterization and prediction of the thermo-mechanical behavior of CMCs has become the subject of extensive investigations worldwide. As CMC prototype and trial products reach the marketplace, the relative scarcity of standards (i.e., test methods, classification systems, unified terminology, and reference materials) for CMCs and the lack of design codes and their related data bases for CMCs are limiting factors in commercial diffusion, industrial acceptance and regularity approval[4] of these advanced materials.

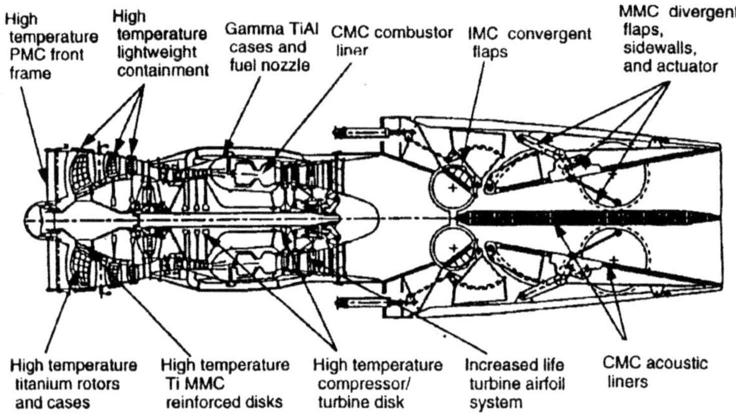

Figure 1 CMC applications in an advanced gas turbine (Ref 5)

In this paper, the issue of standards for CMCs in Next Generation Nuclear Power (NGNP) Gen IV applications and their extension to design codes and data bases is addressed. Of particular concern are component geometries (such as tube-shaped control rod guides) that may require specific test methods. First standards for CMCs are defined, followed by discussion of the current status of standards and codes for CMCs. Finally, some conclusions on the future of standards and codes for CMCs are presented.

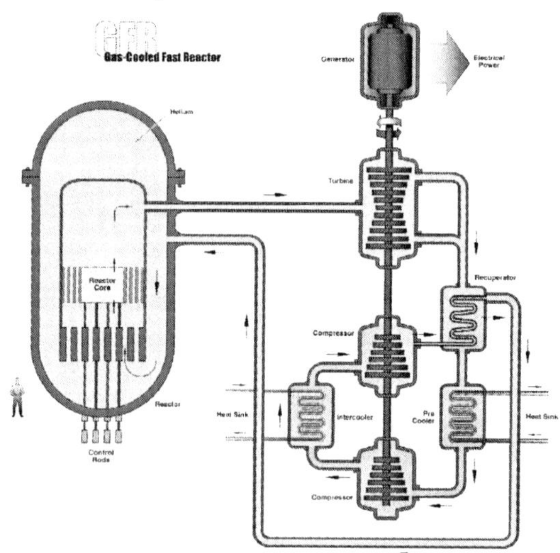

Figure 2 CMC tube as control rod guides in advanced (GenIV) gas cooled fast reactors (Ref 3)

STANDARDS

As a technical term, "standards" has multiple interpretations. For example, for the researcher or for those in the technical community "standards" may be units of measure and/or fundamental test methodologies. For the manufacturer or end-product user, "standards" may be materials specifications and tests to meet requirements. Commercial standards equate to the rules/terms of information transfer among designers, manufacturers and product users.[4] There are even differences between levels of standards: company (internal use and consensus); industry (trade use with limited consensus); government (wide use and various levels of consensus); full-consensus (broadest use and greatest consensus).

Currently there are few --national or international-- full-consensus standards[4] for evaluating advanced ceramics and CMCs in particular (see Table 1). Technical and pragmatic issues related to standardization efforts for CMCs have been discussed[4] for full consensus standards [that is, American Society for Testing and Materials (ASTM) Subcommittee C28.07 on Ceramic Matrix Composites, Comité Européen de Normalisation (CEN) Subcommittee TC184/SC1 on Ceramic Composites, and International Organization for Standardization (ISO) Technical Committee TC206 on Fine (Advanced, Technical) Ceramics]. In addition, other noteworthy, non full-consensus standards for CMCs have been developed by NASA High Speed Research/Enabling Propulsion Program (HSR/EPM) in the United States and Petroleum Energy Center (PEC) in Japan.

However, despite these widespread efforts, the lack of standards has limited the ability to evaluate CMCs on a common-denominator basis and may be hampering further material development[4]. It is noteworthy that the rate of standards development for CMCs has been increasing. Equally important, new efforts within ISO TC206 on Fine Ceramics are aimed at harmonizing existing standards rather than initiating new efforts, thus assisting the introduction of CMC standards.

HARMONIZATION OF TESTING

Uniaxial tensile testing is the most fundamental and, therefore, most common test for CMC. As a result, there are currently five tensile test methods for CMCs at room temperature as shown in Table 1. Of these five standards, three (ASTM, CEN, and EPM) were developed independently. Two others (ISO and PEC) were developed by harmonizing (i.e., choosing the "best" aspects) of the pre-exisiting standards.

For example, common terminology and nomenclature allows communication between testers without confusion. In ISO 15733, such common terms as fine ceramic and ceramic matrix composite are defined as follows:

fine (advanced, technical) ceramic, n - A highly engineered, high performance predominately non-metallic, inorganic, ceramic material having specific functional attributes.

ceramic matrix composite, n - A material consisting of two or more materials (insoluble in one another), in which the major, continuous component (matrix component) is a ceramic, while the secondary component/s (reinforcing component) may be ceramic, glass-ceramic, glass, metal or organic in nature. These components are combined on a macroscale to form a useful engineering material possessing certain properties or behavior not possessed by the individual constituents.

Table 1 Some Test Standards for Ceramic Matrix Composites (Ref 5)

Requirement	ASTM	CEN	ISO	EPM (HSR/EPM)	PEC
Bulk CMC, R.T. in Air					
Tension ··· In-plane ··· Transverse	C1275 C1468	EN658-1 ---	ISO15733 ---	D-001-93 ---	TS-CMC01 ----
Compression ··· In-plane	C1358	ENV658-2	DIS20504		
Shear ··· In-plane ··· Interlaminar	C1292 C1292	ENV12289 ENV658-4	DIS20506 DIS20505		---- TS-CMC06
Flexure ··· Bending ··· Shear	C1341 ----	ENV658-3 ENV658-5		D-003-93 ----	TS-CMC04 -----
Mechanical Fatigue ··· Tensile ··· Flexural	C1360 ----			D-002-93 ----	TS-CMC10 ---
Density		ENV 1389			
Elastic Constants	C1258				TS-CMC13
Fracture		ENV13234			TS-CMC08 TS-CMC09
Bulk CMC, H.T. in Air/Inert					
Tension ··· In-plane ··· Transverse	C1359 ----	ENV1892 ENV1893		D-001-93 ----	TS-CMC01 ---
Compression ··· In-plane		ENV12290 ENV12291			
Shear ··· In-plane ··· Interlaminar	---- C1425	ENV1894			---- TS-CMC06
Flexure ··· Bending ··· Shear	C1341 ----			D-003-93 ----	TS-CMC04 ----
Mechanical Fatigue ··· Tensile ··· Flexural				D-002-93 -----	TS-CMC10 -----
Thermal Fatigue					TS-CMC14
Creep	C1337	ENV13235		D-004-93	TS-CMC11
Stress Rupture				D-004-93	
Specific Heat		ENV1159-3			
Thermal Diffusivity		ENV1159-2			
CTE		ENV1159-1			
Ceramic Fiber, R.T. in Air					
Tension	C1557	ENV1007-4			
Diameter		ENV1007-1 ENV1007-3			
Density		ENV1007-2			

ASTM = American Society for Testing and Materials. CEN = Comité Européen de Normalisation. ISO = International Organization for Standardization. EPM = Enabling Propulsion Materials. PEC = Petroleum Energy Center

ISO 15733 also serves to show the diversity of test methods. In this case, no single tensile test specimen geometry has been forthcoming as the "best." Therefore, a range of successful test specimen geometries is illustrated in the standard to illustrate to users a variety of possibilities (see Figure 3).

The number of properties extractable from a uniaxial tensile tests of CMCs is illustrated in ISO15733 (see Figure 4). Each of these properties has explicit formulae for determining it from the stress strain curve.

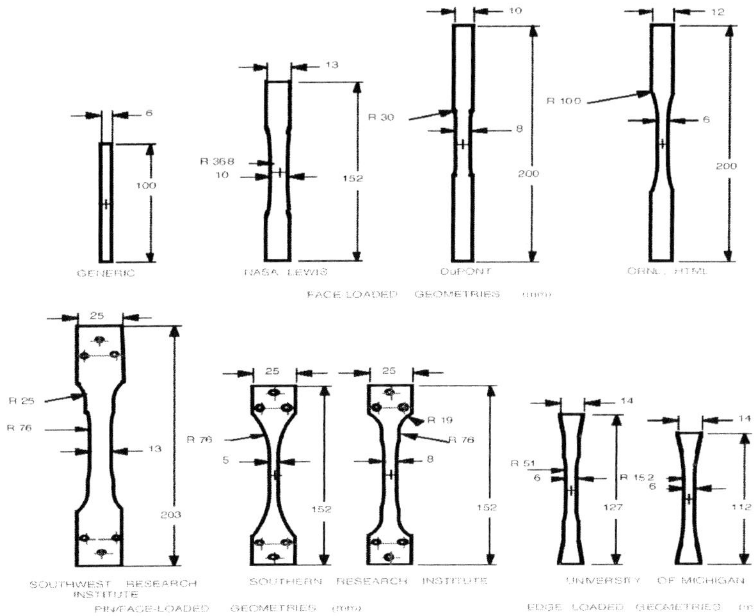

Figure 3 Examples of various tensile test piece geometries (see ISO15733)

For Gen IV gas cooled fast reactor (GFR) applications, a targeted application of CMCs is tubes for control rods. These CMC would be subjected to temperatures of 490 to 850°C at pressures up to 9 MPa of gaseous helium in addition to fast neutron fluence. Standards for mechanical behaviour of CMC tubular components have been addressed only in passing discussions. Standards for irradiation effects of CMCs have not even been proposed. Therefore, to address the specific requirements for NGNP applications, CMC standardization need to be extended to include test methods for tubes in adverse conditions.

DESIGN CODES

Interpretation of the term "design code" is not always clear. In this discussion, "design code" is not a design manual (that is, a "cookbook" design procedure that results in a desired component or system). Rather, "design codes" are widely-accepted, but general rules for the construction of components or systems with emphasis on safety. A main objective is the reasonably-long safe-life of the design while providing the reasonably certain protection of life and property. While recognizing the needs of the users, manufacturers and inspectors, the safety of the design can never be compromised.

Because it does not impose specific rules for design, the "design code" allows flexibility for introducing new designs as required for performance, efficiency, usability, or manufacturabilty while still providing constraints for safety. Such a "design code" is wide ranging and incorporates links between materials, general design (formulae, loads, allowable stress, permitted

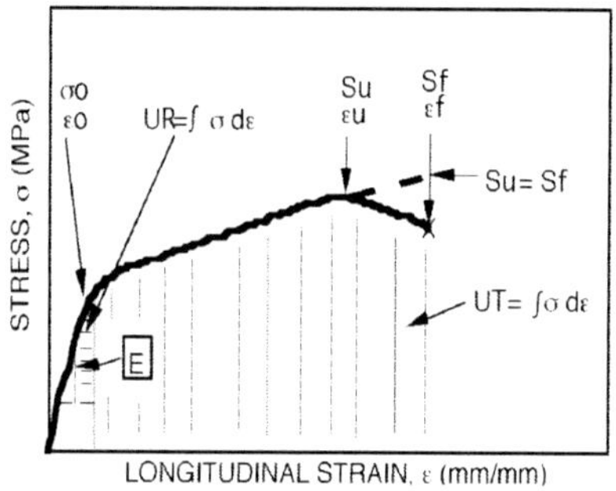

Figure 4 Some explicit parameters extractable from stress-strain curves (ISO15733)

details), fabrication methods, inspection, testing, certification, and data reports, and finally quality control to insure that the "design code" has been followed. Thus, implicit in the "design code" are many of the standards for materials testing, characterization, and quality control. Additionally, unlike standards that do not provide for compliance or accountability, "design codes" require compliance through documentation, and certification through inspection and quality control.

Two efforts are currently underway to establish such design codes for CMCs: Military Handbook 17 (Mil-Hdbk-17) CMC effort and ASME Task Group on Ceramic and Graphite Pressure Equipment. Test standards are incorporated in both these efforts to provide common and consensus methods for determining the properties and performance of materials "approved" for use under the design code. Figure 5 is a schematic illustration of the use a "design by analysis" aspect of a design code in which the long-term performance of a component is predicted using an iterative algorithm that requires as input, the long-term performance of materials determined using standards.

The extension of standards to industry-approved and embraced design codes is particularly important for GenIV nuclear applications because of the oversight provided by the U.S. Nuclear Regulatory Agency (NRC). The NRC is the "watchdog" agency of the nuclear power sector and as such is insistent on documentation, best practices and rigorous procedures. Design codes along with the standards incorporated in the design code represent debated, balloted and approved consensus documents that can go a long way toward meeting the approval of the NRC.

ROUND ROBIN

Round robin test programs using existing test standards can provide two primary outcomes: 1) determination of weaknesses and strengths of consensus standards and 2) statistically-significant set of data for properties and performance. In the year 1999 a well conducted round robin on a "mature" fibre-reinforced CMC was conducted by 21 different participants on 400 test specimens using ASTM Test Methods C1273, C1292, C1341.[6] Tests were conducted in in-plane and trans-thickness tension, in-plane and interlaminar shear, and in-plane flexure, all at room temperature. Statistically significant round robin results are shown in Figure 6.

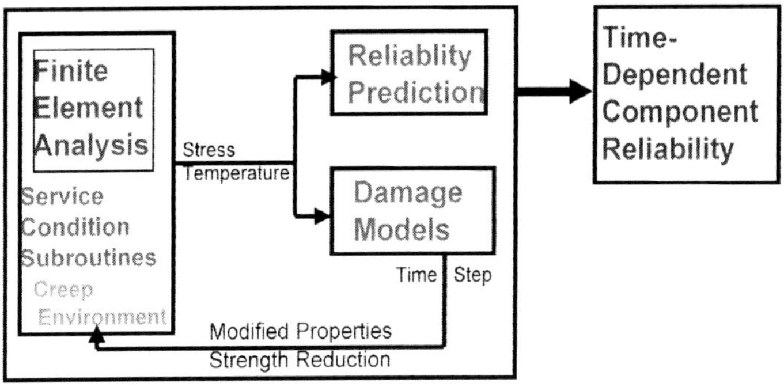

Figure 5 Schematic illustration of "design by analysis" approach for CMCs where time-dependent performance of components is predicted using long term performance of materials measured using standards. (Ref 5)

Physical Characteristics

Dow Corning Sylramic™ S200 (CG Nicalon™)

8 plies of 8HSW, 0/90 symmetric layup

V_f=45±1.5%, ρ=2.20±0.01 g/cm^3, V_p=2.73±0.99% (n=100)

Elastic Constants (in-plane)

Modulus of elasticity, E = 93 ± 6 GPa (n=80)

Shear modulus, G = 33 ± 2 GPa (n=50)

Poisson's ratio, ν = 0.12 ± 0.01 (n=10)

Elastic Limits

Tensile Proportional limit stress, σ_o = 85 ± 3 MPa (n=80)

Tensile Proportional limit strain, ε_o = 1000 ± 0 μm/m (n=80)

Shear proportional limit stress, τ_o = 64 ± 4 MPa (n=50)

Shear proportional limit strain, γ_o = 2000 ± 0 μm/m (n=50)

Flexure proportional limit stress, σ_{of} = 83 ± 6 MPa (n=100)

Flexure proportional limit strain, ε_{of} = 1000 ± 0.0 μm (n=100)

Ultimate Strengths (interlaminar)

Interlaminar shear strength, τ_i = 32 ± 5.5 MPa (n=50)

Trans thickness tensile strength, S_{uTT} = 24 ± 6 MPa (n=50)

Ultimate Strengths (in-plane)

Ultimate tensile strength, S_{uts} = 250 ± 18 MPa (n=80)

Shear strength, τ_u = 111 ± 7 MPa (n=50)

Flexural strength, S_{uF} = 339 ± 37 MPa (n=100)

Ductility (in-plane)

Tensile fracture strain, ε_f = 4400 ± 400 μm (n=80)

Modulus of toughness, U_T = 623500 ± 102405 J/m^3 (n=80)

Figure 6 Statistically-significant properties determined in round robin for Dow Corning Sylramic S200 CMC

CONCLUSIONS

Standards for CMCs now number about 60 since 1991. As indicated in this paper, these standards have numerous uses: harmonize international standards, establish the basis for determining properties and performance for "qualified materials" within design codes, and provide common methods for cooperative round robin programs for developing statistically significant data sets.

However, while the development of standards, codes and data bases for CMCs to date, for CMC applications in GenIV nuclear power generation, the current state of standards and codes for CMCs is inadequate. Not only do new standards need to be developed for non-standard tubular shapes under a variety of load (e.g., tension, flexure, internal pressurization) and environmental (e.g., elevated temperature, high pressure gas and fast neutron fluence) but new design codes must be developed to integrate CMCs into evolving GenIV applications.

REFERENCES

[1] Jenkins, M. G., Piccola, J. P. Jr., Mello, M. D., Lara-Curzio, E., Wereszczak, A. A. (1993) "Mechanical Behaviour of a 3-D Braided, Continuous SiC Fiber-Reinforced / CVI SiC Matrix Composite at Ambient and Elevated Temperatures," *Ceramic Engineering and Science Proceedings*, **14** , 9-10. 991-997 (1993)

[2] Karnitz, M. A., Craig, D. A. and Richlen, S. L. (1991) "Continuous Fiber Ceramic Composite Program," *Ceramic Bulletin*, **70**, 3, 430-435 (1991).

[3] U.S. DOE Nuclear Energy Research Advisory Committee, "A Technology Roadmap for Generation IV Nuclear Energy Systems," GIF-002-00, U.S. Department of Energy, Washington, DC (2002)

[4] Schneider, S. J. and Bradley (1988) "The Standardization of Advanced Ceramics," Advanced Ceramic Materials, Vol 3, No. 5, pp. 442-449.

[5] Jenkins, M. G., "Standards for CMCs: What are They Good for?" HT-CMC5 Proceedings, American Ceramic Society, Westerville, OH (2005)

[6] Jenkins, M. G. , Lara-Curzio, E., Gonczy, S.T. and Zawada, L.P. "Multiple-Laboratory Round-Robin Study of the Flexural, Shear and Tensile Behaviour of a Two-Dimensionally Woven Nicalon™/Sylramic™ Ceramic Matrix Composite," pp. 15-30 in ASTM STP 1392, ASTM International, West Conshohocken, PA. (2000)

Component Development and Industrial Applications

DEVELOPMENT OF STRONG PHOTOCATALYTIC FIBER WITH SURFACE GRADIENT STRUCTURE

Toshihiro Ishikawa
Ube Research Laboratory, Ube Industries, Ltd.,
1978-5 Kogushi, Ube City, Yamaguchi Prefecture, 755-8633, Japan

ABSTRACT

In order to avoid large problems regarding peering of the titania layer coated on the substrate, we developed an epoch-making "strong titania fiber" consisting of photoactive surface layer with a nanometer-scale compositional gradient, which can effectively oxidize any kind of organic materials. An effective water-purification system using this fiber has been also developed. The basis of this technology is to incorporate a selected low-molecular-mass additive ($Ti(OC_4H_9)_4$) into a precursor polymer from which the ceramic forms. After melt-spinning the resulting precursor polymer, thermal treatment of the spun fiber leads to controlled phase separation ("bleed-out") of the additive; subsequent calcination stabilizes the compositionally changed surface region, generating a functional surface layer. This fiber consists of the silica-based core-structure and the gradient-like surface titania layer, which are strongly sintered. We also developed a water-purifier using this fiber (felt material). Any bacteria (common bacterium, legionera pneumophila, colon bacillus, heterotrophic bacteria, and so forth) and organic chemicals (dioxin, PCB, and so forth) were effectively decomposed into CO_2 and H_2O passing through the above purifier.

INTRODUCTION

Many types of polymer-derived ceramic fibers have been developed using a polycarbosilane $(-SiH(CH_3)-CH_2-)_n$ as the starting material [1-4]. Through research, many modifications have been made to obtain much higher heat-resistant fibers [2-4]. Of these, we developed the highest heat-resistant SiC polycrystalline fiber (Tyranno SA fiber) possessed of the excellent heat-resistance up to 2000°C in 1998 [3]. This fiber was synthesized from polyaluminocarbosilane prepared by addition of very small amount of organic compound of aluminum, which serves as a sintering aid. Using a polyaluminocarbosilane precursor made it easy to control the aluminum concentration less than an upper concentration limit of the solid-soluble aluminum in the SiC crystal. This led Tyranno SA fiber to having excellent mechanical properties (3 GPa) and stability up to very high temperatures (~2000 °C). This fiber has very high thermal conductivity (64 W/mK), and some applications, which require good thermal shock resistance and high thermal conductivity, have been examined. In this time, we also developed unique fiber-bonded ceramic (SA-Tyrannohex), which was composed of highly ordered, close-packed structure of very fine hexagonal columnar fibers with very thin interfacial carbon layer between fibers for creating high fracture toughness. This ceramic is one of biomimetic ceramics aiming for natural wood structure or shell structure. Each fiber element consists of the same SiC polycrystalline structure as the above-mentioned Tyranno SA fiber. Accordingly, SA-Tyrannohex shows steady strength and very high fracture energy up to 1700°C in air.

Recently, using this base technology, we developed a new type of functional ceramic fiber with gradient surface structure [6, 7]. This fiber was produced from a polycarbosilane containing an excess amount of selected low-molecular-mass additive, which can be converted into a functional ceramic by heat-treatment. Thermal treatment of the precursor fiber leads to controlled phase separation ("bleed out") of the low-molecular-mass additives from inside to outside of the precursor fibers. After that, subsequent calcination generates a functional surface layer during the production of bulk ceramic components. As the embodied functional material of our new process, we developed a strong photocatalytic fiber composed of anatase-TiO_2 surface structure and silica core structure. The above-mentioned technical background is summarized in Fig.1.

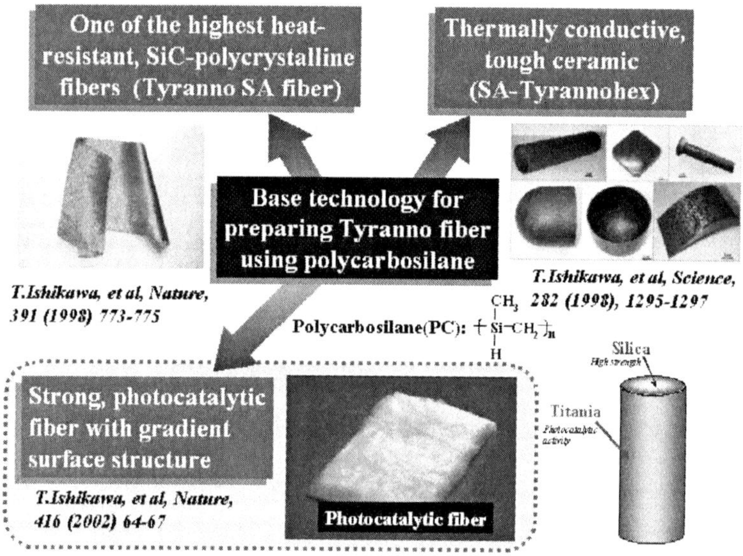

Fig.1 Background of our technology

Anatase-TiO_2 is well known as a semiconductor catalyst, which exhibits a better photocatalytic activity by irradiation of a light with energy greater than the band gap (3.2eV) [8]. The photocatalytic activity appears by irradiation of an ultraviolet (UV) light with wavelength shorter than 387nm (=3.2eV). The decomposition of harmful substances using the photocatalytic activity of anatase-TiO_2 has attracted a great deal of attention [9-16]. This effect is attributed to the strong oxidant (hydroxyl radical), which generates at the surface of the TiO_2 crystals by the irradiation of UV light. The fundamental mechanism of the photocatalytic activity of titania is shown in Fig. 2.

At present, most research have been performed using powder material or coated material on the substrate. Of these, powdery photocatalysts have some difficulties in practical use [9]. For example, they have to be filtrated from treated water. Coated photocatalysts on the substrate cannot provide sufficient contact area with harmful substances [9]. In addition, coated layer is easily peeled off from the substrate during usage. In order to avoid these problems, other types of

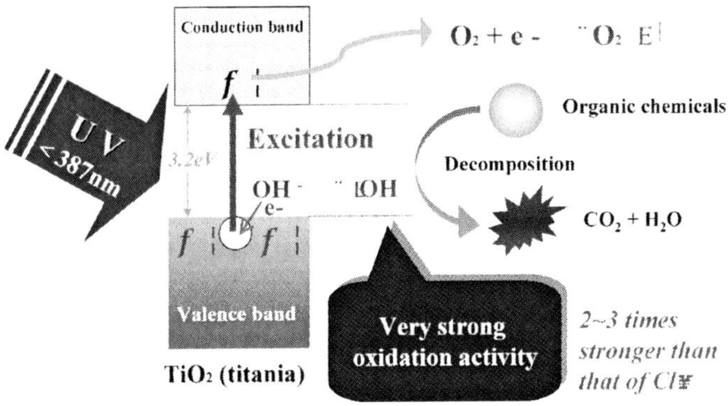

Fig.2 Fundamental mechanism of the photocatalytic activity of titania

research concerning fibrous photocatalysts have been conducted [10]. However, up to the present, a combination of excellent photocatalytic activity and high fiber strength has not been achieved using sol-gel method or simple polymer-blend [17, 18]. According to the aforementioned new process, we achieved the development of a new photocatalytic, strong (2.5 GPa) and continuous fiber with small diameter (5~7μm), namely, a type of titania-dispersed silica-based fiber with a sintered anatase-TiO_2 layer on the surface. The surface gradient layer composed of nanoscale TiO_2 crystals (8 nm) was strongly sintered and exhibited excellent photocatalytic activity, which can lead to the efficient decomposition of harmful substances and any bacterium contained in air and/or water by irradiation of UV light. In this paper, the abovementioned photocatalytic fiber produced by our new *in situ* process and its actual applications are described.

SYNTHESIS OF OUR NEW PHOTOCATALYTIC FIBER

Polytitanocarbosilane containing an excess amount of titanium alkoxide was synthesized by the mild reaction of polycarbosilane $(\text{-SiH(CH}_3\text{)-CH}_2\text{-})_n$ (20kg) with titanium (IV) tetra-n-butoxide (20kg) at 220 °C in nitrogen atmosphere. The obtained precursor polymer was melt-spun at 150 °C continuously using melt-spinning equipment. The spun fiber, which contained excess amount of unreacted titanium alkoxide, was pre-heat-treated at 100 °C and subsequently fired up to 1200 °C in air to obtain continuous, transparent fiber (diameter: 5~7μm). In the initial stage of the pre-heat-treatment, effective bleeding of the excess amount of unreacted titanium (IV) tetra-n-butoxide from the spun fiber occurred to form the surface gradient layer containing large amount of titanium (IV) tetra-n-butoxide. During the next firing process, the pre-heat-treated precursor fiber was converted into a titania-dispersed, silica-based fiber covered with gradient titania (our photocatalytic fiber). The fundamental concept of the new production process for our photocatalytic fiber is shown in Fig.3.

Figure 4 shows the surface appearance and the cross section of our photocatalytic fiber. As can be seen from this figure, the surface of the fiber is densely covered with nanoscale anatase-TiO_2 particles (8 nm), which are strongly sintered with each other directly or through

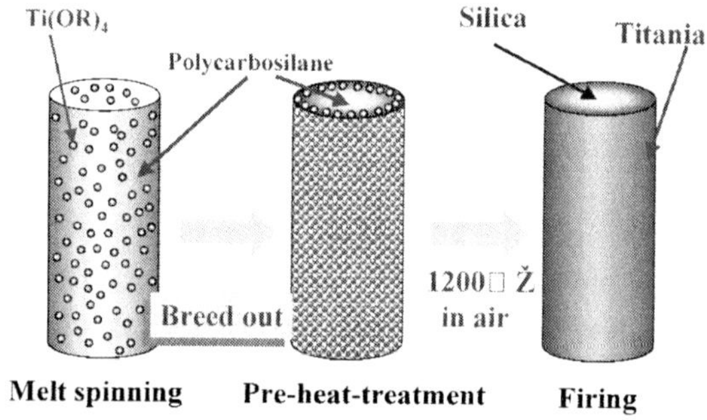

Fig.3 Fundamental concept of our photocatalytic fiber

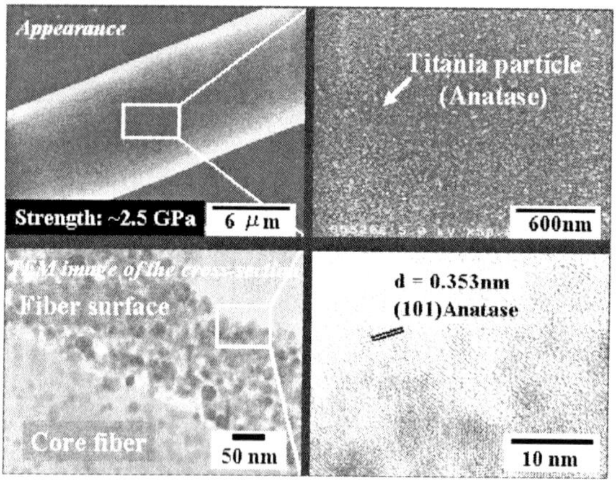

Fig.4 The surface and cross-section of our photocatalytic fiber

with amorphous silica phase. The thickness of the surface TiO_2 layer is approximate 100~200 nm. The tensile strength of this fiber as measured by a single filament method was 2.5GPa on the average using an Orientec UTM-20 with a gauge length of 25 mm and cross-head speed of 2 mm/min. This mechanical strength is markedly superior to that of existing photo-catalytic TiO_2 fibers (<1GPa), which were produced by a sol-gel method [17] or using polytitanosiloxanes [18]. The high strength of our photocatalytic fiber is closely related to the dense structure without pores, which is caused by its higher firing temperature compared with other TiO_2 fibers.

The photocatalytic activity of our fiber is caused by the anatase-titania on the surface of each fiber. It is well known that anatase-titania converts to rutile at temperatures ranging from 700°C to 1000°C [16]. In particular, pure nanocrystalline anatase easily converts to rutile at lower temperature (~500°C) [20]. In our case (firing temperature: 1200°C), it is thought that the surrounding silica phase caused the stabilization of the anatase phase. At the interface between titania and silica, atoms constructing titania are substituted into the tetrahedral silica lattice forming tetrahedral Ti sites [21]. The interaction between the tetrahedral Si species and the tetrahedral Ti sites in the anatase is thought to prevent the transformation to rutile. Figure 5 and 6 show the crystalline structure of titania along with the relationship between silica and titania, and phase diagram for the titania-silica phases, respectively.

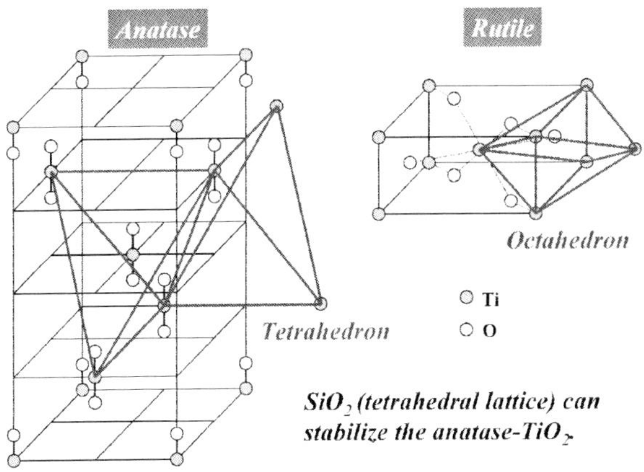

Fig.5 The crystalline structure of titania along with the relationship between titania and silica

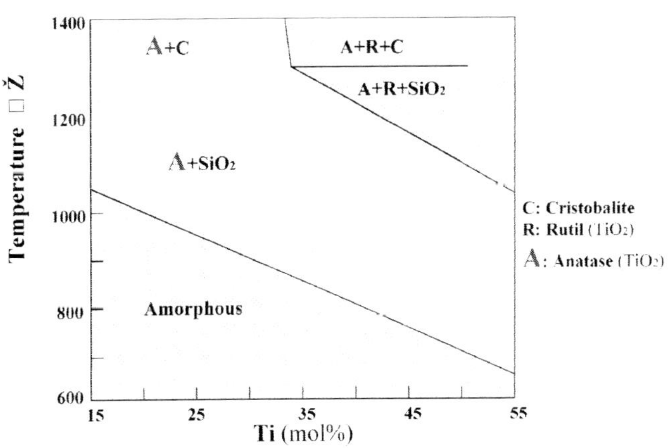

Fig.6 The phase diagram for titania-silica phases

APPLICATION of OUR PHOTOCATALYTIC FIBER

We developed a water-purifier for pollutants using a felt material made of the aforementioned photocatalytic fiber (Fig.7). The average intensity of UV light on the photocatalytic fibers should be $10mW/cm^2$ for obtaining good decomposition activity.

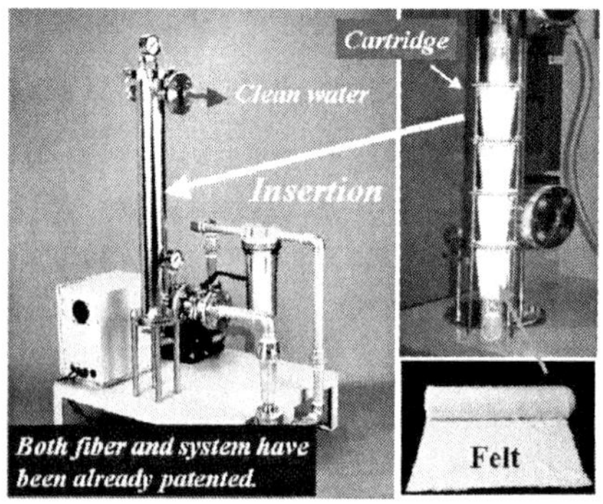

Fig.7 Water-purifier using our photocatalytic fiber

This is a very simple purifier with a module composed of the cone-shaped felt material (made of our photocatalytic fiber with a very high quantum efficiency (~40%) [19]) and UV lamp. Purification of the water of collective bathtubs and swimming pools was performed using this purifier. The muddiness of the pool water was remarkably improved (Fig.8) by the passage through the purifier. Organic filth and chloramines also decreased after passage through the purifier.

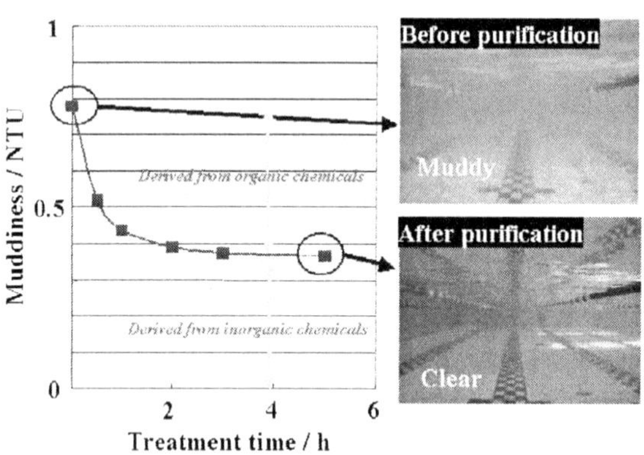

Fig.8 Purification of the pool-water by the purifier using our photocatalytic fiber

In addition, many bacteria (common bacterium, legionera pneumophila and coliform), which existed in the initial bath water, were effectively decomposed into CO_2 and H_2O. The experimental data on the sterilization of legionera pneumophila is shown in Fig.9.

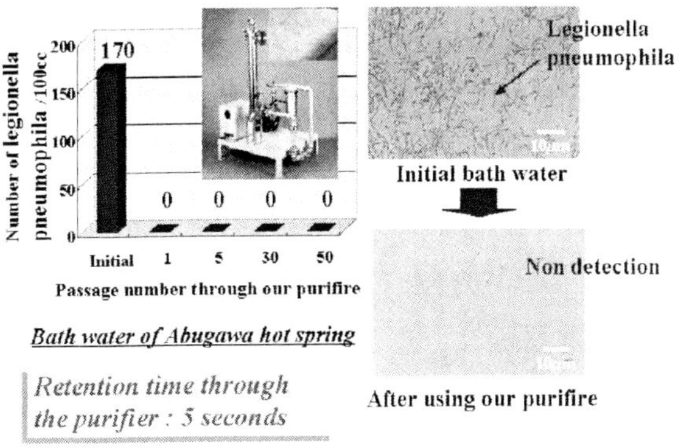

Fig.9 Sterilization of legionera pneumophila by the purifier using our photocatalytic fiber

As shown in figure 9, the legionera pneumophila were perfectly decomposed by single passage through the water-purifier using our photocatalytic fiber. In this case, the retention time for the single passage through the purifier was only 5 s. Although this time may seem too short, it is sufficiently long compared with the lifetime of hydroxyl radical (10^{-6} s). Each oxidation reaction ought to proceed within the aforementioned lifetime of hydroxyl radical. Accordingly, the decomposition reaction can be accomplished like this, as long as the number of photons is sufficient during the passage.

The result on the decomposition of colon bacillus by our photocatalytic fiber irradiated by UV light is shown in Fig.10 along with the comparison data obtained by only UV irradiation.

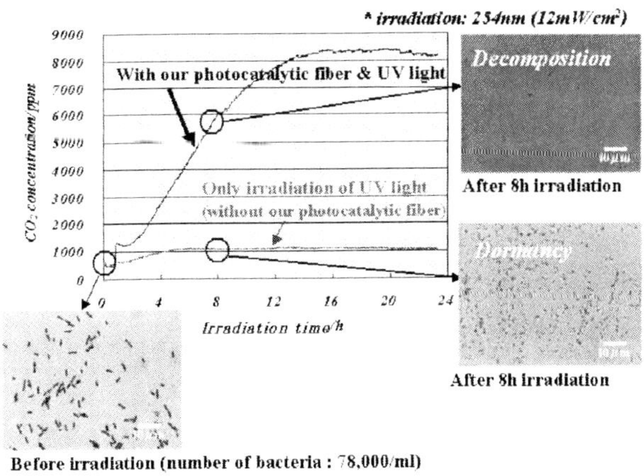

Fig.10 Decomposition of colon bacillus using our photocatalytic fiber

As can be seen from this figure, use of our photocatalytic fiber led to effective decomposition of colon bacillus accompanied by the generation of carbon dioxide. On the other hand, irradiation of UV light alone resulted in the many dead bodies of colon bacillus with no apparent decomposition. A damaged colon bacillus is shown in Fig.11. Although almost all of colon bacillus were perfectly decomposed into carbon dioxide and water, some residual bacteria were also remarkably damaged. This type of decomposition reaction proceeds on the surface of the photocatalytic fiber when colon bacillus contacts each fiber irradiated by UV light.

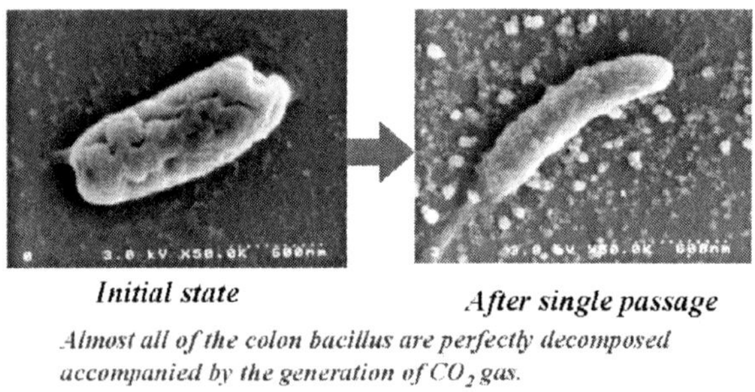

Fig.11 Damaged colon bacillus after single passage through the purifier using our photocatalytic fiber

Fundamentally, this can be applied to purification system of any kind of organic chemicals. Figure 12 shows the result of the field test on decomposition of dioxin. Dioxin is one of many persistent organic pollutants (POPs). It has been recognized that the perfect decomposition of POPs is very difficult. However, the use of our photocatalytic water-purifier enables the oxidation of dioxin into carbon dioxide and water. This result is caused by the oxidation activity of hydroxyl radical generated on the surface of our photocatalytic fiber irradiated by the UV light.

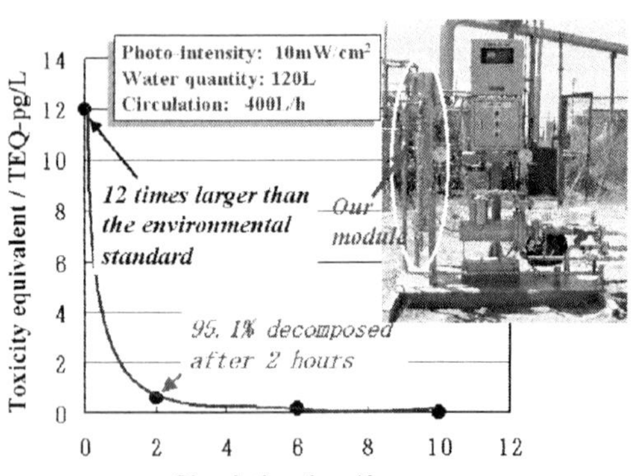

Fig.12 The result of the field test on decomposition of dioxin

By the effective formation of the anatase-titania on the surface of each fiber, our photocatalytic fiber can show the excellent decomposition activity of any kind of organic chemicals. Figure 13 shows the result of a purification test of the water containing bacillus subtilis, which is covered with hard shell out of the body. Our purifier enabled to decompose the shell structure by only single passage. From the result obtained without UV irradiation, it is found that the decrease in the number of bacillus subtilis was not caused by filtration.

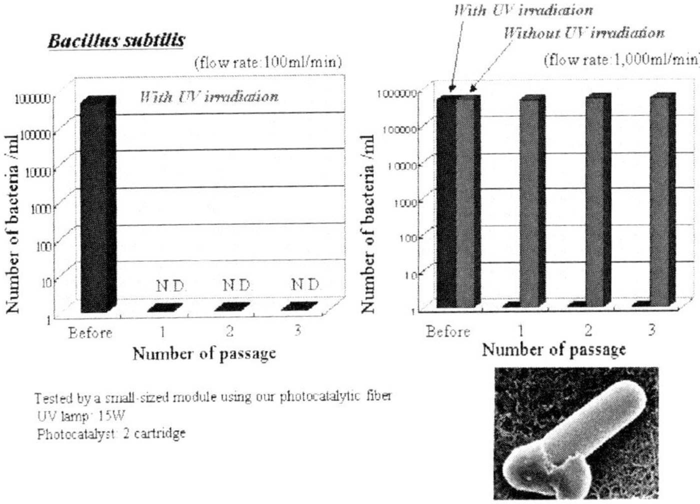

Fig.13 The result of the purification test on bacillus subtilis

Figure 14 shows evidence that the hard shell of the bacillus subtilis was actually damaged by contacting our photocatalytic fiber irradiated by UV light.

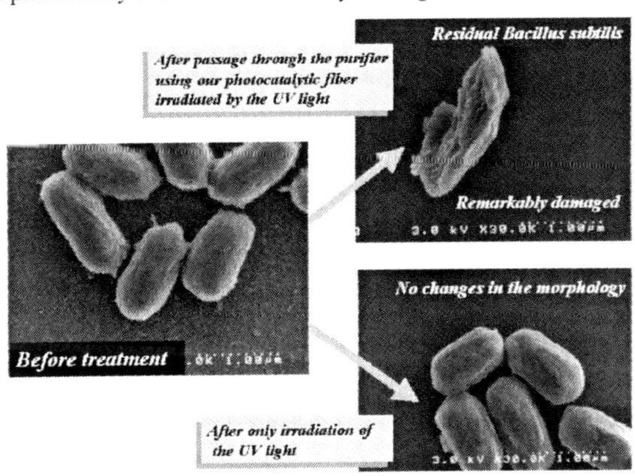

Fig.14 Damaged hard shell of the bacillus subtilis by contact with our photocatalytic fiber

It is well known that botulinus and anthrax are a variety of the aforementioned bacillus subtilis. And this type of bacteria hardly diminished by boiling water or chlorine. Accordingly, the water-purifier using our photocatalytic fiber is found to be very effective for avoidance of this type of hazard.

Regarding drinking water, you are also endangered if you use common water-purifier containing some active carbon filter. In the active carbon filter, lots of bacteria can propagate. We detected the number of heterotrophic bacteria in the outlet water from the active carbon filter (Fig.15). As can be seen from this result, there are found to be tremendous number of heterotrophic bacteria in the outlet water from the active carbon filter. This is very dangerous for you. Even such water could be perfectly purified by the use of our photocatalytic purifier (Aquasolution: AQ in Fig.15).

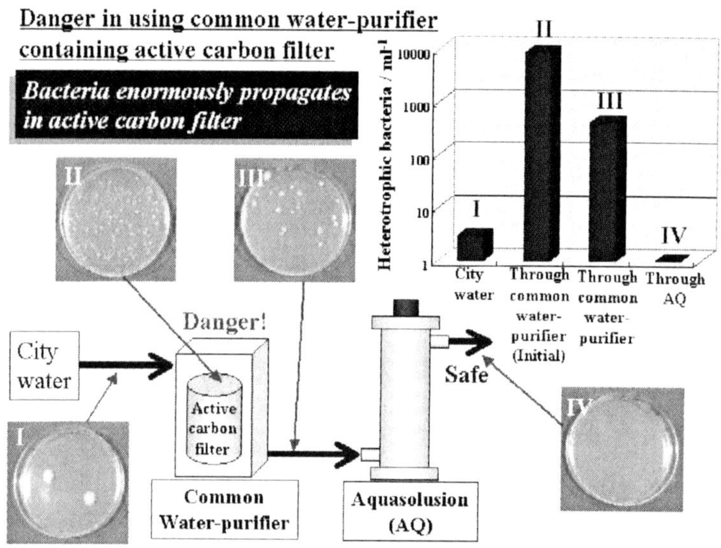

Fig.15 Danger of the outlet water from the active carbon filter

This dangerous phenomenon was also confirmed in the field of medical water. Figure 16 shows the changes in the number of bacteria in the purification system for preparing the water for dialysis. Active carbon filter is used in this system. In this case, a tremendous number of bacteria were detected in the outlet water from the active carbon filter. However, after a single passage through our photocatalytic purifier behind the active carbon filter, all of bacteria were decomposed.

As can be seen from the aforementioned information, we need to reconsider the use of active carbon filter, which is often believed to be a safe purification system. In fact, active carbon filter should be recognized as a kind of breeding zone of bacteria. In order to avoid such dangerous situation, the photocatalytic purification system must be very effective.

Furthermore, we have also confirmed the effective purification of the water for rinsing mouth or tooth-grinder in the dental systems.

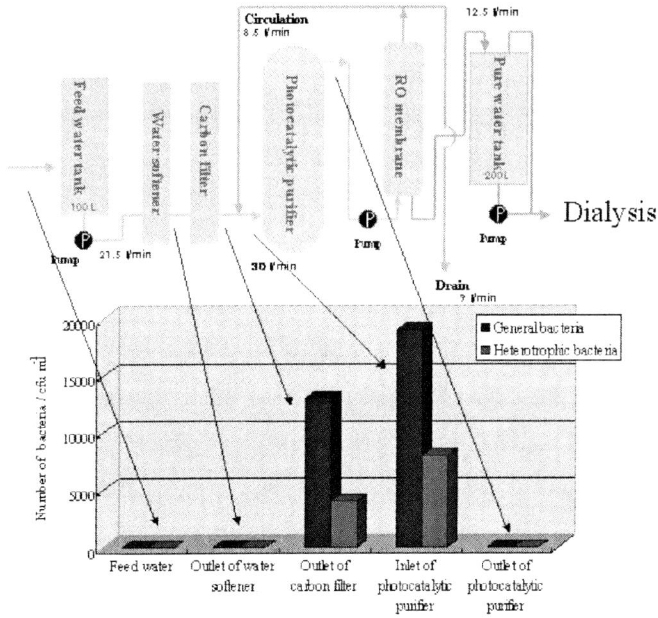

Fig.16 Purification of the medical water using photocatalytic purifier

SUMMARY AND CONCLUSION

On the basis of a precursor method using a polycarbosilane, we developed a new *in situ* process for preparing functional ceramic fibers with a gradient surface layer. This process treated a polycarbosilane containing an excess amount of selected low-molecular-mass additives, which can be converted into functional ceramics by heat-treatment. Thermal treatment of the precursor fiber leads to controlled phase separation ("bleed out") of the low-molecular-mass additives from inside to outside of the fiber. After that, subsequent calcination generates a functional ceramic fiber with a gradient surface structure during the production of bulk ceramic components. As the embodied functional ceramic fiber, we developed a strong photocatalytic fiber composed of a surface gradient titania-crystalline layer and a silica-based core structure. This fiber showed an excellent photocatalytic activity, that is to say, the very high quantum efficiency (~40%) and a decomposition ability of organic chemicals and any kind of bacteria, and so forth, by irradiation of UV light. Very tough bacteria such as bacillus subtillis and anthrax, which are covered with hard shell and hardly decomposed by chlorine or boiled water, can be also effectively decomposed by our photocatalytic purifier. Our system has been successfully applied to purification of many types of water in wide fields.

REFERENCES

[1]S. Yajima, Y. Hasegawa, K. Okamura, T. Matsuzawa, "Development of High Tensile Strength Silicon Carbide Fiber Using an Organosilicon Polymer Precursor", *Nature*, 273, 525-527 (1978).

[2]M. Takeda, J. Sakamoto, A. Saeki, Y. Imai, and H. Ichikawa, "High Performance Silicon Carbide Fiber Hi-Nicalon for Ceramic Matrix Composites", *Ceram.Eng.Sci.Proc.*, 16[4], 37-44 (1995).

[3]T. Ishikawa, Y. Kohtoku, K. Kumagawa, T. Yamamura & T. Nagasawa, "High-Strength Alkali-Resistant Sintered SiC Fibre Stable to 2200 ^{o}C", *Nature*, 391, 773-775 (1998).

[4]J. Lipowitz, J. A. Rabe, G. A. Zank, A. Zangvil, Y. Xu, "Structure and Prooperties of Sylramic TM Silicon Carbide Fiber – A Polycrystalline, Stoichiometric SiC Composition", *Ceram.Eng.Sci.Proc.*, 18[3], 147-157 (1997).

[5]T. Ishikawa, S. Kajii, K. Matsunaga, T. Hogami, Y. Kohtoku & T. Nagasawa, "A Tough, Thermally Conductive Silicon Carbide Composite with High Strength up to 1600 ^{0}C in Air", *Science*, 282, 1295-1297 (1998).

[6]T. Ishikawa, H. Yamaoka, Y. Harada, T. Fujii, T. Nagasawa, "A general process for in situ formation of functional surface layers on ceramics", *Nature*, 416, 64-67 (2002).

[7]T. Ishikawa, Y. Harada, H. Hayashi, S. Kajii, "Silica-Group Composite Oxide Fiber and Process for the Production", *US Patent 6,541,416 B2* (Foreign Application Priority Date: June 13, 2000).

[8]I. Sopyan, S. Murasawa, K. Hashimoto, and A. Fujishima, "Highly Efficient TiO2 Film Photocatalyst, Degradation of Acetaldehyde", *Chemistry Letters (The Chemical Society of Japan)*, pp.723-726 (1994).

[9]T. Ishikawa, "Ceramic Fiber with Decomposition Ability of Dioxine", *Miraizairyo, 3(2), 26-33 (2003).*

[10]H. Koike, Y. Oki, and Y. Takeuchi, *Mater.Res.Soc.Sym.Proc., 549, 141 (1999).*

[11]A. Matsuda, Y. Kotani, T. Kogure, M. Tatsumisago, T. Minami, "Transparent Anatase Nanocomposite Films by the Sol-Gel Process at Low Temperatures", *J.Am.Ceram.Soc.*, 83[1], 229-231 (2000).

[12]D. R. Park, J. Zhang, K. Ikeue, H. Yamashita, M. Anpo, "Photocatalytic Oxidation of Ethylene to CO_2 and H_2O on Ultrafine TiO_2 Photocatalysts in the Presence of O_2 and H_2O", *J.Catal.*, 185, 114-119 (1999).

[13]T. Ishikawa, "Photocatalytic Fiber with Gradient Surface Produced from a Polycarbosilane and its Applications", *International Journal of Applied Ceramic Technology*, 1[1], 49-55 (2004).

[14]T. Ishikawa, "Advances in Inorganic Fibers", *Adv. Polym. Sci.*, 178, 109-144 (2005).

[15]N. Takeda, M. Ohtani, T. Torimoto, S. Kuwabata, H. Yoneyama, "Evaluation of Diffusibility of Adsorbed Propionaldehyde on Titanium Dioxide-Loaded Adsorbent Photocatalyst Films from Its Photocomposition Rate", *J.Phys.Chem.B*, 101, 2644-2649 (1997).

[16]C. K. Chan, J. F. Porter, Y. G. Li, W. Guo, C. M. Chan, "Effects of Calcination on the Microstructures and Photocatalytic Properties of Nanosized Titanium Dioxide Powders Prepared by Vapor Hydrolysis", *J.Am.Ceram.Soc.*, 82[3], 566-572 (1999).

[17]H. Koike, Y. Oki, Y. Takeuchi, "Preparing Titania fibers and their Photo-catalytic Activity", *Mater.Res.soc.Sym.Proc.*, 549, 141-146 (1999).

[18]T. Gunji, I. Sopyan, Y. Abe, "Synthesis of Polytitanosiloxanes and their Transformation to SiO_2-TiO_2 Ceramic Fibers", *J.Polym.Sci., Part A Polym.Chem.*, 32, 3133-3139 (1991).

[19]T. Ishikawa, "Photocatalytic Fiber with Gradient Surface Structure Produced from a Polycarbosilane and Its Application", *Int.J.Appl.Chem.Technol.*, 1[1], 49-55 (2004).

[20]P. I. Gouma, P. K. Dutta, M. J. Mills, "Structural Stability of Titania Thin Films", *Nano Structured Materials*, 11(8), 1231-1237 (1999).

[21]C. Anderson, A. J. Bard, "Improved Photocatalytic Activity and Characterization of Mixed TiO_2/SiO_2 and TiO_2/Al_2O_3", *J.Phys.Chem.B*, 101, 2611-2616 (1997).

HOT SECTION Si_3N_4 MATERIALS DEVELOPMENT FOR ADVANCED MICROTURBINES APPLICATIONS

Vimal K. Pujari, Ara M. Vartabedian, William T. Collins
Saint-Gobain Ceramics & Plastics, Inc.
Northboro Research & Development Center
9 Goddard Road, Northboro, MA 01532
508-351-7929, E-mail: vimal.k.pujari@saint-gobain.com

INTRODUCTION

Silicon nitride offers excellent thermo-mechanical properties such as high strength and creep resistance at elevated temperatures, low co-efficient of thermal expansion and high thermal shock resistance. This makes it a suitable candidate material for high temperature gas turbine components such as integrally bladed rotors (IBR), vanes and combustor liners. At Saint-Gobain in the last decade, extensive effort was directed towards the development of high temperature silicon nitride based ceramic materials (NT154), net shape forming processes and prototype component field evaluations[1]. For example, under the Department of Energy Ceramic Technology Project, a highly reliable closed loop processing (CLP) methodology[2] was developed leading to significant improvements in the properties of NT154 silicon nitride.

The objective of the research effort described here is to re establish the NT154 material and the related processes as they relate to high temperature (1100-1300°C) microturbine applications for the distributed generation of power (approximately 25-500 kW). The successful use of ceramics as an enabling material for achieving higher temperature microturbine operation will result in improved efficiency, reduced fuel consumption, and lower emissions.

The technical effort involved a three-pronged approach: i) NT154 material development and improvement, ii) net-shape forming development of microturbine components such as rotors and vanes and iii) development of recession resistant solution for silicon nitride in a high temperature (1100-1300°C), high gas velocity (20-50m/sec), humid (10% H_2O) gas turbine environment. The progress towards each of these three areas is highlighted.

APPROACH

Material development effort was focused on re-establishing NT154 silicon nitride process and properties. NT154 silicon nitride is a glass encapsulated HIPed composition containing 4wt% Y_2O_3. The NT154 composition, which was originally developed in the late 1980's, had been dormant for approximately 8 years.

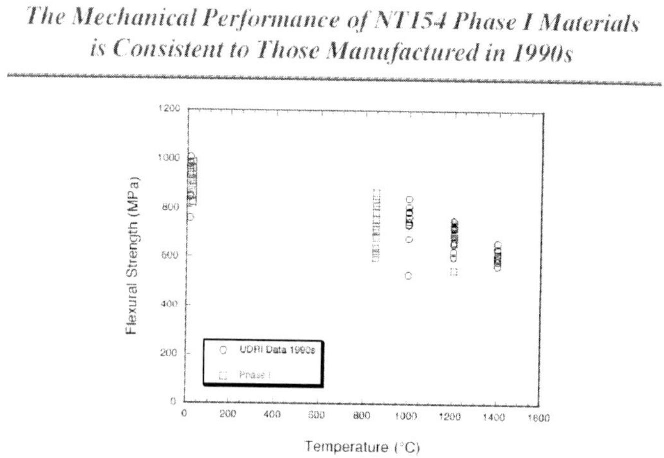

Figure 1: Re-established NT154 (Courtesy of HT Lin, ORNL)

Under the on-going activities for the improvement of NT154 properties a major effort was directed towards the enhancement of the as-processed (AP) strength of NT154 since all the aerofoil sections of the net shape formed components are not subjected to finish machining.

For the net shape forming of complex turbine components, a novel green machining process was developed and optimized utilizing Design of Experiment (DOE) approach. The DOE involved variables related to the green blank properties, cutting tools and machining parameters.

A new focus in the area of material development is recession (weight loss) control. Material recession, due to volatilization in humid gas turbine environments, is an issue for silicon nitride and has been reported in the literature extensively.

Through detailed thermodynamic computations, novel EBC (Environmental Barrier Coating) compositions were defined, synthesized and evaluated for suitability in CTE match with silicon nitride and durability (recession resistance) in high temperature (1204°C), high humidity environment. Recession testing was performed at Oak Ridge National Laboratory (ORNL) in the Keiser Rig test equipment.

RESULTS

Material Development and Improvement

As mentioned previously, the NT154 process was re-established during the current DOE funded program. In the first set of tiles delivered to ORNL, the fast fracture strength of NT154 agreed with historical values for the material (Figure 1).

The second set of tiles delivered to ORNL for mechanical property evaluation examined the effect of the binder system being used for the net shape green machining development.

The room temperature bulk and AP fast fracture and slow crack growth (SCG) properties from that testing is shown in Figure 2. The results are very similar to those measured on the first set of delivered tiles that had minimal binder content. In essence, the green machining binder system seems to have had a minimal

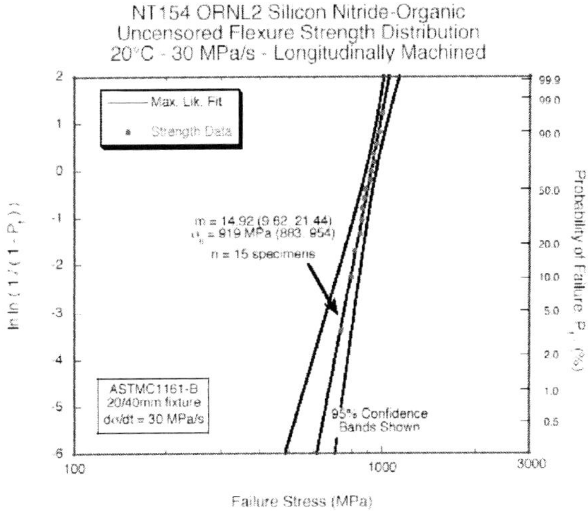

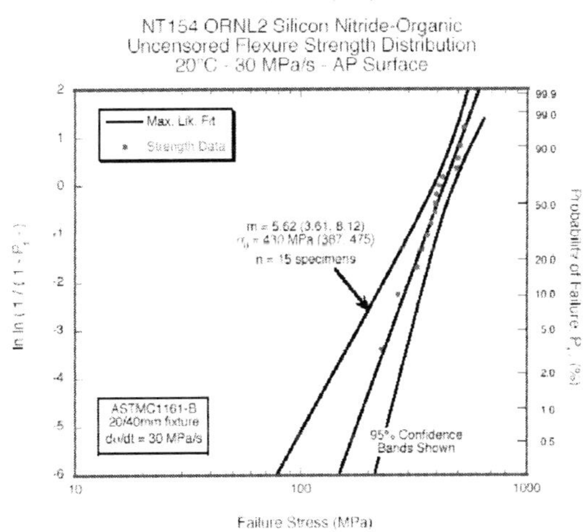

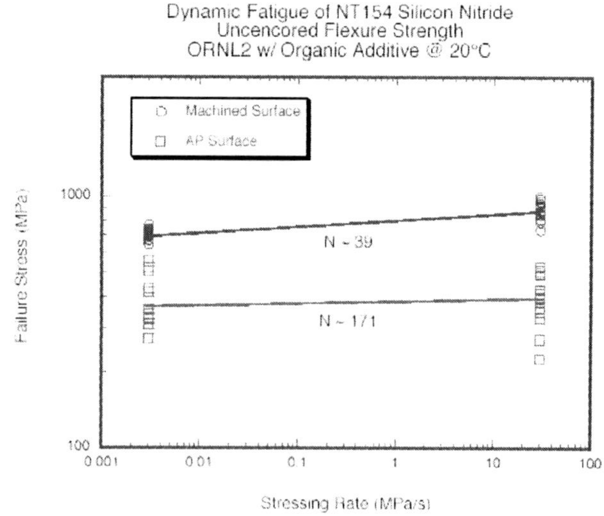

Figure 2: Mechanical Property Data for the Green Machining Composition of NT154 (Courtesy of HT Lin, ORNL). Dynamic Fatigue exponent, N

impact on the room temperature properties (strength and Weibull modulus m). These results have been further confirmed internally by Saint-Gobain laboratory at the Northboro R&D Center. In fact, internal testing indicates that the high temperature fast fracture strength actually increases with the binder system addition. The strength increase is as much as 21% at 1200°C with a high value of 766 MPa.

The main area identified for improvement continued to be as-processed (AP) strength, which has historically been a concern for silicon nitride. The glass encapsulated HIP process used to densify NT154 creates a reaction layer in the silicon nitride due to its interaction with the glass encapsulant. In addition to the decreased toughness present in the reaction layer, the glass / silicon nitride interaction can also create surface defects. Combined, these two factors lead to an AP strength that is less than the bulk, machined specimen, strength of NT154. Therefore the objective was to minimize the glass / silicon nitride interaction. This objective was accomplished through a new proprietary HIP process. The process was developed in a lab scale HIP during 2003 and optimized for the production HIP in 2004. When optimized, the process results in a near surface toughness that is comparable to the bulk material toughness and a surface roughness that is similar to the roughness of the sample in the green state.

The AP flexural strength of NT154 samples from various HIP runs is shown in Figure 3. It can clearly be seen that the new process always results in an improved AP strength. An average improvement of 46% has been seen with a maximum AP strength of 717MPa.

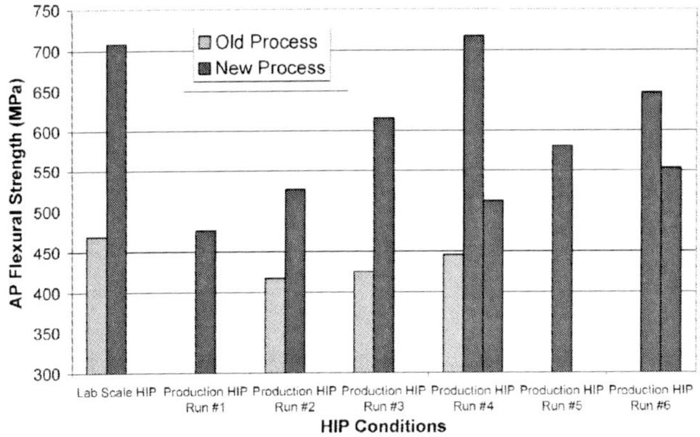

Figure 3: Comparing the AP Strength of the Old and New HIP Processes

NET SHAPE FORMING

Four radial rotors of an experimental Ingersoll-Rand microturbine rotor design were green machined using an optimized procedure that was established. The average surface roughness of the green rotors is ~30□in (~0.8 μm Ra). Coordinate Measuring Machine (CMM) measurements on the green rotors indicate that measurements were within 0.004" of the nominal which is an indication of the consistency and accuracy of the green forming process.

The green-machined radial design rotors (Figure 4) were densified with the proprietary new HIP process explained earlier. The rotors have an excellent average surface roughness of 30-38□in on the as-processed (AP) surfaces. Based on the CMM measurements taken, the rotors exhibited an average part to part variation of 0.003", which is an indication of the consistency of the process. Also, a measured concentricity within 0.003" for the rotors is an indication of the uniform shrinkage seen with this process.

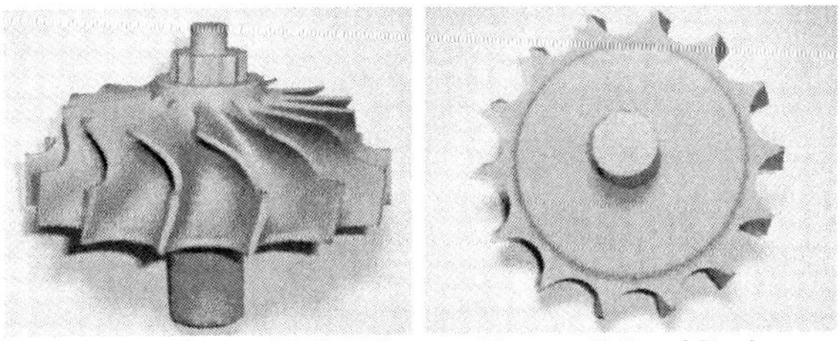

Figure 4: Dense NT154 Rotor of Ingersoll-Rand Design

Two of the rotors were delivered to ORNL for mechanical property evaluation. Disks core-drilled from the blade and hub regions were tested in biaxial flexure to establish the AP and bulk strength of actual components. The biaxial flexure AP strengths, were measured to be 575 and 1088 MPa (Figure 5a), for the blade and hub region respectively. Machined biaxial flexure discs cut from hub section of the rotors were also tested (Figure 5b). A characteristic strength and Weibull modulus of 1070 MPa and 12.7 were measured. These AP and machined biaxial strength values from the rotors compare well with the data generated using MOR bars machined from tiles. This series of tests suggest that the mechanical property data generated from test bars have been reproduced in large net-shape components.

The additional two rotors were also densified successfully. One of the two rotors was spin tested at room temperature approaching 97,000 rpm without failure. Spin testing above 100,000 rpm is planned as soon as a new hub assembly design is completed.

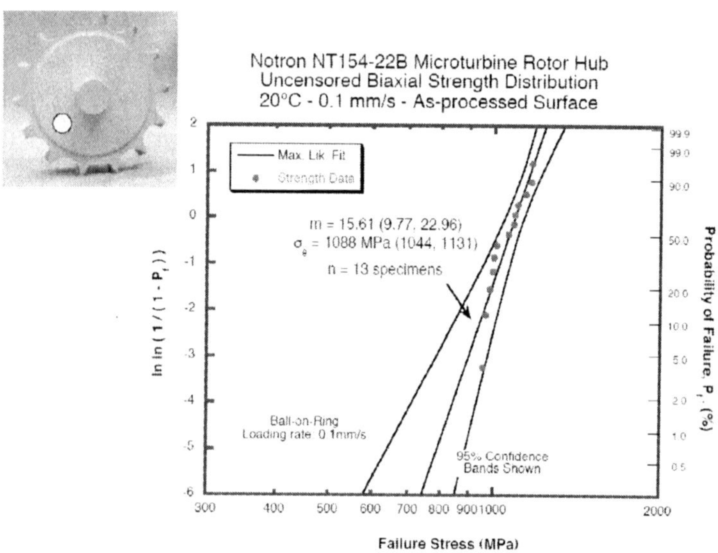

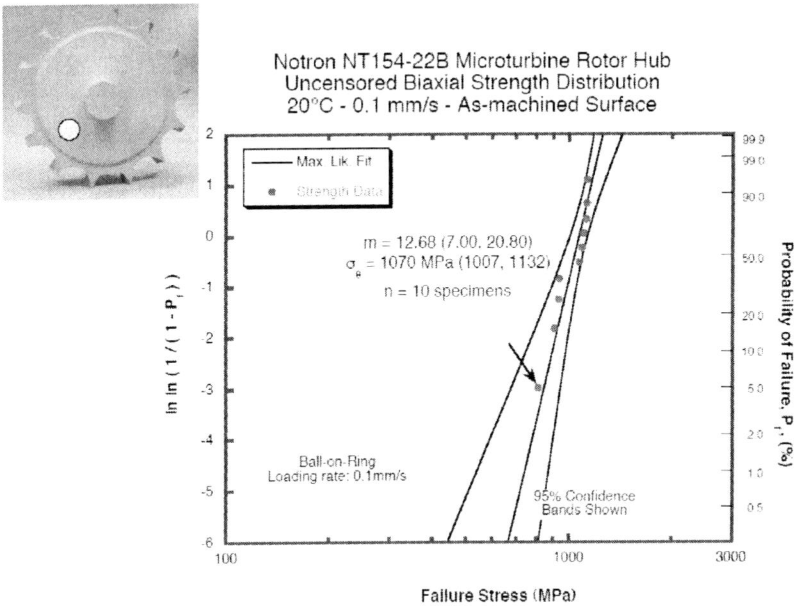

Figure 5: a) AP and b) machined biaxial flexure strength of hub region of NT154 rotor (courtesy of HT Lin, ORNL)

RECESSION CONTROL

Material recession, due to volatilization in humid gas turbine environments, is an issue for silicon nitride and silicon carbide, and has been reported in the literature[3]. Recession resistant solutions for NT154 are being investigated as part of Phase II of this Dept. of Energy funded program.

Baseline (uncoated) NT154 tiles were supplied to ORNL for Keiser Rig testing to evaluate the environmental stability in the Microturbine environment. Silica scale thickness data for 1000 hours of testing is shown in Figure 6. Micrographs showing the silica scale after 500 hours of testing seem to indicate that microstructural differences between the samples appear to impact the silica scale thickness, with the coarser microstructure resulting in a decreased silica scale thickness. In general, the measured thickness of the silica scale formed on NT154 is very similar to that measured on competitor's silicon nitride, which verifies the need for an Environmental Barrier Coating (EBC) or suitable surface modification.

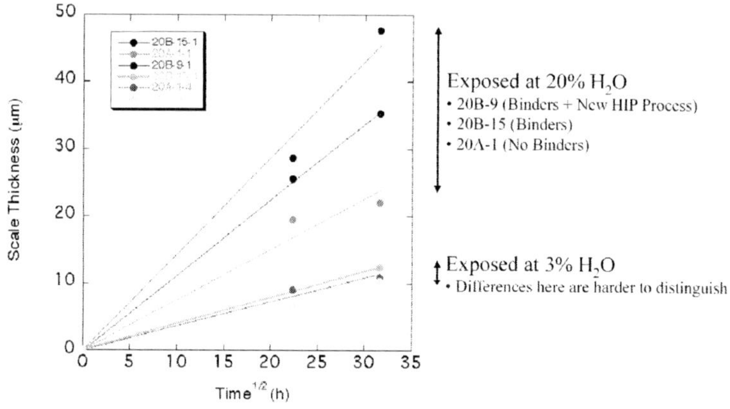

Figure 6: Keiser Rig Data for NT154 (Courtesy of K. More, ORNL)

The development of a recession resistant solution for NT154 consists of three key areas, i) Identification of candidate materials, ii) Application of recession resistant materials on to NT154, and iii) Densification of recession resistant material.

Candidate materials have been chosen based on literature data, thermodynamic calculations and available CTE data. A close CTE match to NT154 (3.3e-6/°C for 20-1370°C) is critical to result in crack free EBC coatings of required thickness (100- 150μm).

Three application / densification approaches were evaluated internally during 2004, i) Pack cementation and ii) HIP Engineered Environmental Protection Surface (HEEPS) and iii) Dip coating of rare earth based composite systems.

The pack cementation (PC) approach consists of imbedding dense NT154 in a reactive powder bed and firing at a specified temperature and environment. The intent is to completely transform the surface of the silicon nitride into a candidate recession resistant material.

The internal PC work concentrated on oxide powder beds. The majority of the work was done in air at 1500-1600°C. The intent was to form continuous layers of rare earth silicates and/or disilicates, which have been shown to have low volatility in the turbine environment. In general it appears that, for the conditions used, only minor modifications to the silicon nitride surface were seen. The issues include: 1) Incomplete transformation of the surface, 2) Low reaction depths (typically <10□m), 3) Near surface cracking, and 4) Creep deformation of the silicon nitride during 1600°C pack cementation runs. As a result, the samples showed only a modest improvement over baseline NT154 in an internal recession test. SEM images of a representative collection of PC samples are shown in Figure 7.

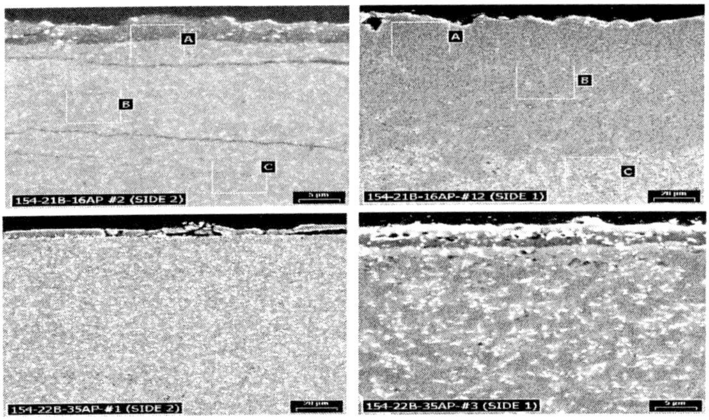

Figure 7: SEM images of cross-section of PC NT154 samples

HIP Engineered Environmental Protection Surface, or HEEPS, is a patented approach to co-densify an EBC coating and NT154 during the typical NT154 HIP process (high temperature HEEPS). Therefore the EBC coating is applied to green NT154 prior to the HIP step. Various rare earth oxide and silicate coatings were explored with limited success.

It appears that, for the coatings applied, there is an issue with the coatings diffusing into the silicon nitride and possibly also into the HIP glass encapsulant. Microprobe scans of cross-sections indicate the coating penetrates as much as 5mm into the sample. Also, there is no evidence of a coating in the samples examined. To date, the HEEPS samples tested in the internal recession rig have shown no improvement over baseline NT154.

Under the dip coating effort, a composite two layer (bond coat and top coat) proprietary EBC system has been designed based upon high temperature thermodynamic properties, CTE match and chemical reactivity. As shown in Figure 8, this two layer EBC coating (100-200μm) has been successfully dip coated and sintered on a NT154 test coupon without cracking or delamination.

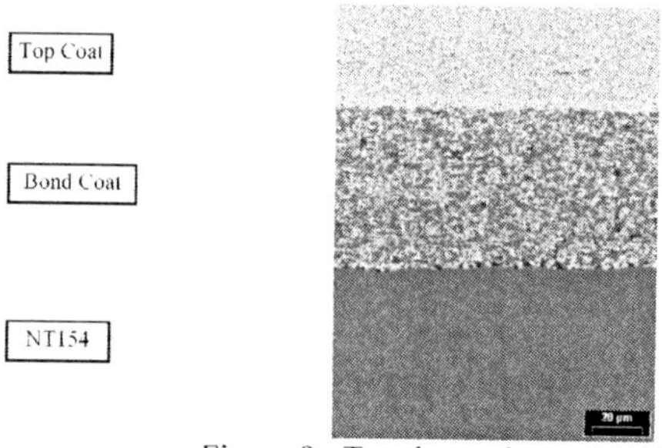

Figure 8: Two layer sintered EBC coating

A few hot pressed test coupons of the potential EBC compositions have also been tested at ORNL in the high pressure Keiser rig (1250°C / 18atm H2O pressure / 20 atm total pressure) exhibiting a stable microstructure and negligible recession or weight loss up to 500 hours accumulated so far. A cross section of 500 hour tested sample is shown in figure 9. Additional 500 hours of Keiser rig testing is in progress at ORNL.

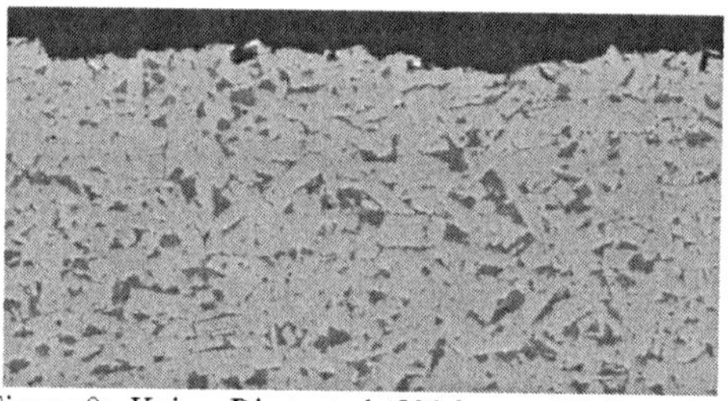

Figure 9: Keiser Rig tested (500 hours) EBC coating

CONCLUSIONS

After an eight year hiatus, Saint-Gobain has re-introduced and re-established NT154 high temperature silicon nitride for use in turbine applications. Further improvements to the HIP process resulted in a 46% improvement in as-processed strength.

Net shape forming by green machining has been developed and demonstrated on several rotors. Excellent dimensional control and surface finish have been achieved. Test specimens machined from rotors have been tested in biaxial flexure at ORNL. Machined strengths up to 1070 MPa and as-processed strengths up to 1088 MPa have been measured. The series of tests suggest that the mechanical property data generated from test bars have been reproduced in large net-shape components. This result is further corroborated by a rotor spin test up to 97,000 rpm without failure.

Work is on-going on the development of recession resistant solutions for NT154. Three different recession resistant solutions namely Pack Cementation, HEEPS and Dip Coating of EBC compositions were evaluated. The first two techniques showed limited success. Dip coating, on the other hand shows great promise. A composite rare earth based EBC composition has been developed and successfully dip coated and sintered, to desired thickness, crack-free on NT154 test coupons. Limited recession tests (500 hours) performed in the Keiser rig at ORNL show no measurable weight loss. Additional testing is underway.

Future work will involve optimization of EBC dip coating and sintering for thicknesses in the range 100-200μm on test coupons for long term recession testing both in the Keiser rig as well as the high velocity NASA rig. Once proven, the dip coating procedure will be further optimized for complex shape turbine components.

REFERENCES

[1]B.J. McEntire, et al, "Ceramic Component Processing Development for Advanced Gas-Turbine Engines", ASME Technical Paper, 91-GT-120, Presented at the International Gas Turbine and Aeroengine Congress and Exposition, Orlando, June 3-6, 1991.

[2]VKPujari et al, *Development of Improved Processing and Evaluation Methods for High Reliability Structural Ceramics for Advanced Heat Engine Applications, Phase I Final Report*, ORNL/Sub/98-SB182/1 ORNL Ceramic Technology Project, U.S. DOE Office of Transportation Technologies, August 1993.

[3]Robinson and Smialek, *SiC recession caused by SiO2 scale volatility under combustion conditions: I, experimental results and empirical model*, J. Am. Ceram. Soc., 1999, 82(7), 1817-1825.

DEVELOPMENT OF SOFC STACK AT KIST USING 10X10 cm^2 ANODE SUPPORTED CELLS

Hae-Weon Lee, Hwa-Young Jung, Ji-Won Son, Joosun Kim and Jong-Ho Lee
Nano-Materials Research Center, Korea Institute of Science and Technology
Seoul 136-791, Korea

ABSTRACT

This paper presents the development of planar-type solid oxide fuel cell (SOFC) stacks based on an internal gas manifold and a cross-flow type gas channel design. In this study, a single-columned 12-cell SOFC stack was assembled using 10 x 10 cm^2 anode-supported unit cells, metallic interconnects, and glass-based compression-seal gaskets. We characterized the power-generating characteristics of the unit cell and 12-cell stack via DC current-voltage load test. According to the performance evaluation, the 12-cell stack produced about 300 W in total power at an average cell voltage of around 0.7 V, while it showed some scale-up problems related to multi-cell stacking.

INTRODUCTION

The solid oxide fuel cell (SOFC) has been developed for next-generation power systems, which combine heat and electrical power. Recently, SOFC has received more attention, due to its high efficiency, high waste-heat utilization, and high fuel flexibility, all of which are the major motives behind replacing the conventional combination of a central power station and decentralized domestic heating systems. Nonetheless, SOFC still has a number of drawbacks in its application to an actual power system. Above all, the reliability and cost competitiveness of the SOFC system are recognized as the key technical barriers blocking SOFC entrance to the commercial market. Therefore, most research and development activities on SOFCs are mainly focused on the development of a commercially viable SOFC technology with high electrochemical performance and long-term stability[1-3].

In general, a high operating temperature of SOFC is favorable for a heat and power co-generation system but this high temperature normally has seriously adverse effects on the stability and reliability of a cell stack, due to the degradation of cell stack constituent materials and components over a long operation time. Therefore, the current focal point of SOFC research and development is to enhance the SOFC performance and, thereby, to reduce its operating temperature to a level at which reliability and cost competitiveness can be guaranteed.

In our previous studies[4-5], we successfully fabricated high-performance anode-supported 5 x 5 cm^2 unit cells with optimized electrode microstructures. In particular, we successfully constructed a novel current collecting layer on the cathode, and we verified its current collecting capability during a unit cell operation[6]. As a result, the current collecting layer with a controlled microstructure could enhance the unit cell performance by reducing the ohmic and/or polarization resistance of the cathode.

Based on our previous results, we fabricated larger sized (10 x 10 cm^2) anode-supported unit cell with improved cathode, and applied this toward the manufacturing of SOFC stacks. In this paper, we introduce our development of SOFC stacks based on large sized 10x10 cm^2 anode-supported unit cells as a plausible means to achieve low or intermediate operating temperature. The fabrication procedure of our planar type stack and its performance results are reported.

UNIT CELL AND STACK FABRICATION

In our previous investigations[4-5], we developed high-performance anode-supported type SOFCs for intermediate operating temperatures (700-800 °C). Fine yttria-stabilized zirconia (Tosoh, Japan), coarse yttria-stabilized zirconia (Unitec, USA), and nickel oxide (J.T. Baker, USA) powders were used to prepare the NiO/YSZ anode substrate. The granulated NiO/YSZ composite powders were compacted by means of uni-axial pressing into a green anode substrate. An anode functional layer, which has the same composition as the anode substrate while containing only fine YSZ powder, was deposited on NiO/YSZ substrate via screen-printing. A very thin (less than 10 μm) YSZ electrolyte layer was also screen-printed onto the anode functional layer and co-fired at 1400 °C in air for 3 hours.

The raw powders of $(La_{0.7}Sr_{0.3})_{0.95}MnO_3$ (LSM) and $(La_{0.7}Sr_{0.3})_{0.95}CoO_3$ (LSCo) for the cathode and the current collecting layer, respectively, were synthesized via a modified glycine-nitrate-process. To improve the cathode performance and to prevent any interfacial reaction between the LSCo current collecting layer and the YSZ electrolyte, we applied a bi-layered cathode consisting of LSM/YSZ (60:40 wt%) and LSM before we applied the current collecting layer. The bi-layered cathode and the additional LSCo layer were also screen-printed onto the sintered anode/electrolyte substrate and were then co-fired at 1150 °C in air for 3 hours. A photograph of the final 10 × 10 cm^2 unit cells of 1 mm thickness is shown in Fig. 1.

Fig. 1. photograph of the anode supported type 10x10 cm^2 unit cells.

Figure 2 and 3 show photographs of the sealing gasket and the interconnector for the SOFC stacks used in this study. For the planar SOFC stack, proper designs for the sealing and the interconnect are very important to ensure proper fuel and oxidant supply through the gas flowing paths without any leak or buildup of gas pressure, both of which can deteriorate the performance and efficiency of SOFCs during power generation. In general, glass based materials are highly attractive for sealing purposes because of their structural deformability for gas tightness at high temperatures. However, serious technical problems of glass-based sealants in terms of chemical and thermo-mechanical stability are still unsolved.

In this study, we developed a very unique compression-seal gasket design (Fig. 2) based on a glass and ceramic composite. This compression seal gasket showed extremely good sealing capability between neighboring components, i.e., YSZ electrolyte and Inconnel based alloys, in a general operating temperature range of 700-800 °C. Moreover, this gasket type sealant could effectively reduce the thermal stresses between the stack components, allowing relatively rapid thermal cycling of the stack.

On the other hand, we employed a commercial Inconnel-based alloy as a metal interconnect. We chose it due to its rather good oxidation resistive characteristics even though it is not so suitable material due to the great difference in thermal expansion coefficient from other stack components. The final appearance of the Inconnel-based interconnector with an internal gas manifold design is shown in Fig. 3.

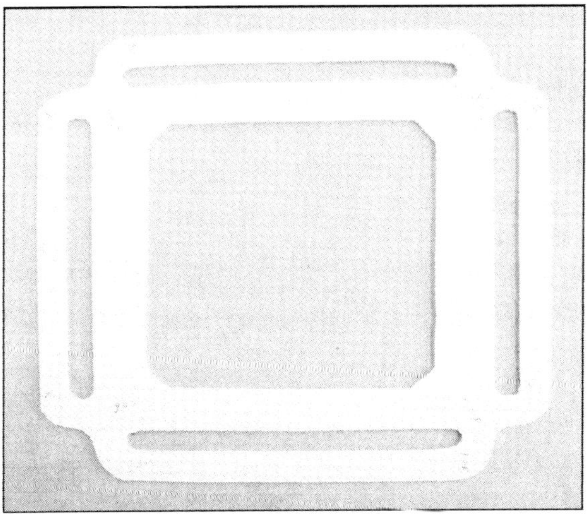

Fig. 2. Photographs of the SOFC Sealing gasket

Fig. 3. Photographs of the SOFC stack interconnector.

Figure 4 shows the configuration of the 300W SOFC stack based on 10x10 cm^2 anode supported cells. As shown in Fig. 4, 12 stack segments were piled up to construct the 300W scaled stack. Each stack segment consists of a ceramic-based 10 × 10 cm^2 unit cell, a glass-based sealing gasket, and metallic interconnects. In both ends of the stack, two planar end plates which have been made of Inconnel alloy are installed to properly align each stack segment.

Porous Ni sponge was used as a current collector and was inserted between the anode and the interconnector for better electrical contacts. Moreover, mechanical load cell was installed to apply a constant pressure to the stack to enhance the electrical contact. For the electrical connection to the outside of the stack, we used Inconnel tube which has been welded to the end plates of the stack. Each stack segment has a separate voltage lead wire to check the voltage of each unit cell during stack operation.

The sealing gasket in each segment has a role not only to prevent mixing of fuel and oxidant gas between each stack segment but also to prevent a direct contact between metallic interconnectors assuring current flow only through the unit cells. Moreover this gasket-type compression sealant can relieve the thermal stresses caused by the mismatches of thermal expansion coefficient between individual components of the SOFC stack during operation.

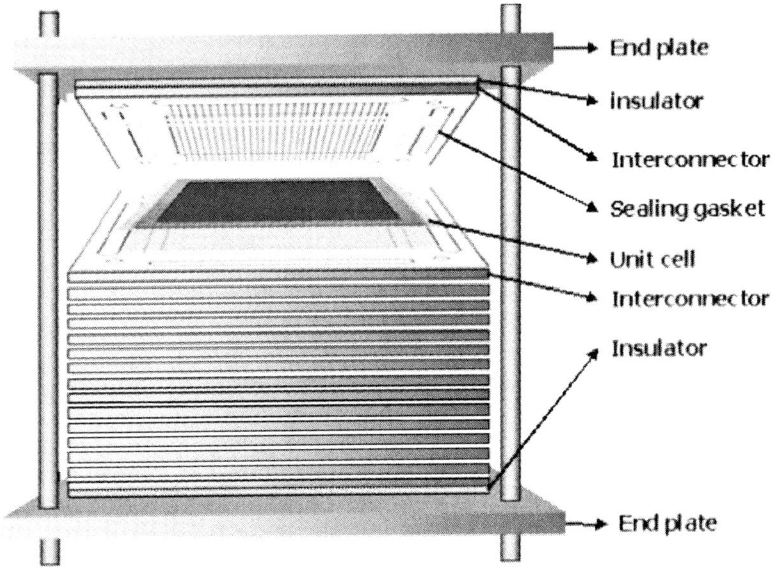

Fig. 4. Configuration of the 300W SOFC stack based on 10x10 cm^2 anode supported cells.

Figure 5 shows the structural schematic diagram of the planar SOFC stacks developed in our laboratory. As shown in Fig. 5, gas flow channels of the interconnector were designed to induce a cross-flowing of fuel and oxidant gas. Considering the difference in the total flow rate between the fuel (H_2) and the oxidant (air) gas, the metallic interconnector (~5 mm thickness) was fabricated to have asymmetric cathode and anode gas manifolds with different cross-sectional areas.

The current-voltage and current-power characteristics of the unit cell and short stacks were measured with a SOFC test station (Toyo, SAT890-100W). Unit cell performance was characterized in a temperature range of 600~800 °C whereas stack performance was characterized at 750°C. In both cases, air was used as an oxidant and moisturized hydrogen (H_2 + 3% H_2O) was used as a fuel.

Performance evaluation was carried out under various operation conditions. For the electrochemical characterization, the current-interruption technique was used to measure the ohmic loss (IR-drop) of the unit cell. The AC impedance spectra of unit cell were collected from using a Solartron SI 1260 FRA (UK) and/or Solartron 1287 potentiostat (UK) employed in conjunction with the 1260 FRA. The AC impedance spectra were acquired in the frequency range from 0.01Hz to 50 kHz under operation temperatures.

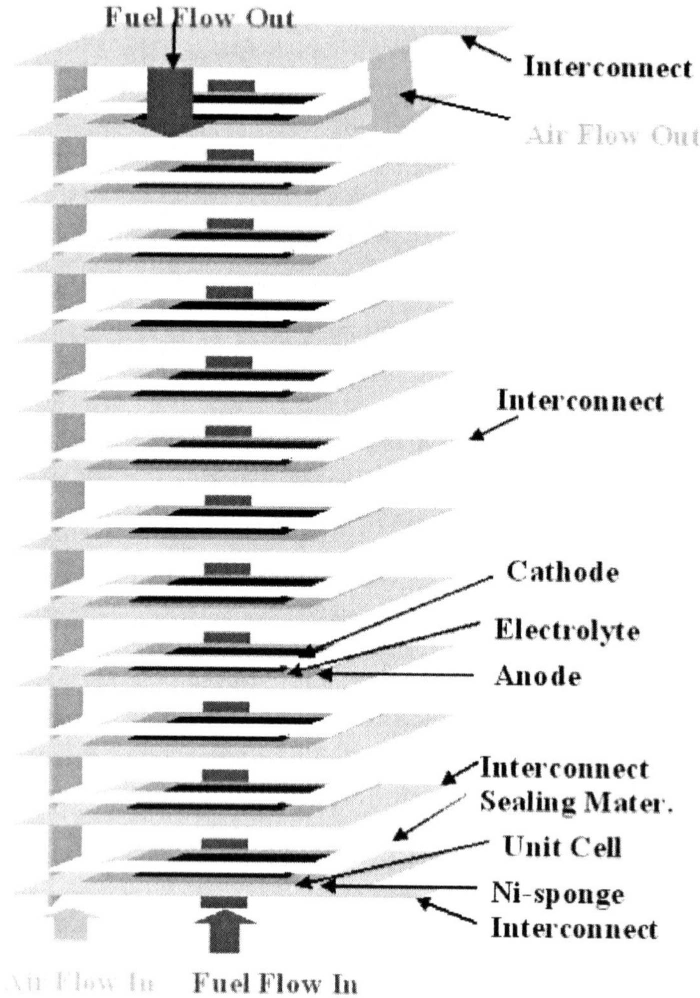

Fig. 5. Schematic diagram of a planar SOFC stack based on anode-supported 10x10 cm^2 unit cells.

UNIT CELL AND STACK PERFORMANCE

According to our previous investigations[6-8] on the electrode microstructure and its related physical properties, proper manipulation of the processing variables related with the electrode fabrication is essential to obtain optimal unit cell performance. In particular, the electrical and/or electrochemical properties of the electrode, which determine the ohmic and diffusional polarization loss of the unit cells, are normally tailored for the appropriate connection of both the electrical conducting path and the gas diffusion path. Based on these investigations, we extended our fabrication technique to fabricate larger area unit cell (10 x 10 cm^2) with a controlled electrode microstructure. We then characterized the power-generating characteristics of this cell.

Figure 6 shows an SEM micrograph of the 10 x 10 cm^2 unit cell with a current collecting layer. As shown in the cross-sectional view in Fig. 6, the unit cell comprises six different layers, including a single functional layer for both the cathode and anode side, respectively. The thickness of the dense electrolyte layer was about 8 μm and the overall cathode thickness was around 50 μm.

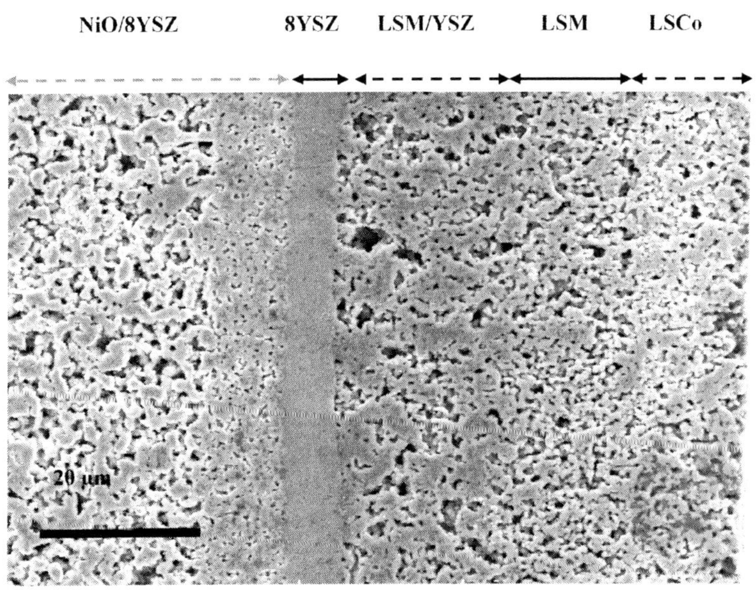

Fig. 6. Cross-sectional view of a 10x10 unit cell with LSCo current collecting layer.

Figure 7 shows the power generating curves of the 10 x 10 cm^2 (effective area 81cm^2) anode-supported cells with an LSCo current collecting layer; the cell was being operated under the conditions of 3% H_2O added hydrogen as a fuel and air as an oxidant. The fuel utilization condition of 60% for the unit cell operation was determined from the criterion of ~ 0.7 V at 750 °C. The airflow was fixed as 2 times the stoichiometric amount ('2 stoichs'). As shown in Fig. 7, the open circuit voltages of the cells were in a range of 1.15~1.20 V in all temperature conditions. The power density of the cell was over 0.5 W/cm^2 at 0.8 V at 750 °C. Furthermore, the cell area specific resistance (ASR) from I-V characteristics and the ohmic ASR from the DC current interruption technique were about 0.362 Ωcm^2 (at 0.8 A/cm^2) and 0.174 Ωcm^2 at 750 °C, respectively.

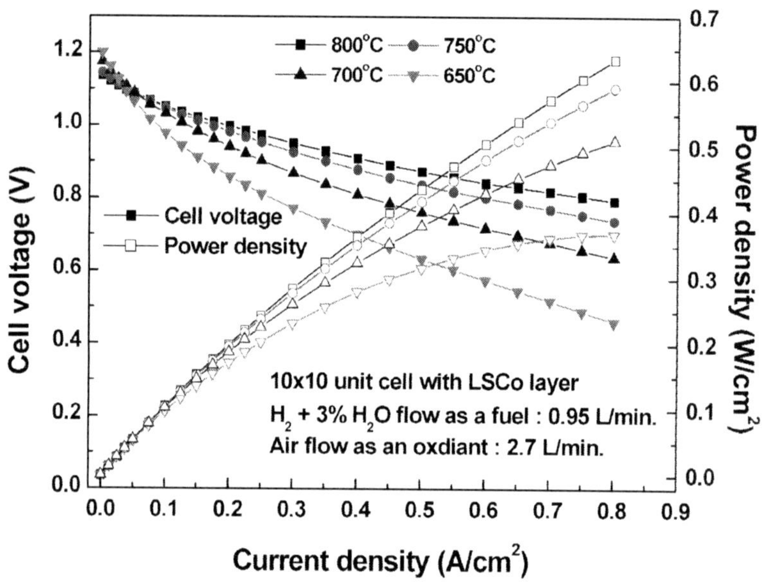

Fig. 7. Power generating characteristics of 10x10 cm^2 unit cell with LSCo layer under various operating temperatures. (Gas flowing conditions of fuel and air were 60L/h and 160L/h, respectively.)

The current-power characteristics were measured at the temperature of 750°C. Air was used as an oxidant, and 3% moisturized hydrogen gas (H_2 + 3% H_2O) was used as a fuel. For the performance test of stack, it was operated under similar operating conditions to that of unit cell whereas the fuel and oxidant utilizations were kept around 40%. According to the performance evaluation, the open circuit voltages (OCV) of each cell in 12-cell stack are around of 1.1 V which is comparable with that from unit cell test. The obtained OCV value was almost equal to the theoretical value, indicating that all the stack components were working properly without gas leakage with dense electrolyte, optimal electrical potential through the cells, and tight stack housing. The overall stack output was about 300W at the condition of 40A and average cell voltage of around 0.7V. (Fig. 8) Such level of power density was approximately one half of unit cell, which indicating that some scale-up problems had occurred during the multi-cell stacking.

Such a scale-up problem is suspected to arise mainly from the deterioration of electrical contact between a cathode and an interconnect. In this study, we used bare Inconnel metal without any surface modification, such as oxidation resistive coating or high conductive coating. Hence rather high resistive surface layer can be formed in oxidative atmosphere during the high temperature operation even though it has relatively high resistivity against oxidation. Moreover the cathode layer can be easily damaged due to considerably higher thermal expansion coefficient of Inconnel alloy, which might become another source of contact problem. Such kinds of contact problems at cathode-interconnect interface can degrade not only the power generating characteristics but also the long-term stability of the stack. Further postmortem characterization of the stack to verify our interpretation will be addressed in a forthcoming paper.

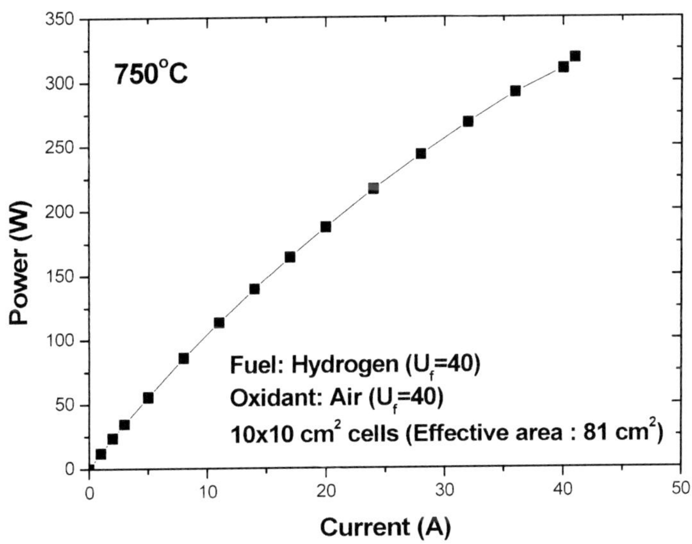

Fig. 8. Power generating characteristics of 12-cell stack with 10x10 cm^2 unit cells at 750°C.

CONCLUSION

We fabricated anode-supported 10x10 cm^2 unit cells with the optimized electrode microstructure, which shows excellent power density over 0.5W/cm^2 at 0.8V (750°C). We developed planar type stack by using this 10x10 cm^2 anode-supported cells, Inconnel based interconnects and glass-base compression-seal gaskets. According to the performance evaluation, the stack output was about 300W at the condition of 40A and average cell voltage around 0.7V.

REFERENCES

[1]N. Q. Minh, "Ceramic Fuel Cells," *J. Am. Ceram. Soc.*, **76**, 563-588 (1993).

[2]N. Q. Minh, T. Takahashi, Chap. 1-5 in Science and Technology of Ceramic Fuel Cells, Elsevier, Amsterdam, 1995.

[3]S. C. Singhal, K. Kendall, Chap. 1 in High Temperature Solid Oxide Fuel Cells, Elsevier, UK, 2003.

[4]J.-H. Lee, H. Moon, H.-W. Lee, J. Kim, J.-D. Kim, K.-H. Yoon, "Quantitative Analysis of Microstructure and Its Related Electrical Property of SOFC Anode, Ni-YSZ Cermet," *Solid State Ionics*, **148**, 15-26 (2002).

[5]D.-S. Lee, J.-H. Lee, J. Kim, H.-W. Lee, H.-S. Song, "Tuning of the Microstructure and Electrical Properties of SOFC Anode via Compaction Pressure Control during Forming," *Solid State Ionics*, **166**, 13-17 (2004).

[6]H. Y. Jung, W.-S. Kim, S.-H. Choi, H.-C. Kim, J. Kim, H.-W. Lee, J.-H. Lee, "Effect of Cathode Current Collecting Layer on the Unit Cell Performance of Anode-supported SOFCs," *J. Power Sources* (2005) in press.

[7]J.-H. Lee, J.-W. Heo, D.-S. Lee, J. Kim, G.-H. Kim, H.-W. Lee, H.-S. Song, J.-H. Moon, "The Impact of Anode Microstructure on the Power Generating Characteristics of SOFC," *Solid State Ionics*, **158**, 225-232 (2003).

[8]K.-R. Lee, S. H. Choi, J. Kim, H.-W. Lee, J.-H. Lee, "Viable Image Analyzing Method to Characterize the Microstructure and the Properties of the Ni/YSZ Cermet-Anode of SOFC," *J. Power Sources*, **140**, 226-234 (2005).

ENGINEERED REFRACTORIES FOR SLAGGING GASIFIERS

Kyei-Sing Kwong, James Bennett, Rick Krabbe, Hugh Thomas, and Cynthia Powell
Albany Research Center, DOE
1450 Queen Ave. S.W.
Albany, OR 97321

ABSTRACT

The widespread commercial adaptation of slagging gasifier technology to produce power, liquid fuels, and/or chemicals from coal and other fossil-based feedstocks and from biomass, will depend in large measure on the technology's ability to prove itself both economic and reliable. Improvements in gasifier reliability, availability, and maintainability will in part depend on the development of improved structural materials with longer service life in this application. Current generation refractory materials used to line the gasifier vessel, and contain the gasification reaction, may last no more than three months in commercial applications. The downtime required for tear-out and replacement of these critical materials results in gasifier on-line availabilities that fall short of targeted goals. In this talk we will discuss the development of improved refractory materials engineered specifically for longer service life in this application, and present results from recent field trials in commercial systems.

INTRODUCTION

Gasification offers one of the cleanest ways to convert carbon based materials into electricity, hydrogen, and other chemical products. In addition, gasification is one of the best ways to produce clean-burning hydrogen for tomorrow's automobiles and power-generating fuel cells. The Department of Energy's (DOE) Office of Fossil Energy is working on gasifier advances that enhance efficiency, environmental performance, and reliability, as well as expanding the gasifier's flexibility to process a variety of coals and low-cost, widely available feedstocks. The capability to produce electricity, hydrogen, chemicals, or various combinations while eliminating nearly all air pollutants and potential greenhouse gas emissions makes gasification one of the most promising technologies for the energy plants of tomorrow[1], especially as oil and natural gas become more expensive as supplies are depleted.

In general, gasification technology, such as integrated gasification combined cycle (IGCC), includes three major processes: carbon gasification, gas purification and separation, and power generation (or chemical production) (Figure 1). Carbon gasification occurs by using high temperatures to break down virtually any carbon-based feedstock into its basic constituents. A slagging gasifier is a high pressure/high temperature reaction vessel fed by large quantities of carbon-rich fuels that produce highly corrosive slags as a by-product. Refractory linings are used in slagging gasifiers to protect the steel vessel from the attack of molten slag and to insulate it from energy loss.

Gasifiers are the heart of IGCC power systems, and refractory lining service life is one of the most important variables in defining gasifier reliability and economics.[2] In order to move gasification technology forward, researchers at the DOE's Albany Research Center (ARC), have conducted fundamental studies on gasification operational conditions, refractory properties, post-mortem analyses, material development, and field trials to find a solution for extending the service life of the refractory. An ARC-developed refractory material is currently in field tests, and shows promise as a potential lining material for use in slagging gasifiers.

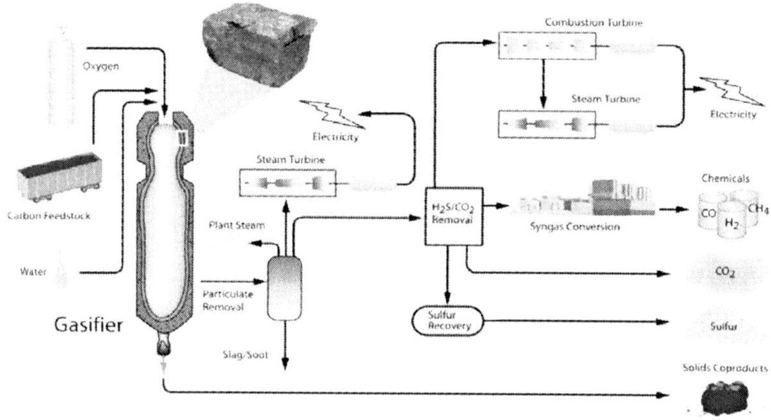

Figure 1 The configuration of integrated gasification combined cycle (IGCC) processes.

REFRACTORY LINING AND GASIFICATION OPERATIONAL CONDITIONS

The slagging gasifier is a pressurized vessel lined with multiple refractory layers that contain up to 10 tons/hour of flowing, molten slag. The slags originate from the coal, petcoke or other carbon-based feedstocks injected as a high speed slurry into the gasifier[3]. Depending on the melting point and viscosity characteristics of the slag, the typical slagging gasifier operates at temperatures between 1300 and 1600 °C and pressures up to 1000 psi.

The hot-face refractory lining inside a slagging gasifier must be able to withstand the following failure inducing factors:

- High temperature operation
- Aggressive low melting, low viscosity and high acidic coal slag attack
- Variation in feedstock
- Particle erosion from injected slurry
- Alternating oxidizing and reducing atmosphere
- Erratic operating conditions such as frequent of shutdowns and aberrant temperature control
- Load at high temperatures (creep)
- Flame impingement.

Material research in the mid 70's to 80's suggested that a high chromia content in a refractory (75 wt pct or higher) is required for these severe gasifier environments[4]. Currently, refractory lifetime ranges between 3 and 24 months. The actual service life of the refractory depends on the chemistry of the feedstock, operational conditions and gasifier design. In addition, different parts of a gasifier may have different wear mechanisms, resulting in a frequency of refractory lining replacement that varies with location.

The high material costs and the downtime associated with replacing the refractory liner may impact the future of gasification technology in power generation when competing against natural gas and pulverized coal power plants. A gasifier availability of 85-95% for utility applications and more than 95% for other applications is required to move IGCC technology forward in the marketplace[2]. To achieve these kinds of availabilities, refractory service life must be improved.

POST-MORTEM ANALYSES

The Albany Research Center has performed extensive post-mortem analyses of spent refractories from commercial gasifiers which utilize coal as the major feedstock. Figure 2 shows a spent refractory with a 3 centimeter layer of slag penetration, and a crack about 10 centimeters wide and 12 centimeters deep within the penetration zone. Detailed analyses found that calcium silicates from the slag penetrated deeply into the refractory through pores, while ferrous iron oxide (FeO) reacted with Cr_2O_3 to form a dense spinel layer on the surface. Detailed fundamental studies on slag and refractory reactions in the laboratory support these post mortem analyses[5]. Laboratory studies also found that on its own, neither SiO_2 nor CaO will penetrate deeply into the refractory under gasifier conditions.

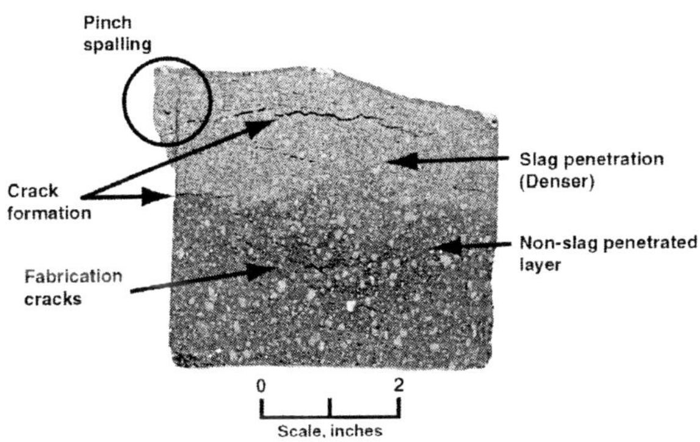

Figure 2. A cross section of a spent refractory showing a 3 centimeter slag penetration zone

Chromium oxide does not have good thermal stability, and suffers thermal shock from fast thermal cycling. Gasifiers can shut down as often as every 30-45 days for non-refractory issues such as system maintenance or down-stream problems. Every time a gasifier cycles, it increases the chance of crack formation and growth in the slag-penetrated regions, leading to structural spalling. When material is lost, a fresh refractory surface is exposed and the cycle of slag attack and penetration begins again. Figure 3 presents a graphical representation of this cycle of material loss by structural spalling.

Figure 4 shows that the amount of refractory material loss during a commercial gasifier operation[6]. This figure indicates sudden material loss occurring every 1300-1500 hours. The service life of a refractory could be doubled if structural spalling could be prevented.

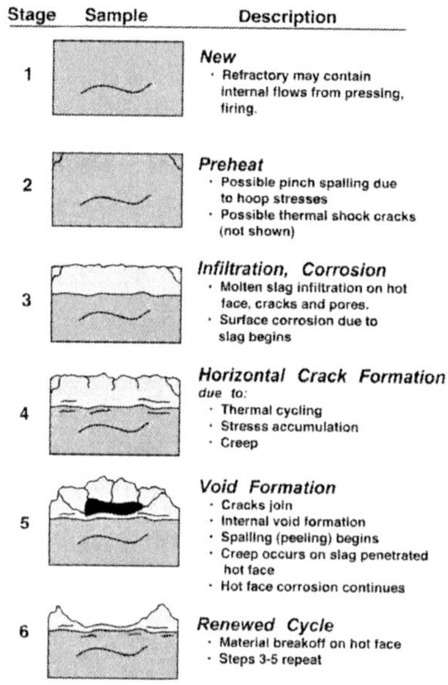

Figure 3 A model showing the cycle of material loss by structural spalling.

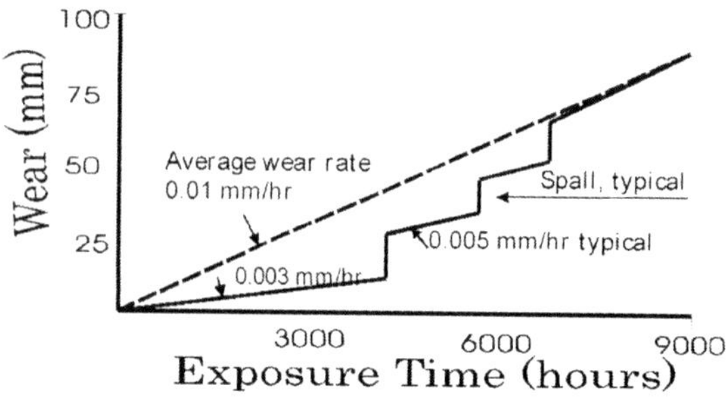

Figure 4 The amount refractory material loss during a commercial gasifier operation.[6]

LABORATORY RESEARCH AND MATERIAL DEVELOPMENT

Based on the previously discussed failure mechanisms, it is obvious that refractory service life in slagging gasifiers is defined at least in part by the gasifier design and its operation, the nature and limitation of materials, and business decisions that can not be directly impacted by refractory research. However, post-mortem analyses showed that structural spalling is a major cause of the premature refractory failure, something that can be improved by engineering refractory materials through laboratory research.

Researchers at ARC have conducted numerous trials in their search to solve the spalling problem, including improving the microstructure of the hot face refractory, incorporating non-wetting materials, promoting sintering, and adding phosphates. ARC researchers found that phosphate additives in the Cr_2O_3 refractory were particularly effective in decreasing slag penetration in laboratory tests. Phosphate additives have been commonly used by the refractory industry, especially in ramming mixes or castable, for enhancing the bonding strength from room to high temperatures. Phosphate additives in Cr_2O_3 refractories were designed with the goal of reacting with the calcium silicates in the molten slag, to form highly siliceous melts that stop or slow slag penetration. In addition, phosphate additives were found to provide other benefits, including decreasing the porosity of the Cr_2O_3 refractory and improving its thermal shock resistance. Extensive laboratory tests, from two-inch cubes to brick size samples, confirmed the benefits of phosphate additives.

Figure 5 shows cross sections of commercial and phosphate containing high Cr_2O_3 refractory samples after high temperature static exposure tests. It shows that phosphate additives to a high Cr_2O_3 refractory reduced slag penetration versus the commercial refractory. Rotary slag drum tests were performed to compare the corrosion and slag penetration resistance of ARC-developed refractories (brick C) with other similar high chromia commercial gasifier refractories (brick A&B) in a more dynamic, flowing slag environment (Figure 6). These bricks all contain about 90% Cr_2O_3. The tests indicated that the ARC developed refractory had the best resistance to slag penetration and thermal shock. Table 1 lists the comparison of physical properties among ARC-developed refractory and commercial bricks. Cup, rotary slag drum, and physical tests all indicate that the ARC-developed refractory should provide better performance than other commercial bricks in slagging gasifiers.

(A) (B)

Figure 5 Cross sections of (a) a commercial 90% chrome refractory, and (b) ARC improved refractory following exposure to a coal slag at 1600°C. The depth of slag penetration is marked with a dotted line in each case.

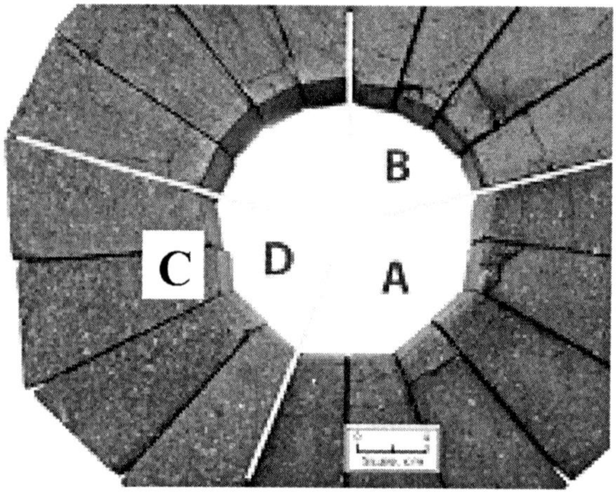

Figure 6 A section of refractory brick after a rotary slag drum test (A: commercial brick 1, B: commercial brick 2, C: ARC-developed brick)

Table I. Physical Properties of High Cr_2O_3 Gasifier Refractories.

	A	B	C
Bulk Density (g/cc)	4.27	4.07	4.20
Apparent Porosity (%)	14.8	16.5	12.7
CCS (MPa)	65.5	66.9	63.1
Reheat expansion *1	+0.64	-0.08	+0.11
Creep deformation*2	+0.18	-1.98	- 0.24
Rotary slag *3			
- % area change	+2.3	+2.3	+6.5
mm slag penetration	3.5	6.0	1.8

*1 = 1550°C, 50 hr hold
*2 = 1550°C, 345 kPa, 50 hr hold
*3 = 1657°C, 5 hrs of slag feed, 2 ½ rpm.

Detailed studies by SEM and XRD have been conducted with a goal of defining how phosphate additives improve slag penetration resistance. Because of the complexity of the slag and refractory reactions, it is likely that multiple mechanisms combine to make the ARC-developed brick more slag resistant. Possible mechanisms are as follows:

1) Microstructure improvement: Because the additive phosphate is a liquid binder in the formulated mixture, the green porosity of the refractory decreased up to 3%, as indicated by sample C in Table 1. Evidence from sintering studies of slurry coatings made from refractory powders indicate that the ARC-developed refractory sinters better than other commercial chromia refractories[7] (figure 7). Enhanced sinterability can ensure better binding among matrix materials and aggregates in the refractory to limit slag penetration.

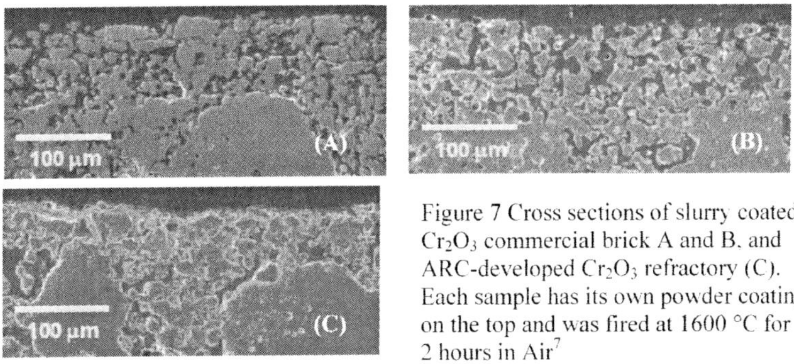

Figure 7 Cross sections of slurry coated Cr_2O_3 commercial brick A and B, and ARC-developed Cr_2O_3 refractory (C). Each sample has its own powder coating on the top and was fired at 1600 °C for 2 hours in Air[7]

2) Oxygen potential: The oxygen potential can affect the reaction of Cr_2O_3 with FeO_n. The three phase diagrams in Figure 8 demonstrate that the amount of liquid phase varies tremendously under different oxidation/reduction environments. For example, if we assumed that 60 wt% FeO reacts with 40 wt% Cr_2O_3 at 1600 °C, then 47% liquid can be formed in a reduced/neutral environment (figure 8a), while 13 % liquid exists in an atmosphere consisting of equal parts CO_2 and H_2 (figure 8b). No liquid is formed in air environment[8] (figure 8c). $AlPO_4$ is an oxidant and may increase the oxygen potential at local sites, decreasing the formation of liquid phase. Thermodynamic calculations, based on the gasifier environments, indicate that the iron in the coal slag should occur predominantly as FeO, with trace amounts of Fe_2O_3 or Fe. SEM analyses indicate that trace Fe or FeS can exist on the surface of slag grains. Wet chemistry analyses detect trace amounts of Fe_2O_3 in some coal slags.

3) $AlPO_4$-Cr_2O_3 reaction: SEM with EDAX microstructural analysis indicate the possibility of an $AlPO_4$-Cr_2O_3 reaction to form a compound of Al-Cr-P-O. It is known that Al_2O_3 can react with Cr_2O_3 to form a complete solid solution. However, no definite evidence from laboratory tests proves the possibility of a solid solution between $AlPO_4$ and $CrPO_4$. More tests will be performed to verify a compound or a solid solution consisting of Al_2O_3-Cr_2O_3-P_2O_5.

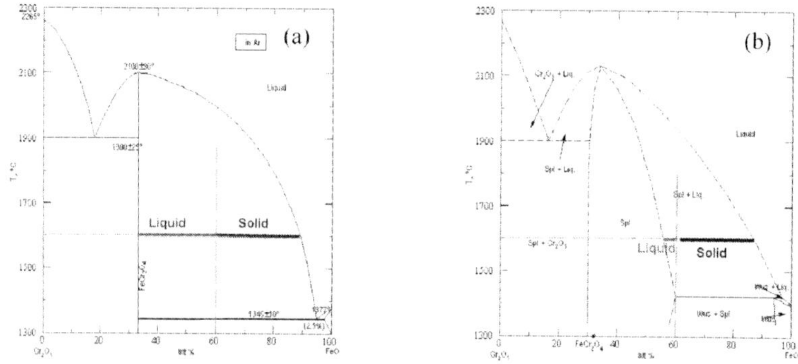

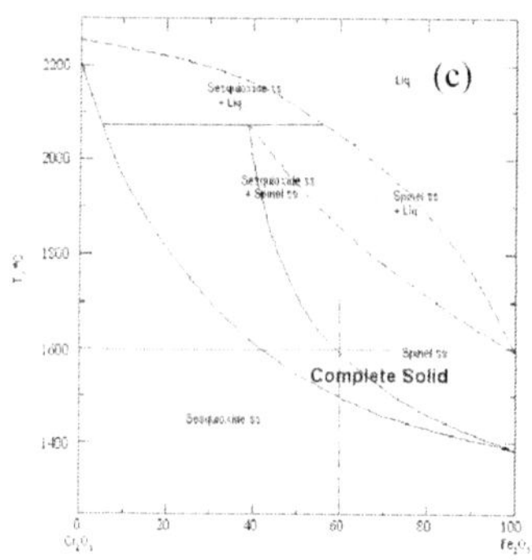

Figure 8 Phase diagrams of Cr_2O_3-FeO_x under different oxidation/reduction environments: (a) Ar, (b) equal amount of CO_2 and H_2, and (c) Air.

4) Calcium silicates and $AlPO_4$ reactions: Post-mortem analyses indicate that penetrated slag consists of calcium silicates (or calcium aluminosilicates). Thermodynamic calculations predict that calcium phosphates can be formed by the reaction of calcium silicates with $AlPO_4$. X-ray diffraction studies showed indirect evidence of calcium phosphate formation in a mixture of calcium silicates and $AlPO_4$. This reaction is expected to create a viscous melting material because of the formation of calcium phosphates and the glass former of P_2O_5.

The Potential of Non-Chromia Refractories

While chromia refractories currently offer the best available service life to the gasifier operator, there are many reasons why a non-chromia refractory might be preferable.[10] Researchers at ARC are working on developing new non-chromia refractory bricks with the goal of partly or completely replacing the hot face of Cr_2O_3 gasifier refractory lining. Fundamental scientific methods have been adopted to identify potential candidates using thermodynamic calculations, phase diagram studies, material property investigation, and literature reviews. Laboratory analyses are being performed to verify potential candidates by visual inspection, optical microscopy, X-ray diffraction (XRD), thermogravimetric analysis (TGA), differential thermal analysis (DTA), and scanning electron microscopy (SEM). Issues related to chemical corrosion resistance, age deterioration, steam reaction, hardness, vapor pressure, sinterability, cost, environment, and others have been considered. Each component's function in the refractory has been designed and engineered from fundamental studies and laboratory tests.

Figure 9 shows improved corrosion and penetration resistance in a non-chromia cup test made from one of ARC's newly engineered formula. However, more dynamic and physical property tests must be conducted on these potential compositions to confirm their improved performance, including rotary slag tests, creep tests, hot and cold strength, and thermal expansion.

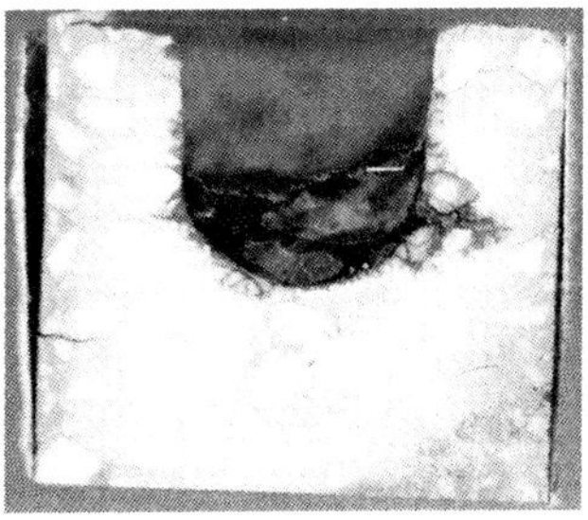

Figure 9 A cross section of non-chromia refractory after exposure coal slag 1 hour at 1500 °C.

FIELD TRIALS

Researchers at ARC have worked with refractory producers, and gasifier manufacturers and operators with a goal of improving the service life of refractory linings. Laboratory studies of an ARC-designed refractory for this application have been followed by a scale-up to full-sized brick production and test panel field trials in several commercial gasifiers (Figure 10). Although post mortem studies of the field trial brick are not yet completed, visual inspection suggests that the ARC-designed material does in fact perform better in slagging gasifier environments than the currently used high-chromia refractories.

A patent on phosphates added to chromium refractory for gasifier applications has been issued to ARC/DOE[9]

Figure 10 Field trial of test material after 17 days service

CONCLUSIONS

Coal gasification offers one of the cleanest ways to convert coal into electricity, hydrogen, and other chemical products. The service life of refractory liners for gasifiers has been identified by users as a critical barrier to IGCC system economics, reliability, and on-line availability. In order to move gasification forward, researchers at Albany Research Center, DOE, conducted fundamental studies on gasification operational conditions and refractory properties, post-mortem analysis, material development, and field trials to find a solution for extending the service life of refractory. Structural spalling has been identified as one of major failure mechanism for the premature service life of refractory. A model has been built to explain stages of wear mechanisms for gasifier refractory. Using phosphate additives in Cr_2O_3 refractory has proven an effect way to reduce the slag penetration in refractory bricks. Based on research results, field trial bricks were produced by a refractory company and were installed in slagging gasifiers. Results from field tests shows that the ARC developed refractory has promise as a potential lining material for use in slagging gasifiers. A patent on phosphates added to chromium refractory for gasifier applications has been obtained.

REFERENCES

[1]"Gasfication Technologies "A Program to Deliver Clean, Secure, and Affordable Energy"" NETL, DOE, November, 2001

[2]G.Stiegel, and S. Clayton, "DOE Gasification Industry R& D Survey: A Perspective of Long Term Market Trends and R&D needs" in Proceedings from the Gasification Technologies 2001 Annual Meeting, San Francisco, CA.

[3]"Tampa Electric Integrated Gasification Combined Cycle Project" Project Performance Summary Clean Coal Technology Demonstration Program June 2004 DOE/FE-0469

[4]J. Bonar, C. Kennedy, and R. Swaroop, "Coal-Ash Slag Attack and Corrosion of Refractories" Am Ceram Soc Bull. 59 473-478 (1980)

[5]J. Rawers, J. Bennett, and K-S. Kwong, "Slag-Refractory Interactions" in Proceedings from 13th Annual Conference on Fossil Energy Materials, 1999, Knoxville, TN

[6] W. T. Bakker "Materials Guidelines for Gasification Plants" EPRI Report TR-110507, 1998, Palo Alto, CA.

[7]B. L. Armstrong, G. H. Kirby, K. M. Cooley, K. S. Kwong, and C. A. Powell "Corrosion Resistant Ceramic Coatings" FEARM 2005 conference, Oak Ridge National Laboratory, Oak Ridge, TN.

[8] "Phase Equilibria Diagrams" CD ROM database version 3, American Ceramic Society and National Institute of Standards and Technology

[9]K. S. Kwong, C.P. Powell, J.P. Bennett, R.E. Chinn, and A.V. Petty "Use of Phosphates to Reduce Slag Penetration in Cr_2O_3-based Refractories" US patent 6815386.

[10] K.S. Kwong, J.P Benett, C.A. Powell, H. Thomas, and E. Krabbe, "Development of Improved Performance Refractory Liner Materials for Slagging Gasifiers," Proceedings of the 22nd Annual International Pittsburgh Coal Conference, 2005,Pittsburgh, PA

ACKNOWLEDGMENTS

The work was supported by the DOE's National Energy Technology Laboratory. The authors are very grateful to Beth L. Armstrong, Glen H. Kirby, and Kevin M. Cooley at Oak Ridge National Laboratory for the work of slurry coating on refractory samples.

Author Index